ELEMENTS OF ALGEBRA
AND ALGEBRAIC COMPUTING

ELEMENTS OF
ALGEBRA
AND
ALGEBRAIC
COMPUTING

John D. Lipson
University of Toronto

1981

Addison-Wesley Publishing Company
Advanced Book Program
Reading, Massachusetts

London · Amsterdam · Don Mills, Ontario · Sydney · Tokyo

Library of Congress Cataloging in Publication Data

Lipson, John D
    Elements of algebra and algebraic computing.

    Bibliography: p.
    Includes index.
    1. Algebra. 2. Algebra—Data Processing. I. Title.
QA154.2.L56   512   80-22161
ISBN 0-201-04115-4

Manufactured in the United States of America
ABCDEFGHIJ—MA—8987654321

To
Pamela *and* Nancy

CONTENTS

PREFACE

This book is an outgrowth of a two-semester course in Applied Algebra given over the past several years in the Department of Computer Science, University of Toronto. The goal of this book, like that of the course, is to present some of the great ideas of modern and universal algebra, and to show the applicability of those ideas to computing. The book, like the course, is intended to offer students of computer science, applied mathematics, and engineering an alternative viewpoint to that of the traditional "pure" treatments of modern algebra.

Some justification for Part One, Mathematical Foundations, is perhaps in order. I have long believed that algebra is sufficiently close to the foundations of mathematics to make some intuitive set theory worthwhile. The progression from set theory, with its concepts of set, subset, function, and equivalence relation, to algebra, with its corresponding concepts of algebraic system, subalgebra, morphism, and congruence relation is a most natural one. This is the idea behind Chapter I.

Chapter II, The Integers, besides gathering some vital facts about the integers, serves as a warmup for algebra (which is appropriate, since so much of algebra has been motivated by the integers). A high level axiomatization of the integers (as an ordered integral domain with well-ordering of the positive elements) affords an opportunity to illustrate the postulational approach of modern algebra in a familiar concrete setting. Out of this axiomatization comes the inductive property of the integers, which in turn gives an indispensable tool for reasoning about integers: proof by induction.

An important ancillary purpose of Part One is to bring the student on board by giving him an appreciation, mainly by example, of what constitutes a mathematical proof. In a typical classroom setting, with students having diverse backgrounds and preparation, this goal might be the most significant of all.

The Prologue to Part Two, Algebraic Systems, advances a principal thesis of the book: that algebra can be profitably studied from a number of levels of abstraction (three in particular). Of course the *concrete* algebraic systems of everyday mathematics motivate the *abstract* systems of modern algebra; e.g., integers, rationals, reals, polynomials, matrices collectively lead to the unifying concept of a ring. But there is a yet higher level of abstraction, one that unifies analogous concepts of modern algebra—this is the viewpoint of *universal algebra*, where, for example, the subgroups, subrings, sublattices of modern algebra lead to the universal notion of a subalgebra. Perhaps the most distinctive aspect of our treatment of algebra is the systematic use of this universal viewpoint. (While many students find the transition from concrete, manipulative algebra to modern algebra a difficult one, the transition from modern to universal algebra seems to be an easy and natural one.)

In Chapter III, Semigroups, Monoids, and Groups, we put our levels-of-abstraction principle into practice. Functions under composition serve as the prime concrete motivation for studying abstract algebraic systems based on a single binary operation. With semigroups, monoids, and groups on the table, a measure of unity is achieved by the universal concepts of subalgebra and morphism.

Chapter IV, Rings, Integral Domains, and Fields, is of paramount importance for computing, motivated as it is by the familiar number systems. The specialization from rings to integral domains hinges on the notion of a zerodivisor, from integral domains to fields on the notion of a unit (invertible element). Accordingly, these two kinds of elements are singled out for special attention. (In particular it is hoped that the student will acquire a new respect for the laws of "ordinary" algebra when shown that a quadratic equation may have more than two roots over a ring with zerodivisors.) The properties of the integers under quotient-remainder division (established in Chapter II) are shown to hold more generally for Euclidean domains, hence for polynomials over a field.

In Chapter V, Quotient Algebras, we return to the viewpoint of universal algebra, which is appropriate for a chapter that is concerned with the technical content of abstraction. The highlight of this chapter is the result that the homomorphic images (= abstractions) of any algebraic system A are *all* given up to isomorphism by quotient algebras of $A—A/E$ for some congruence relation E. (In view of its importance to algebra generally, this result might well be called the fundamental theorem of universal algebra.) When specialized to rings, the "fundamental theorem" yields the intimate connection between quotient rings and ideals. We use this theorem to define into existence a number of useful rings, in particular the integers modulo m, modular polynomial rings, and truncated power series rings.

In Chapter VI, Elements of Field Theory, the "fundamental theorem" of the previous chapter plays an important supporting role, where we show that a variety of important fields, including the complex field and all finite fields, can be defined as quotient rings by appropriate ideals. The material on finite fields is perhaps of most interest for algebraic coding theory (an application which is not covered in this book).

Part Three, Algebraic Computing, is the *raison d'être* for this book. The underlying theme here is that the computer manipulation of algebraically structured data can be a useful and intriguing enterprise, made all the more interesting by the fact that algebraic concepts can be exploited to achieve efficient algorithms.

In Chapter VII, Arithmetic in Euclidean Algorithms, we present and analyze a number of classical algorithms for arithmetic in Euclidean domains. The highlight of this chapter is the application of the Invariant Relation Theorem—a result from computer programming—to derive Euclid's extended algorithm for computing greatest common divisors in abstract Euclidean domains.

In Chapter VIII, Computation by Homomorphic Images, the homomorphic image concept finds powerful application in an algorithmic setting. The idea is a simple one: to carry out a computation over an algebraic system A, perform corresponding computations over a number of homomorphic images A' of A, and use these computed A' values to ascertain the desired A value. We focus on the A = Euclidean domain case and the role played by the Chinese Remainder Theorem.

The concluding chapter, Chapter IX, is devoted to one of the "super" algorithms of computer science, the fast Fourier transform. In particular we focus on its role as an algebraic algorithm for speeding up operations on integers, polynomials, and power series. The highlight of this chapter is an algebraic application of Newton's method of numerical computing fame.

* * *

Most (but not all) of the exercises have been classroom tested. I have tried to include interesting exercises that avoid the extremes of triviality and excessive difficulty.

The notes occasionally introduce supplementary material but are mainly intended for historical and bibliographical purposes.

Within each section (Section 1.1, 1.2, etc.), lemmas and theorems are numbered independently. To help the reader distinguish between the major and minor results, I have arbitrarily called the former "theorems" and the latter "propositions." No distinction is made between theorems and propositions for numbering purposes; e.g., we might have Theorem 1 followed by Proposition 2 followed by Theorem 3, etc.

I have profited greatly from comments and suggestions of several generations of Applied Algebra students at the University of Toronto. It is a pleasure also to acknowledge the advice and encouragement that I received from students and colleagues—Dennis Arnon, Allan Borodin, Stan Cabay, David Elliott, John Guttag, Kent Laver, Juliette Sandorfi, Martin Tompa, George Tourlakis, Chris Wilson, and Uriel Wittenberg, to name but a few. Crucial to the completion of this book was my 1978 sabbatical leave spent in the inspiring environs of Oxford University's Programming Research Group. I thank Tony Hoare, Joe Stoy, and my other colleagues at PRG for their warm hospitality. Some of the results of the book depend on current research; in this regard I gratefully acknowledge the support received by the Natural Sciences and Engineering Research Council of Canada. Finally my heartfelt thanks goes to Mrs. Vicky Shum for her skillful typing of the manuscript throughout its various incarnations.

JOHN D. LIPSON

ELEMENTS OF ALGEBRA
AND ALGEBRAIC COMPUTING

ONE

MATHEMATICAL FOUNDATIONS
Sets and Integers

I

SETS, RELATIONS, AND FUNCTIONS

Sets, relations, and functions are fundamental to algebra, indeed to mathematics generally. In this chapter we gather together the notions and notation pertaining to these concepts that we shall need throughout the text.

1. TERMINOLOGY OF SETS

The concept of a set—a collection of things—stands as one of the most significant ideas of twentieth century mathematics. This is quite remarkable, because there is nothing especially quantitative or geometric about the concept of a set that would mark it as being intrinsically mathematical. Indeed, just the opposite might seem to be the case. The large number of synonyms, each with its own nuance, for the word *set—collection, class, family, flock, herd, pack,* to name a few—underscores the everyday nature of the concept. Nevertheless, starting with the pioneering work of the German mathematician Georg Cantor (1845–1918) in the late nineteenth century, the simple notion of a set has profoundly influenced all aspects and levels of mathematical inquiry, from the foundations of mathematics to the "new math" curricula of grade schools.

1.1 Set Membership

The feature that most visibly differentiates the mathematical from the everyday treatment of sets is the use of symbolic (= shorthand) notation. An important

John D. Lipson, Elements of Algebra and Algebraic Computing. ISBN 0-201-04115-4.

example of such notation is the *membership*, or *elementhood*, relation $\in$: We use the notation $x \in A$ to mean x is an element of the set A (also read: x is a member of A; x is contained in A; x belongs to A). We write $x \notin A$ for the negation of $x \in A$.

To specify a set we must designate what its elements are:

(1) The roster approach: The constituent elements are enclosed in braces, $\{\dots\}$. Thus $\{a, b, c\}$ designates the set with elements a, b, and c; $\{1, 2, 3, \dots\}$ designates the set of all positive integers.

(2) The defining property approach: If $p(x)$ is an assertion (a statement that is either true or false) about x, then $\{x \mid p(x)\}$ designates the set of all x such that $p(x)$ is true. Thus $\{x \mid x \text{ is a prime integer } \leq 10\}$ designates the set $\{2, 3, 5, 7\}$; $\{x \mid x = 2y \text{ for some integer } y\}$ designates the set of all even integers $\{0, \pm 2, \pm 4, \dots\}$.

(3) Use of special symbols: We set aside special symbols to designate commonly recurring mathematical sets. The following are the symbols that we shall use for the familiar number sets:

Z: the set $\{0, \pm 1, \pm 2, \dots\}$ of integers;

P: the set $\{1, 2, 3, \dots\}$ of positive integers;

N: the set $\{0, 1, 2, \dots\}$ of nonnegative integers, or "natural numbers";

n: the set $\{1, 2, \dots, n\}$ of positive integers $\leq n$;

Q: the set $\{m/n \mid m, n \in \mathbf{Z}, n \neq 0\}$ of rational numbers;

R: the set of real numbers;

C: the set $\{a + ib \mid a, b \in \mathbf{R}\}$ of complex numbers $(i = \sqrt{-1})$;

Z*   (**Q***, **R***, **C***): the set of *nonzero* integers (rational, real, complex numbers);

Q$^+$ (**R$^+$**): the set of *positive* rational (real) numbers.

1.2 Equality and Inclusion

The most fundamental question one can ask about two sets is, when are they the same? We define two sets A and B to be equal when they have precisely the same elements. Thus,

(1) $A = B$ means $x \in A \Rightarrow x \in B$, and
$$x \in B \Rightarrow x \in A.$$

Remarks on logic and notation.   "$p \Rightarrow q$" denotes the implication "p implies q," or "if p then q." "$p \Leftrightarrow q$" denotes the logical equivalence of p and q: "$p \Rightarrow q$ and conversely $q \Rightarrow p$," or "p if and only if q." (Thus the above definition of equality of sets could be stated "$A = B$ means $x \in A \Leftrightarrow x \in B$.")

An important logical equivalence is the one that exists between any implication "$p \Rightarrow q$" and its contrapositive, "(not q) $\Rightarrow$ (not p)." To assert either one of "$p \Rightarrow q$" or "(not q) $\Rightarrow$ (not p)" is to assert the falsity of "p and (not q)." This logical equivalence (in symbols: $[p \Rightarrow q] \Leftrightarrow [(\text{not } q) \Rightarrow (\text{not } p)]$) is the basis of so-called indirect proofs. To establish the implication $p \Rightarrow q$, the direct method is to assume p and prove q as a consequence; the indirect method is to assume (not *q*) and prove (not *p*) as a consequence. From the viewpoint of logic, both methods are equally valid; it is a matter of taste as to which approach one chooses.

According to our definition of equality, a set is determined solely by its members—i.e., by its *extensional* character—and not by how it is designated—i.e., not by its *intensional* character. (In axiomatic treatments, this definition of equality is adopted as an axiom—the *Axiom of Extension*.) Thus, the sets $\{2, 3, 5\}$, $\{x \mid x$ is a prime and $x \leq 6\}$, and $\{x \mid x$ is a root of $x^3 - 10x^2 + 31x - 30\}$ are all equal in spite of obvious differences in manner (and clarity) of description. As another case in point, the sets $\{a, b\}$, $\{b, a\}$, and $\{a, a, b, b, b, b\}$ are all equal (you don't change a team by listing its players in a different order, nor do you change a team by listing its players more than once).

Another important relation between sets is the *subset*, or *inclusion*, relation. We write $A \subseteq B$ (A is a *subset* of B, or A is *included* in B) to denote that every element of A is also an element of B. In symbols,

$$(2) \qquad\qquad A \subseteq B \quad \text{means} \quad x \in A \Rightarrow x \in B.$$

As examples, $\mathbf{n} \subseteq \mathbf{P}$ and $\mathbf{Z} \subseteq \mathbf{Q}$, but $\{0, 1\} \not\subseteq \mathbf{P}$ (because $0 \notin \mathbf{P}$). We write $A \subset B$ (A is a *proper* subset of B, or A is *strictly* included in B) to denote that $A \subseteq B$ and $A \neq B$.

Comparing (1) and (2) we see that set equality can be readily expressed in terms of inclusion:

$$(3) \qquad\qquad A = B \quad \text{whenever} \quad A \subseteq B \text{ and } B \subseteq A.$$

Sermon. The reader should let (3) serve as a constant reminder of how to go about showing that two sets are equal: first establish inclusion in one direction, and then in the other. In practice, the argument required to establish inclusion in one direction is usually different from the argument required for the other direction, and in such cases the temptation to try and prove equality in one triumphant step invariably meets with frustration.

We now introduce the set $\varnothing$ with no elements; we refer to $\varnothing$ as the *empty*, or *null*, set. This set, in spite of its utter triviality, manages to have a couple of noteworthy properties.

Proposition 1. The empty set is a subset of every set.

Proof. Let A be any set. Either $\varnothing \subseteq A$ or $\varnothing \not\subseteq A$. Suppose, to the contrary, that $\varnothing \not\subseteq A$. Then there must be some element x that is in $\varnothing$ but is not in A. But such an element x cannot exist because $\varnothing$ is empty. Therefore our assumption that $\varnothing \not\subseteq A$ is untenable, and we are forced to conclude that $\varnothing \subseteq A$, as required. $\square$

Remark. The above proof is perhaps a little unsettling on first exposure, but is typical of arguments involving the empty set. A theorem that the empty set has such-and-such a property is established (so it seems) by default: Since the empty set is empty, it cannot provide an element to refute the assertion of the theorem;

hence the theorem must be true (albeit vacuously). The logic is impeccable—so accept it!

Our reference to *the* empty set is justified by the following

Corollary.   The empty set is unique.

Proof.   Let $\varnothing$ and $\varnothing'$ both be empty sets. (Our goal is to prove them equal.) Since $\varnothing$ is empty, we have $\varnothing \subseteq \varnothing'$ by Proposition 1. Since $\varnothing'$ is empty, we have $\varnothing' \subseteq \varnothing$ again by Proposition 1. The two inclusions $\varnothing \subseteq \varnothing'$ and $\varnothing' \subseteq \varnothing$ give $\varnothing = \varnothing'$, as required. $\square$

2. OPERATIONS ON SETS

In this section we catalogue a number of set-theoretic methods for manufacturing new sets from old (already existing) ones.

2.1 Subset Specification; the Power Set

If A is a given set and $p(x)$ is an assertion about x, then $\{x \in A \mid p(x)\}$ denotes the set (subset of A) of all $x \in A$ such that $p(x)$ is true. Thus, $\{x \in \mathbf{Z} \mid x > 0\}$ specifies the set $\mathbf{P}$ of positive integers; $\{x \in \mathbf{P} \mid x \text{ is a prime} < 10\}$ denotes the set $\{2, 3, 5, 7\}$; $\{x \in \mathbf{R} \mid x^2 + 1 = 0\}$ specifies the empty set, $\varnothing$.

Remark.   *Subset* specification, $\{x \in A \mid p(x)\}$, produces a set in terms of an already existing set, whereas *set* specification, $\{x \mid p(x)\}$, manufactures a set by fiat out of thin air at least in its full generality. Is this distinction significant? The answer turns out to be yes. We refer the interested reader to Note 1.

Whereas subset specification yields a *particular* subset of a given set, our next construct yields *all* subsets of a given set. The *power set* $P(A)$ of a set A is defined as the set of all subsets of A:

$$P(A) = \{X \mid X \subseteq A\}.$$

Example 1.
(a) $P(\{1\}) = \{\varnothing, \{1\}\}$. (Remember: $\varnothing$ is a subset of every set, hence $\varnothing \in P(A)$ for any set A.)
(b) $P(\{1, 2\}) = \{\varnothing, \{1\}, \{2\}, \{1, 2\}\}$.
(c) $P(\{1, 2, 3\}) = \{\varnothing, \{1\}, \{2\}, \{3\}, \{1, 2\}, \{1, 3\}, \{2, 3\}, \{1, 2, 3\}\}$. $\square$

2.2 Union, Intersection, and Complement

Let A and B be sets. The *union* $A \cup B$ of A and B is the set of all elements that lie either in A or in B (or in both—"or" is used in its inclusive sense). The *intersection* $A \cap B$ of A and B is the set of all elements that simultaneously lie both in A and B. In symbols,

$$A \cup B = \{x \mid x \in A \text{ or } x \in B\},$$
$$A \cap B = \{x \mid x \in A \text{ and } x \in B\}.$$

Example 1. Let $A = \{a, b, c\}$, $B = \{b, c, d\}$, and $C = \{d, e\}$. Then $A \cup B = \{a, b, c, d\}$, $A \cap B = \{b, c\}$, $A \cap C = \varnothing$. □

Two sets that have an empty intersection (such as A and C in the above) are said to be *disjoint*.

A convenient pictorial tool for visualizing relationships among sets is provided by *Venn diagrams*, in which sets are represented by regions. In Figs. 1(a) and 1(b), A is represented by the left-hand circle and B by the right-hand circle; $A \cup B$ is then given by the hatched area of (a), $A \cap B$ by the hatched area of (b). The Venn diagram of Fig. 1(c) represents three sets A, B, and C that satisfy $C \subseteq A$ and $B \cap C = \varnothing$.

The definitions of union and intersection can be naturally extended from a pair of sets to an arbitrary collection ($=$ set) $\mathcal{C}$ of sets. The union $\cup \mathcal{C}$ and the intersection $\cap \mathcal{C}$ of $\mathcal{C}$ are defined by

$$\cup \mathcal{C} = \{x \mid x \in A \text{ for some set } A \in \mathcal{C}\},$$

$$\cap \mathcal{C} = \{x \mid x \in A \text{ for all sets } A \in \mathcal{C}\}.$$

Thus, $A \cup B = \cup \{A, B\}$ and $A \cap B = \cap \{A, B\}$. If $\mathcal{C} = \{A, B, C\}$, where $A = \{a, b, c\}$, $B = \{b, c, d\}$, and $C = \{d, e\}$ as in Example 1, then $\cup \mathcal{C} = \{a, b, c, d, e\}$ and $\cap \mathcal{C} = \varnothing$.

Frequently the collection $\mathcal{C}$ is an *indexed* set $\{A_i \mid i \in I\}$; corresponding to each *index* $i \in I$ (I is called the *index* set) there is a set A_i. Another notation for $\{A_i \mid i \in I\}$ is $\{A_i\}_{i \in I}$ or simply $\{A_i\}_i$ when I is understood. For the case of an indexed set $\{A_i \mid i \in I\}$, we usually write $\cup_{i \in I} A_i$, $\cap_{i \in I} A_i$ rather than $\cup \{A_i \mid i \in I\}$, $\cap \{A_i \mid i \in I\}$.

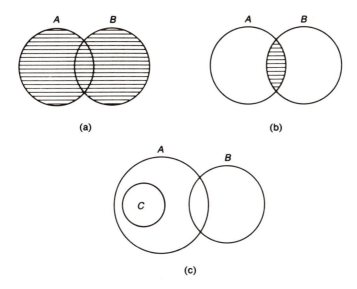

(a)   (b)

(c)

Fig. 1 Venn diagrams.

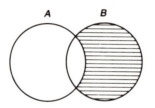

Fig. 2   Venn diagram for $B - A$.

From union and intersection, which correspond to the familiar logical connectives "or" and "and," we pass to the notion of *complement*, which corresponds to the logical connective "not."

The *complement* A' of a set A is defined with respect to some fixed *universe of discourse*. This universe of discourse, call it U, is a set that contains all objects of interest. A' is then defined as the set of all elements in U that do not lie in A:

$$A' = \{x \in U \mid x \notin A\}.$$

The *relative complement* $B - A$ of A in B, also called the *set difference* of B and A, is defined as the set of all elements that lie in B but not in A:

$$B - A = \{x \in B \mid x \notin A\}$$

(Fig. 2). The relative complement $B - A$ becomes the (absolute) complement A' when B is the universe of discourse.

Example 2.   Let $A = \{a, b, c\}$, $B = \{b, c, d\}$, and $C = \{d, e\}$, as in the previous example. If $U = \bigcup \{A, B, C\} = \{a, b, c, d, e\}$, then $A' = \{d, e\}$, $B' = \{a, e\}$, $C' = \{a, b, c\}$. Also, $A - B = \{a\}$, $B - A = \{d\}$, $A - C = \{a, b, c\}$. $\square$

We now record a number of general properties of the operations of union, intersection, and complement. These properties take the form of *laws*, or *identities*, which hold for arbitrary sets. The idea of investigating objects (in this case sets) by looking at the general properties that hold for operations defined on those objects (in this case union, intersection, and complement) is the hallmark of modern algebra.

Theorem 1   (*Algebra of sets*). Let I be an arbitrary set (universe of discourse). Then for any sets $A, B, C \subseteq I$ the following laws are satisfied:

1. *Idempotent:* $A \cap A = A$, $A \cup A = A$;
2. *Commutative:* $A \cap B = B \cap A$, $A \cup B = B \cup A$;
3. *Associative:* $A \cap (B \cap C) = (A \cap B) \cap C$,
                              $A \cup (B \cup C) = (A \cup B) \cup C$;
4. *Absorption:* $A \cap (A \cup B) = A$, $A \cup (A \cap B) = A$;
5. *Distributive:* $A \cap (B \cup C) = (A \cap B) \cup (A \cap C)$ ($\cap$ over $\cup$),
                              $A \cup (B \cap C) = (A \cup B) \cap (A \cup C)$ ($\cup$ over $\cap$);
6. *Complementarity:* $A \cap A' = \varnothing$, $A \cup A' = I$;
7. *Consistency:* $A \cap B = A$ $\Leftrightarrow$ $A \subseteq B$ $\Leftrightarrow$ $A \cup B = B$.

Remark. An analogy between union and intersection on one hand and ordinary (numerical) addition and multiplication on the other is suggested by the associative and commutative laws and the first of the distributive laws. However, the idempotent laws and the second distributive law (whose numerical analog would be $a + bc = (a + b)(a + c)$) are clearly not shared by numbers. Indeed, these latter properties are what make *Boolean algebra*, as the algebra of sets is called, so different from the algebra of numbers. (Note 2.)

The proofs of the various properties of sets asserted in Theorem 1 are straightforward consequences of the definitions of the operations involved. For example, let us prove the first of the absorption properties. First we show that $A \subseteq A \cap (A \cup B)$. Assume $x \in A$. Then $x \in A$ or $x \in B$ trivially, whence $x \in A \cup B$ by the definition of $\cup$. Thus we have $x \in A$ and $x \in A \cup B$, which means $x \in A \cap (A \cup B)$ by the definition of $\cap$, and we have established $A \subseteq A \cap (A \cup B)$. The reverse inclusion holds trivially. The other properties are proved just as easily (Exercise 3).

2.3 The Cartesian Product

We denote by (a, b) the *ordered pair* containing elements ("components") a and b in that order. Equality of ordered pairs is declared according to

$$(a, b) = (c, d) \quad \Leftrightarrow \quad a = c \text{ and } b = d.$$

This is to be contrasted with equality of ordinary (unordered) pairs, where according to the definition of set equality, $\{a, b\} = \{c, d\}$ means that either (1) $a = c$ and $b = d$ or (2) $a = d$ and $b = c$. Thus $\{1, 2\} = \{2, 1\}$, whereas $(1, 2) \neq (2, 1)$. (Interestingly enough, the concept of ordered pair can be defined solely in terms of unordered pairs—Exercise 7.)

The *Cartesian product* $A \times B$ of sets A and B is the set of all ordered pairs with first component from A and second component from B:

$$A \times B = \{(a, b) | a \in A, b \in B\}.$$

Example 1. If $A = B = \mathbf{R}$, then we obtain the set $\mathbf{R} \times \mathbf{R}$ of ordered pairs of real numbers. Each $(a, b) \in \mathbf{R} \times \mathbf{R}$ can be interpreted as the *Cartesian coordinates* of a point in the plane relative to given x-y coordinate axes (Fig. 3) (which accounts for the name Cartesian product). $\square$

Example 2. If $A = \{0, 1\}$ and $B = \{a, b, c\}$, then

$$A \times B = \{(0, a), (0, b), (0, c), (1, a), (1, b), (1, c)\}. \quad \square$$

The Cartesian product $A \times B$ of two sets can be generalized to the Cartesian product $A_1 \times \cdots \times A_n$ of n sets $A_1, \ldots, A_n$: This n-fold Cartesian product, which is also denoted by $\prod_{i=1}^{n} A_i$, consists of all *n-tuples*, or *n-vectors*, $(a_1, \ldots, a_n)$

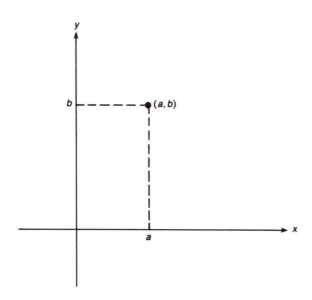

Fig. 3 Cartesian coordinates.

where $a_i \in A_i$. Equality of n-vectors is declared according to

$$(a_1, \ldots, a_n) = (b_1, \ldots, b_n) \quad \Leftrightarrow \quad a_i = b_i \text{ for } i = 1, \ldots, n.$$

When each A_i is the same set A (a frequent case in practice), we denote the Cartesian product $\prod_{i=1}^{n} A_i$ by A^n and refer to it as the (n-fold) Cartesian *power* of A. Thus A^n is the set of all n-vectors $(a_1, \ldots, a_n)$ with components $a_i \in A$.

The possibilities $n = 1$ and $n = 0$ are not excluded. A^1 is the set $\{(a) \mid a \in A\}$ of all 1-vectors. A^0 is the singleton set $\{(\ )\}$; $(\ )$ is called the *empty vector*.

3. RELATIONS

3.1 The Relation Concept

By a *relation* ρ from a set A to a set B we mean a subset of $A \times B$; i.e., $\rho \subseteq A \times B$. When $A = B$ (the most frequent case in practice) we call ρ a relation on A. (Note 1.)

If $(a, b) \in \rho$, then we write $a \rho b$ ("a is ρ-related to b"). If $(a, b) \notin \rho$, then we write $a \not{\rho} b$ ("a is not ρ-related to b"). This "$a \rho b$" notation for relations reflects standard practice, where, for example, we write $1 \leq 2$, not $(1, 2) \in \leq$.

Examples of relations abound: $\leq$ on the set **P**, **Q**, or **R**; $\subseteq$ on $P(I)$, the power set of some set I; geometrical relations such as $\{(x, y) \mid x^2 + y^2 = 1\} \subseteq$ **R** $\times$ **R**, the set of all points in the plane lying on a circle of radius 1 with center

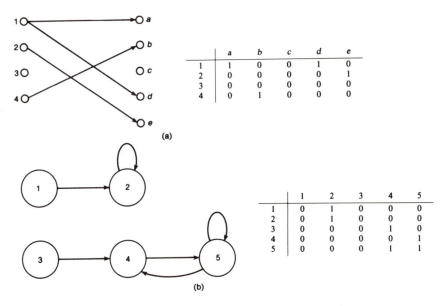

Fig. 4   Representation of relations by graphs and matrices.

at the origin; genealogical relations such as brotherhood—$x\,b\,y$ whenever x is a brother of y, defined on some set of persons.

Some kinds of relations, especially ones not readily specified by a simple rule, can be conveniently represented a graph or a matrix. In Fig. 4(a) at the left, we have represented a relation ρ from $A = \{1, 2, 3, 4\}$ to $B = \{a, b, c, d, e\}$ by a graph in which A and B are represented by two disjoint sets of nodes (such a graph is called *bipartite*), and there is an edge from node a to node b whenever $a\,\rho\,b$. Thus $\rho = \{(1, a), (1, d), (2, e), (4, b)\}$. To the right we have displayed the corresponding *relation matrix* $(r_{a,b}^{\rho})$ of ρ, the zero-one matrix defined in general for $\rho \subseteq A \times B$ as follows: for $a \in A$, $b \in B$,

$$r_{a,b}^{\rho} = 1 \quad \text{if } a\,\rho\,b,$$

$$= 0 \quad \text{otherwise.}$$

For the case of a relation ρ from a set A to itself, a different graphical representation is convenient. There is a single node set corresponding to A, with an edge from node a to node b whenever $a\,\rho\,b$. The graph of the left of Fig. 4(b) thus represents the relation

$$\rho = \{(1, 2), (2, 2), (3, 4), (4, 5), (5, 4), (5, 5)\}$$

on the set $A = \{1, 2, 3, 4, 5\}$, as does the matrix to the right.

Operations on relations. Since relations are sets, all of our set-theoretic operations are applicable. Thus we have the union, intersection, and complement of relations:

$$(a,b) \in \rho \cup \sigma \quad \Leftrightarrow \quad a \rho b \quad \text{or} \quad a \sigma b;$$
$$(a,b) \in \rho \cap \sigma \quad \Leftrightarrow \quad a \rho b \quad \text{and} \quad a \sigma b;$$
$$(a,b) \in \rho' \qquad \Leftrightarrow \quad a \not\rho b.$$

In addition to the above purely set-theoretic operations on relations, we have two new operations on relations, which use the fact that relations are not any old sets, but sets of ordered pairs.

Let ρ be a relation from A to B. The *converse* of ρ, denoted by $\check{\rho}$, is the relation from B to A given by

$$b \check{\rho} a \quad \Leftrightarrow \quad a \rho b.$$

Thus, the converse of $\leq$ on $\mathbf{P}$ is $\geq$. If $a \rho b$ on $\mathbf{P}$ means that a is a divisor of b, then $a \check{\rho} b$ means that a is a multiple of b.

Let ρ be a relation from A to B and σ a relation from B to C. The *composition* of ρ and σ, denoted by $\rho \circ \sigma$ or simply by $\rho \sigma$, is the relation from A to C given by

$$a \rho \sigma c \quad \Leftrightarrow \quad a \rho b \text{ and } b \sigma c \text{ for some } b \in B.$$

The set B in the composition of $\rho \subseteq A \times B$ and $\sigma \subseteq B \times C$ serves as an intermediary for establishing a correspondence between the sets A and C, as illustrated in Fig. 5.

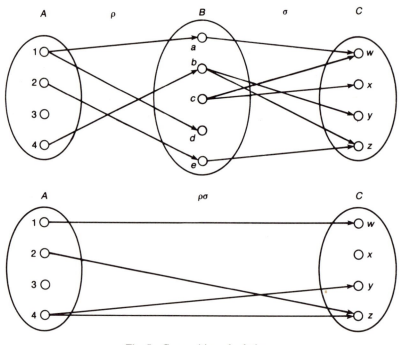

Fig. 5 Composition of relations.

Converse and composition are both very intuitive notions for genealogical relations—Exercise 1.

The operations of converse and composition are related by

Proposition 1. Let ρ be a relation from A to B, and σ a relation from B to C. Then $\widecheck{\rho\sigma} = \widecheck{\sigma}\widecheck{\rho}$.

Proof. (Reminder: Relations are sets, and we know how to go about showing that two sets are equal—establish a two-way inclusion.) Noting first that $\rho\sigma$ and $\widecheck{\sigma}\widecheck{\rho}$ are both relations from C to A, we have for any $c \in C$ and $a \in A$:

$$c \, \widecheck{\rho\sigma} \, a \Rightarrow a \, \rho\sigma \, c \qquad \text{defn. of converse}$$

$$\Rightarrow a \, \rho \, b \text{ and } b \, \sigma \, c$$
$$\text{for some } b \in B \qquad \text{defn. of composition}$$

$$\Rightarrow b \, \widecheck{\rho} \, a \text{ and } c \, \widecheck{\sigma} \, b \qquad \text{defn. of converse}$$

$$\Rightarrow c \, \widecheck{\sigma} \, b \text{ and } b \, \widecheck{\rho} \, a \qquad \text{logic}$$

$$\Rightarrow c \, \widecheck{\sigma}\widecheck{\rho} \, a \qquad \text{defn. of composition,}$$

proving that $\rho\sigma \subseteq \widecheck{\sigma}\widecheck{\rho}$. The inclusion in the other direction is just as easily established. $\square$

Remark. The above result may remind the reader of the matrix relationship $(AB)^{-1} = B^{-1}A^{-1}$; Proposition 1 is actually a far-reaching generalization of that matrix result.

$$* \quad * \quad *$$

The concept of a relation becomes more interesting and significant when we focus on special classes of relations that satisfy specified properties. In the next two sections we consider two important such classes of relations, *equivalence relations* and *partial orders*.

3.2 Equivalence Relations

An *equivalence relation* E on a set A is a relation on A that for all $a, b, c \in A$ is

1. *Reflexive:* $a \, E \, a$;
2. *Symmetric:* $a \, E \, b \Rightarrow b \, E \, a$;
3. *Transitive:* $a \, E \, b$ and $b \, E \, c \Rightarrow a \, E \, c$.

If $a \in A$, we define $[a]$, the *equivalence class* containing a, as the set of all elements of A that are equivalent to a:

$$[a] = \{b \in A \mid a \, E \, b\};$$

the notation is embellished to $[a]_E$ (E-equivalence class) should it be necessary to indicate the equivalence relation under consideration. We denote the set of all equivalence classes of A by A/E:

$$A/E = \{[a] \mid a \in A\};$$

A/E is variously referred to as the *quotient set* of A by the equivalence relation E, A *modulo* E, or A mod E for short.

We need to know when two equivalence classes $[a]$ and $[b]$ are the same.

Lemma 1. $[a] = [b] \Leftrightarrow a E b$. (Thus, we pass from equivalence to equality when we pass from elements of A to equivalence classes of A.)

Proof. ($\Rightarrow$) Assume $[a] = [b]$. Since E is reflexive, $b E b$, whence $b \in [b]$ by the definition of $[b]$. But then $b \in [a]$, giving $a E b$ by the definition of $[a]$.

($\Leftarrow$) Assume $a E b$. Then,

$$c \in [a] \Rightarrow a E c \qquad\qquad \text{defn. of } [a]$$

$$\Rightarrow b E a \text{ and } a E c \qquad a E b, \text{ symmetry}$$

$$\Rightarrow b E c \qquad\qquad\qquad \text{transitivity}$$

$$\Rightarrow c \in [b] \qquad\qquad\quad \text{defn. of } [b],$$

so that $[a] \subseteq [b]$. Similarly $[b] \subseteq [a]$, giving $[a] = [b]$. $\square$

Now for some examples of equivalence relations.

Example 1. There are two trivial equivalence relations on any set A: (1) equality: $a e b$ when $a = b$ and (2) universality: $a u b$ for all $a, b \in A$. These trivial equivalence relations give rise, in turn, to trivial quotient sets. For example, if $A = \{a, b, c\}$ we have $A/e = \{\{a\}, \{b\}, \{c\}\}$ (each equivalence class contains but a single element) and $A/u = \{\{a, b, c\}\}$ (there is but one equivalence class, which contains all the elements of A). $\square$

Example 2. Let L be the set of lines in the plane, and for $l, m \in L$, define $l \parallel m$ to mean l is parallel to m. The relation $\parallel$ on L is clearly an equivalence relation. The equivalence class containing any line l is the set of all lines parallel to l. $\square$

Example 3 (*Equivalence* mod m). We shall make much use of the *equivalence* mod m ($\equiv_m$) relation on $\mathbf{Z}$: for m a positive integer,

$$a \equiv_m b \quad \Leftrightarrow \quad a - b = km \quad \text{for some } k \in \mathbf{Z}.$$

We now check that $\equiv_m$ is an equivalence relation as claimed. *Reflexivity:* $a - a = 0m$, so that $a \equiv_m a$. *Symmetry:* If $a \equiv_m b$, then $a - b = km$ for some $k \in \mathbf{Z}$, whence $b - a = (-k)m$, and so $b \equiv_m a$. *Transitivity:* If $a \equiv_m b$ and $b \equiv_m c$, then $a - b = km$ and $b - c = lm$ for some $k, l \in \mathbf{Z}$, whence $a - c = (k + l)m$, and so $a \equiv_m c$.

The mod m equivalence class containing $a \in \mathbf{Z}$ is $[a] = \{a + km \mid k \in \mathbf{Z}\}$. For example, if $m = 3$ we have

$$[0] = \{ \ldots, -6, -3, 0, 3, 6, \ldots \},$$

$$[1] = \{ \ldots, -5, -2, 1, 4, 7, \ldots \},$$

$$[2] = \{ \ldots, -4, -1, 2, 5, 8, \ldots \},$$

and we see that $\mathbf{Z}/\equiv_3 = \{[0], [1], [2]\}$. (In Section II.3.1 we shall show in general that $\mathbf{Z}/\equiv_m = \{[0], [1], \ldots, [m-1]\}$ and that these m equivalence classes are all distinct.) $\square$

Example 4 (*Rational numbers as equivalence classes*). Let $X = \mathbf{Z} \times \mathbf{Z}^*$ ($\mathbf{Z}^* = \mathbf{Z} - \{0\}$). Thus X is the set of all ordered pairs (a, b) of integers with $b \neq 0$. Define $\sim$ on X by

$$(a, b) \sim (c, d) \quad \Leftrightarrow \quad ad = bc.$$

Then $\sim$ is easily verified to be an equivalence relation of X.

If we associate the ordered pair (a, b) with the rational number a/b (which we can do since $b \neq 0$), then we see that the equivalence class $[(a, b)]$ contains all ordered pairs whose corresponding rational numbers have the same value, namely a/b; e.g., $[(1, 2)] = \{(1, 2), (2, 4), (3, 6), \ldots\}$. Since equality of equivalence classes corresponds to equality of rational numbers, we can identify the equivalence class $[(a, b)]$ with the rational number a/b; e.g., $[(1, 2)] = [(2, 4)]$ just as $1/2 = 2/4$. $\square$

We now investigate the structure of the quotient set A/E for the case of an arbitrary equivalence relation.

Definition. A *partition* π of a set A is a collection $\{\pi_\alpha\}_\alpha$ of nonempty subsets of A, called *blocks*, that satisfies:

1. $\bigcup_\alpha \pi_\alpha = A$ (the blocks of π exhaust A);
2. $\pi_\alpha \cap \pi_\beta \neq \emptyset \Rightarrow \pi_\alpha = \pi_\beta$ (the blocks of π are disjoint).

For example $\{\{1, 2, 5\}, \{3, 4\}, \{6\}\}$ is a partition of $\mathbf{6}$; so are $\{\{1\}, \{2\}, \{3\}, \{4\}, \{5\}, \{6\}\}$ and $\{\{1, 2, 3, 4, 5, 6\}\}$, the latter two being *trivial* partitions of $\mathbf{6}$ (cf. Example 1).

A convenient notation for partitions is the following: denote each block $\{\ldots\}$ by $\overline{\ldots}$ and use spaces rather than commas as separators. Using this notation, the first of the above partitions would be written $\{\overline{1\,2\,5}\ \overline{3\,4}\ \overline{6}\}$.

The main result about equivalence relations, which is illustrated by Examples 1–4, is given by

Theorem 1. Let E be an equivalence relation on a set A. Then the quotient set $A/E = \{[a] \mid a \in A\}$ is a partition of A ("the partition induced by E"; also denoted by π_E).

Proof. Since E is reflexive, $a E a$, hence $a \in [a]$, for all $a \in A$. It then follows that each $[a]$ is nonempty and that $\bigcup_{a \in A} [a] = A$.

As for disjointness, assume that $[a] \cap [b] \neq \emptyset$, so that $c \in [a]$ and $c \in [b]$ for some $c \in A$. We then have $a E c$ and $b E c$ by the definitions of $[a]$ and $[b]$. But $a E c$ and $b E c$ imply $a E c$ and $c E b$ by symmetry, which imply $a E b$ by transitivity. Hence $[a] = [b]$ by Lemma 1. $\square$

A converse to Theorem 1 is provided by

Theorem 2. If $\pi = \{\pi_\alpha\}_\alpha$ is a partition of A, then π is expressible as A/E for an equivalence relation E on A, namely for the equivalence relation E_π ("the equivalence relation induced by π") on A defined by

$$a E_\pi b \quad \Leftrightarrow \quad a \text{ and } b \text{ are in the same block of } \pi.$$

Proof. We leave the easy proof that E_π is an equivalence relation to the reader (Exercise 6). We now show that $A/E_\pi = \pi$.

For any $[a] \in A/E_\pi$, let π_α be the unique block of π that contains a. The claim is that $[a] = \pi_\alpha$. Proof of claim: If $b \in [a]$, then $a E_\pi b$ by definition of $[a]$, whence b, like a, is in π_α by definition of E_π. Thus $[a] \subseteq \pi_\alpha$. As for the reverse inclusion, if $b \in \pi_\alpha$, then $a E_\pi b$ by definition of E_π, whence $b \in [a]$ by definition of $[a]$. Thus $\pi_\alpha \subseteq [a]$, giving $[a] = \pi_\alpha$ as claimed, and we have shown that $A/E_\pi \subseteq \pi$.

As for the reverse inclusion, $\pi \subseteq A/E_\pi$, let $\pi_\alpha \in \pi$. Then $a \in \pi_\alpha$ for some $a \in A$, since $\pi_\alpha \neq \emptyset$. The claim now is that $\pi_\alpha = [a]$, and the proof is exactly as above. Thus $A/E_\pi = \pi$. $\square$

Summarizing Theorems 1 and 2, equivalence relations and partitions go hand in hand: Any quotient A/E of a set A by an equivalence relation E is a partition of A (Theorem 1). Conversely, any partition π of A is a quotient of A by an equivalence relation, namely by the induced equivalence relation E_π (Theorem 2).

The importance of equivalence relations (or partitions) in practice is tied up with the idea of *abstraction*, or *suppression of detail*. When we deal with equivalence classes of A/E, rather than elements of A, then we effectively identify, or at least agree to ignore differences among, equivalent elements of A. Thus A/E is less complicated than A itself, an *abstract* version of A.

3.3   Partial Orders

A *partial order* $\leq$ ("less than or equal to") on a set P is a relation on P that for all $x, y, z \in P$ is

1. *Reflexive:* $x \leq x$;
2. *Antisymmetric:* $x \leq y$ and $y \leq x \Rightarrow x = y$;
3. *Transitive:* $x \leq y$ and $y \leq z \Rightarrow x \leq z$.

We write $x < y$ to mean $x \leq y$ and $x \neq y$ ("x is [strictly] less than y"), $x \geq y$ to mean $y \leq x$ ("x is greater than or equal to y"), and $x > y$ to mean $y < x$ ("x is [strictly] greater than y"). Thus $\geq$ and $>$ are the converse relations to $\leq$ and $<$.

A set P together with a partial order $\leq$ on P is called a *partially ordered set*, or *poset* for short, and is denoted by $[P; \leq]$. (More generally, a set A together with a collection of relations $\rho, \sigma, \ldots$ defined on A, $[A; \rho, \sigma, \ldots]$, is called a *relational system*.)

Some examples of posets:

Example 1. $[\mathbf{P}; \leq]$, the positive integers under ordinary $\leq$, is a poset. This poset is not only *partially* ordered but *totally* ordered according to the following

> **Definition.** A poset $[P; \leq]$ is said to be *totally ordered* if every pair of elements $x, y \in P$ are comparable: either $x \leq y$ or $y \leq x$. A totally ordered set is also called a *chain*.

Example 2. $[P(I); \subseteq]$, the power set of any set I under inclusion, is a poset. The reflexive and transitive laws are trivial consequences of the definition of inclusion (check), while the antisymmetric property is nothing more than the definition of set equality.

    This poset, unlike $[\mathbf{P}; \leq]$, is not totally ordered; incomparable subsets abound. For example, in $P(3)$ we have $\{1, 2\} \not\subseteq \{2, 3\}$ and $\{2, 3\} \not\subseteq \{1, 2\}$.

Example 3. Another familiar poset involving the positive integers is $[\mathbf{P}; |\,]$, where $a | b$ ("a divides b") means $b = ka$ for some $k \in \mathbf{P}$.

    Let us check this one out. *Reflexivity:* $a | a$ since $a = 1a$. *Antisymmetry:* If $a | b$ and $b | a$, then $b = ka$ and $a = mb$ for $k, m \in \mathbf{P}$, giving $b = kmb$. Thus $km = 1$, so that $k = m = 1$ since $k, m \in \mathbf{P}$. This gives $a = b$, as required. *Transitivity:* If $a | b$ and $b | c$, then $b = ka$ and $c = mb$ for $k, m \in \mathbf{P}$, giving $c = m(ka) = (mk)a$. Thus $a | c$, proving transitivity.

    We note that $|$, unlike $\leq$, only *partially* orders $\mathbf{P}$. For example $2 \nmid 3$ and $3 \nmid 2$. $\square$

Example 4. Let $\Pi(X)$ denote the set of all partitions of a set X. For π, $\sigma \in \Pi(X)$, we write $\pi \leq \sigma$ ("π is a refinement of σ," "π refines σ") to mean that $[x]_\pi \subseteq [x]_\sigma$ for all $x \in X$. For example, in $\Pi(7)$, $\{\overline{1\,2\,3}\ \overline{4\,5\,6\,7}\} \leq \{\overline{1\,2\,4\,5}\ \overline{3\,6\,7}\}$. We leave it as an easy exercise (Exercise 2) to show that $[\Pi(X); \leq]$ is a poset. $\square$

Poset diagrams. The familiar idea of an immediate superior in a hierarchy yields a convenient graphical representation of a poset. In a poset $[P; \leq]$, we say that x *covers* y if $x > y$ but $x > z > y$ for no $z \in P$. The *diagram* of a poset $[P; \leq]$ is an oriented graph with a node for each element of P and an edge from y *up to* x whenever x covers y.

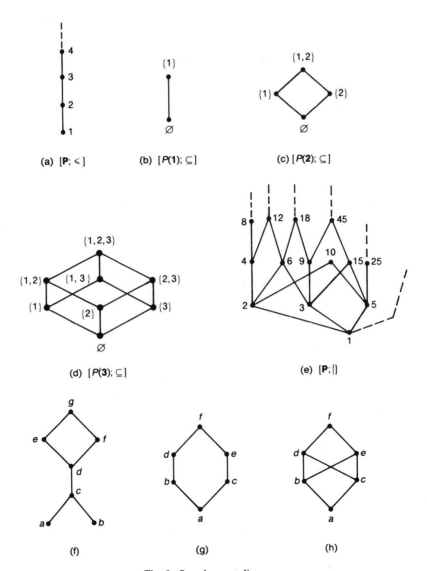

Fig. 6 Sample poset diagrams.

A number of poset diagrams are displayed in Fig. 6.

Special elements in posets. Let $[P; \leq]$ be a poset, and let X be a subset of P. Then

1. $y \in X$ is a *minimal* (*maximal*) element of X if $x < y$ ($x > y$) for no $x \in X$;
2. $y \in X$ is the *least* (*greatest*) element of X if $y \leq x$ ($y \geq x$) for all $x \in X$;
3. $y \in P$ is a *lower* (*upper*) *bound* of X if $y \leq x$ ($y \geq x$) for all $x \in X$;

4. $y \in P$ is the *greatest lower bound*, or g.l.b., (*least upper bound*, or l.u.b.) of X
   if y is the greatest (least) element of the set of all lower (upper) bounds of X.

In a *totally* ordered set, such as $[\mathbf{P}; \leq]$, the concepts of minimal and least
coincide, as do those of maximal and greatest. In a *partially* ordered set, on the
other hand, these concepts are quite distinct. For example, in the poset of Fig.
6(f), the subset $\{d, e, f, g\}$ has a minimal and least element (namely d), whereas
the subset $\{e, f, g\}$ has minimal elements (namely e and f) but no least element.
We further observe that $\{e, f, g\}$ has a g.l.b. (namely d) in contrast to $\{a, b, c\}$
which has no lower bounds hence no g.l.b. And in Fig. 6(h), $\{f, d, e\}$ has lower
bounds a, b, c but still has no g.l.b.

Duality. Corresponding to any poset $[P; \leq]$ is its *dual*, $[P; \geq]$, defined by

$$x \leq y \text{ in } [P; \geq] \quad \Leftrightarrow \quad x \geq y \text{ in } [P; \leq].$$

It is easily verified that $[P; \geq]$ is a poset (Exercise 5).

The diagram of $[P; \geq]$ is drawn with respect to the $\leq$ relation ($x \leq y$
means x is below y) so that the diagram of $[P; \geq]$ is simply that of $[P; \leq]$
turned upside-down. For example, if $[P; \leq]$ is given by the diagram of Fig. 6(f),
then its dual, $[P; \geq]$ is given by

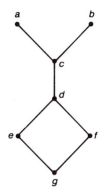

Concepts about posets, like posets themselves, come in dual pairs, as we
have already encountered. Minimal and maximal, least and greatest, lower
bound and upper bound, g.l.b. and l.u.b. are all pairs of dual concepts; each
member of these pairs is obtained from the other by interchanging $\leq$ and $\geq$.

If S is a true statement about a particular poset $[P; \leq]$, then the dual $\check{S}$ of S
—the statement obtained from S by interchanging $\leq$ and $\geq$ in S (hence also
interchanging all order-theoretic concepts in S by their duals)—must be a true
statement about the dual poset $[P; \geq]$. (Why? Because $[P; \leq]$ is obtained from
$[P; \geq]$ in just the same way—by interchanging $\leq$ and $\geq$.) For example, a true
statement about the poset $[P; \leq]$ of Fig. 6(f) is "d is the g.l.b. of e and f";
another is "a and b are minimal elements of $[P; \leq]$." The corresponding dual
statements, "d is the l.u.b. of e and f" and "a and b are maximal elements of
$[P; \geq]$," then apply to the dual poset $[P; \geq]$, drawn above.

From true statements about particular posets, we pass to true statements about arbitrary posets, i.e., *theorems* about posets, for which we have the following *metatheorem* (= theorem about theorems):

Principle of Duality for posets. Let T be a theorem about posets, i.e., a statement that is true for any poset $[P; \leq]$. Then the dual $\check{T}$ of T is also a theorem about posets.

Proof. Suppose, to the contrary, that $\check{T}$ fails to hold in some poset $[P; \leq]$. Then the dual of $\check{T}$, namely $\check{\check{T}} = T$, fails to hold in the dual of $[P; \leq]$, namely the poset $[P; \geq]$, contradicting the assumption that T holds in any poset. Thus $\check{T}$, like T, must be theorem about posets. □

Now for a sample application of the Principle of Duality.

Theorem 1. Let $[P; \leq]$ be a poset. Then a least element of any $X \subseteq P$, whenever it exists, is unique. (Thus we are justified in referring to *the* least element of X and *the* l.u.b. of X.)

Proof. Let l and l' both be least elements of X. Then $l \leq l'$ because l is a least element of X. Also $l' \leq l$ because l' is a least element of X. Hence $l = l'$ by antisymmetry. □

By the Principle of Duality, we conclude the

Dual Theorem. Let $[P; \leq]$ be a poset. Then a greatest element of any $X \subseteq P$, whenever it exists, is unique.

Lattices. Let x and y be elements in a poset $[P; \leq]$. We denote their g.l.b. by $x \wedge y$ ("x *meet* y") and their l.u.b. by $x \vee y$ ("x *join* y"). Thus, if $g = x \wedge y$, then g satisfies

1. $g \leq x$ and $g \leq y$
    (g is a lower bound of $\{x,y\}$),
2. $h \leq x$ and $h \leq y \Rightarrow h \leq g$
    (g is the *greatest* lower bound of $\{x,y\}$);

and dually for $x \vee y$.

Definition. A *lattice* is a poset in which every pair of elements has a meet and join.

Example 5. Any chain (e.g., $[\mathbf{P}; \leq]$) is a lattice in which $a \wedge b = $ minimum of $\{a,b\}$ and $a \vee b = $ maximum of $\{a,b\}$. □

Example 6. $[P(I); \subseteq]$ is a lattice in which $A \wedge B = A \cap B$ and $A \vee B = A \cup B$. (Proof: $A \cap B \subseteq A$ and $A \cap B \subseteq B$. Thus $A \cap B$ is a lower bound of $\{A, B\}$. If $D \subseteq A$ and $D \subseteq B$, then $D \subseteq A \cap B$, trivially. Thus $A \cap B$ is the *greatest* lower bound of $\{A, B\}$. Similarly, $A \cup B$ is the least upper bound of $\{A, B\}$.) □

Example 7. If we assume the fact (which we shall prove later) that every pair of integers has a greatest common divisor (g.c.d.) and least common multiple (l.c.m.), then it follows (check) from the definitions of meet and join that $[\mathbf{P}; |]$ (Fig. 6(e)) is a lattice in which $a \wedge b = $ g.c.d. of a and b, and $a \vee b = $ l.c.m. of a and b. For example, $6 \wedge 9 = 3$, $6 \vee 9 = 18$. $\square$

> **Theorem 2** (*Algebra of lattices*). Let $[L; \leq]$ be a lattice. Then the following laws are satisfied by $\wedge$ and $\vee$:

1. *Idempotent:* $x \wedge x = x$, $x \vee x = x$;
2. *Commutative:* $x \wedge y = y \wedge x$, $x \vee y = y \vee x$;
3. *Associative:* $x \wedge (y \wedge z) = (x \wedge y) \wedge z$,
                   $x \vee (y \vee z) = (x \vee y) \vee z$;
4. *Absorption:* $x \wedge (x \vee y) = x$, $x \vee (x \wedge y) = x$;
5. *Consistency:* $x \leq y \iff x \wedge y = x \iff x \vee y = y$.

The proofs of the above laws follow easily from the definitions of meet and join. As an example, let us show the second of the absorption laws. First, $x \geq x \wedge y$, since $x \wedge y$ is a lower bound of $\{x, y\}$; also $x \geq x$, trivially. Therefore x is an upper bound of $\{x, x \wedge y\}$. But any other upper bound of $\{x, x \wedge y\}$ is $\geq x$, trivially. Thus x is the *least* upper bound of $\{x, x \wedge y\}$, i.e., $x = x \vee (x \wedge y)$, as was to be proved.

The above theorem shows that the algebra of lattices has much in common with the algebra of sets (cf. Theorem 1 of Section 2.2). But see Exercise 8 for an interesting point of distinction.

4. FUNCTIONS

4.1 The Function Concept

Next to the concept of a set, the concept of a function is most fundamental to mathematics. A *function* (also called a *map*) f from a set A (the *domain* of f) to a set B (the *codomain* of f) is a correspondence that assigns to each element a of A a unique element $b = f(a)$ of B (the *image* of a under f, or the *value* of f at a). We denote this situation by $f: A \to B$ or $f: a \mapsto b$ (if A and B are understood). Thus, $x \mapsto \sin x$ denotes the real sine function; $\mathbf{P} \to \mathbf{P}: m \mapsto m^2$ denotes the squaring function defined on $\mathbf{P}$.

As a set-theoretic (static) object, a function $f: A \to B$ is nothing more than a special kind of relation from A to B (i.e., $f \subseteq A \times B$), one that satisfies

1. For each $a \in A$, afb for some $b \in B$
        (f is *totally* defined);
2. afb and $afb' \Rightarrow b = b'$
        (f is *well-defined*, or *single-valued*).

If we relax condition 1, then we obtain what is called a *partial* function. For example, the real square root function, $x \mapsto \sqrt{x}$, is a partial function, undefined for $x < 0$ (to make it well-defined for $x \geq 0$, we conventionally choose the square root to be nonnegative); the complex function $z \mapsto 1/(z^2 + 1)$ is a partial function, undefined for $z = \pm i$. (Of course, these functions become total under the appropriate restrictions of their domain.) When we use the term "function" unqualified, we shall usually mean *total* function.

Two functions $f, g\colon A \to B$ with common domain and codomain are *equal* when $f(a) = g(a)$ for all $a \in A$.

We denote the set of all functions from A to B by B^A; i.e., $B^A = \{f \mid f\colon A \to B\}$.

We say that a function $f\colon A \to B$ is

1. *injective*, or *one-to-one*, if $a \neq b \Rightarrow f(a) \neq f(b)$ (or, equivalently, if $f(a) = f(b) \Rightarrow a = b$);
2. *surjective*, or *onto*, if for every $b \in B$, $b = f(a)$ for some $a \in A$;
3. *bijective* if f is both injective and surjective.

In Fig. 7, f is injective, g is surjective, and h is bijective.

Images and inverse images. Often it is useful to examine the action of a function on subsets (rather than on elements) of the domain or codomain. Let f be a function $A \to B$. For any subset S of A, we define $f(S)$, the *image* of S under f, by

$$f(S) = \{t \in B \mid t = f(s) \quad \text{for some } s \in S\}.$$

For any subset T of B, we define $f^{-1}(T)$, the *inverse image* of T under f, by

$$f^{-1}(T) = \{a \in A \mid f(a) \in T\}.$$

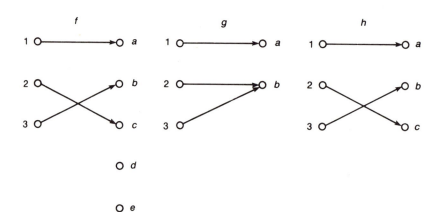

Fig. 7   Injective, surjective, and bijective functions.

The subset $f(A)$ of B is called the *image* of f, $\operatorname{Im} f$ (also called the *range* of f).

Example 1. If f is the integer squaring function and g the real squaring function, then $\operatorname{Im} f = \{0, 1, 4, 9, 16, \dots\}$, $\operatorname{Im} g = \{x \in \mathbf{R} \mid x \geq 0\}$, $f^{-1}(\{4\}) = g^{-1}(\{4\}) = \{2, -2\}$, $f^{-1}(\{5\}) = \varnothing$, $g^{-1}(\{5\}) = \{\sqrt{5}, -\sqrt{5}\}$. $\square$

4.2   Composition and Invertibility

Composition.   The *composition* of functions $f: A \to B$ and $g: B \to C$, denoted by $g \circ f$ or gf, is the function $A \to C$ defined by $gf(a) = g(f(a))$ (apply f first, then g).

(The careful reader has probably already noticed that *functional* composition $g \circ f$ corresponds to the *relational* composition $f \circ g$. We shall remove a possibility of confusion by agreeing that the right-to-left rule always applies when dealing with a composition $g \circ f$ of functions: apply f first, then g.)

Example 1.   For the functions $f(x) = x + 1$ and $g(x) = x^2$ (defined, say, on $\mathbf{N}$), gf is given by $g(f(x)) = g(x + 1) = (x + 1)^2$, and fg by $f(g(x)) = f(x^2) = x^2 + 1$. $\square$

The above example shows that composition is not commutative (i.e., $fg \neq gf$). But it is associative: $[f(gh)](x)$ and $[(fg)h](x)$ are easily checked to be both equal to $f(g(h(x)))$.

A useful tool for visualizing the composition of functions is provided by a *mapping diagram*. In such a diagram the composition gf of functions $f: A \to B$ and $g: B \to C$ is set down thus:

$$A \xrightarrow{f} B \xrightarrow{g} C,$$

where the rule of mapping diagrams is as follows: if two compositions of any number of functions have the same starting and finishing points, then the two compositions are equal. Thus the mapping diagram of Fig. 8(a) conveys that $h = gf$, that of Fig. 8(b) conveys that $gf = \psi\phi$.

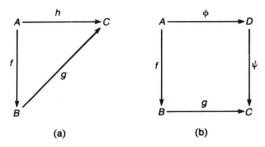

(a)   (b)

Fig. 8   Mapping diagrams.

Invertibility. For any set A, the *identity function* on A is the function $1_A\colon A \to A$ defined by $1_A(a) = a$. For any function $f\colon A \to B$, the identity functions 1_A and 1_B trivially satisfy

$$f1_A = f \quad \text{and} \quad 1_B f = f \quad (\textit{identity laws}).$$

These identity laws constitute the analog of the identity law $x1 = 1x = x$ for numbers. The situation is more complicated for functions because composition is noncommutative; we require the notions of both a *right* identity element and a *left* identity element (1_A and 1_B above).

As with the identity element, the notion of *invertibility* is richer for functions than for numbers, again stemming from the noncommutativity of composition. A function $f\colon A \to B$ is called

1. *left invertible* if there exists a function $g\colon B \to A$ such that $gf = 1_A$;
2. *right invertible* if there exists a function $h\colon B \to A$ such that $fh = 1_B$; and
3. (*two-sided*) *invertible* if both left and right invertible.

In Fig. 9(a), f is left invertible with left inverse g; in Fig. 9(b), f is right invertible with right inverse h.

A function may in general have any number (including none) of left inverses and any number of right inverses (Exercise 1). But interestingly enough,

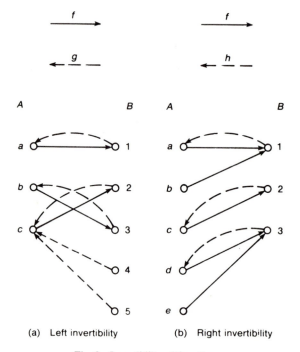

    (a) Left invertibility   (b) Right invertibility

Fig. 9 Invertibility of functions.

if a function has at least one left inverse and at least one right inverse, then every left inverse must equal every right inverse; i.e., in this case the function has a unique *two-sided* inverse. This result is the content of

Theorem 1. If $f: A \to B$ is a function with a left inverse $g: B \to A$ and a right inverse $h: B \to A$ (i.e., $gf = 1_A$ and $fh = 1_B$), then $g = h$.

Proof. The proof depends on purely algebraic properties of functional composition. We have

$$
\begin{aligned}
g &= g1_B & &\text{identity law} \\
  &= g(fh) & &\text{given} \\
  &= (gf)h & &\text{associativity of composition} \\
  &= 1_A h & &\text{given} \\
  &= h & &\text{identity law.} \quad \square
\end{aligned}
$$

The unique two-sided inverse of f is denoted by f^{-1}.

We have met two ways of classifying functions: (1) injective, surjective, or bijective, and (2) left invertible, right invertible, or two-sided invertible. Our next theorem shows that these two ways of classifying functions are equivalent.

Theorem 2. Let $f: A \to B$ be any function.

  1. f is left invertible $\Leftrightarrow$ f is injective;
  2. f is right invertible $\Leftrightarrow$ f is surjective.

Proof of 1. Left invertible $\Rightarrow$ injective: Assume that f has a left inverse g. (We now show that f is injective, i.e., that $f(a) = f(b) \Rightarrow a = b$.) Then

$$
\begin{aligned}
f(a) &= f(b) \\
     &\Rightarrow gf(a) = gf(b) & &\text{applying } g \text{ to both sides} \\
     &\Rightarrow 1_A(a) = 1_A(b) & &g \text{ is a left inverse of } f \\
     &\Rightarrow a = b & &\text{defn. of } 1_A,
\end{aligned}
$$

and so f is injective.

Injective $\Rightarrow$ left invertible: Let $f: A \to B$ be injective. Our goal now is to specify a left inverse $g: B \to A$ for f. For motivation, let us take a look at the injective function f and its left inverse g of Fig. 9(a). We observe that g left inverts f by sending elements in the image of f back where they came from—because of the injectivity of f, an element in the range of f can only come from one place. (Elements outside the range of f play no essential role in its left invertibility; g simply assigns such elements to an arbitrary (fixed) element of the domain of f—c in Fig. 9(a).)

With this illustration at hand, it is an easy matter to show how to construct a left inverse for any injective function. With $f: A \to B$ injective, we define $g: B \to A$ by

$$g(b) = a \text{ if } b = f(a) \text{ for some } a \in A,$$

$$= \alpha \text{ otherwise (i.e., if } b \notin \operatorname{Im} f), \text{ where}$$
$$\alpha \text{ is some fixed element of } A.$$

The claim now is that (i) g is a function and (ii) g is a left inverse for f. As for (i), we note that if $b = f(a)$ for *some* $a \in A$, then $b = f(a)$ for *precisely one* $a \in A$, because f is injective. Hence g is single-valued, hence a function. As for (ii), we have for any $a \in A$,

$$gf(a) = g(f(a)) \qquad \text{defn. of composition}$$

$$= a \qquad \text{defn. of } g,$$

and so $gf = 1_A$. This completes the proof that f is left invertible.

We leave the proof of part 2 (right invertible $\Leftrightarrow$ surjective) to the reader— Exercise 2. (This time use the surjective function g of Fig. 9 as a guide.) $\square$

As an immediate consequence of the definitions of bijective and (two-sided) invertible functions, we have the following

Corollary. A function is bijective if and only if it is invertible.

4.3   Characteristic Functions

Let A be a set. Any subset S of A naturally induces a binary-valued function on A, namely the function c_S defined by

$$c_S(x) = 1 \quad \text{if } x \in S,$$

$$= 0 \quad \text{if } x \notin S.$$

The function c_S is called the *characteristic function* of S (on A).

Example 1. Let $A = \{a, b, c, d, e\}$ and $S = \{a, d, e\}$. Then c_S is given by:

x	a	b	c	d	e
$c_S(x)$	1	0	0	1	1

$\square$

The characteristic function essentially reduces the concept of subset to that of binary-valued function; each uniquely determines the other. This is the content of

Proposition 1. Let A be a given set. Then the function $P(A) \to \{0, 1\}^A$ defined by $S \mapsto c_S$ ($S \subseteq A$)—i.e., the correspondence between a subset and its characteristic function—is a bijection.

Proof. $S \mapsto c_S$ is surjective: Let c be any function $A \to \{0, 1\}$. Then c is the characteristic function of some subset of A, namely the subset $S_c = \{x \in A \mid c(x) = 1\}$.

$S \mapsto c_S$ is injective: Let S and T be subsets of A. If $S \neq T$, there must be one element in one of the sets, say S, that is not in the other (otherwise S and T would be equal). Suppose that $x \in S$ and $x \notin T$. Then $c_S(x) = 1$ and $c_T(x) = 0$, whence $c_S \neq c_T$, proving that $S \mapsto c_S$ is injective. $\square$

Our goal now is to define some useful operations on binary-valued functions—operations which arise frequently in the role of such functions as characteristic functions. To this end we first define three operations on the set $\{0, 1\}$—*meet* ($\wedge$), *join* ($\vee$), and *complement* ('):

x	y	$x \wedge y$	$x \vee y$		x	x'
0	0	0	0		0	1
0	1	0	1		1	0
1	0	0	1			
1	1	1	1			

The most useful and intuitive interpretation of these operations is as logical connectives, with x and y being interpreted as assertions, and 1 and 0 being interpreted as truth-values of assertions, true and false respectively. The meet operation then corresponds to "and," the join operation to "or," and the complement to "not."

We now define the meet, join, and complement of binary-valued functions by elementwise application of the corresponding operations on $\{0, 1\}$: If f and g are functions $A \to \{0, 1\}$, then $f \wedge g, f \vee g$, and f' are functions $A \to \{0, 1\}$ defined by:

$$(f \wedge g)(x) = f(x) \wedge g(x);$$
$$(f \vee g)(x) = f(x) \vee g(x);$$
$$f'(x) = f(x)'.$$

The following proposition highlights the intimate connection between the concepts of subset and characteristic function by displaying the analogy between the operations $\cap$, $\cup$, and ' on sets and the above operations $\wedge$, $\vee$, and ' on binary-valued functions.

Proposition 2. Let S and T be subsets of a given set A. Then

$$c_{S \cap T} = c_S \wedge c_T;$$
$$c_{S \cup T} = c_S \vee c_T;$$
$$c_{S'} = c_S'.$$

We leave the proof of Proposition 2 as an exercise (Exercise 1).

4.4 Functions and Equivalence Relations

Any equivalence relation E on a set A naturally induces a function on A, namely the map $\nu: A \to A/E$ that assigns each element of A to its equivalence class:

$$\nu: a \mapsto [a];$$

ν is called the *natural map* of E.

The converse situation also holds: Any function f defined on A naturally induces an equivalence relation on A, namely the relation E_f defined by

$$a E_f b \iff f(a) = f(b).$$

The equivalence relation E_f induced by f is also called the *kernel relation* of f. Similarly, the induced partition is called the *kernel partition* of f and denoted by π_f.

Example 1. A function f and its kernel partition are given below.

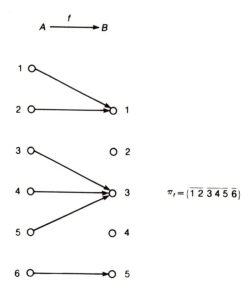

$$\pi_f = \{\overline{1}\;\overline{2\;3\;4\;5}\;\overline{6}\}$$

The importance of the kernel concept is simply this. If we are interested in the value of a function $f: A \to B$ at a particular argument $x \in A$, then we can replace x by any element y that is equivalent modulo the kernel relation—by the definition of E_f, we are guaranteed to get the same result. From the viewpoint of f, the domain A can be treated at the level of abstraction of A/E_f.

Example 2. Find the kernel relation of $f: x \mapsto x^2$, x real. Solution: We have

$$f(x) = f(y) \Rightarrow x^2 = y^2$$

$$\Rightarrow x = \pm y,$$

and so $x E_f y \Leftrightarrow x = \pm y$—a formal statement of the well-known fact that when we square a real number, we can ignore its sign. $\square$

We conclude this section with a theorem that tells us how to "factor" an arbitrary function into the "product" (composition) of two simpler functions.

Theorem 1 (*Decomposition Theorem for Functions*). Any function $f: A \to B$ can be expressed as the composition $\psi\nu$ of a surjective function ν and an injective function ψ, in accordance with the following mapping diagram:

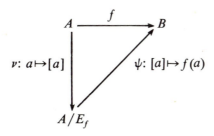

Proof. We must show (i) that ν is surjective, (ii) that ψ is injective, and (iii) that $f = \psi\nu$.

(i) The natural map $\nu: A \to A/E$ corresponding to any equivalence relation E on A is trivially surjective.

(ii) We must first show that ψ is well-defined, i.e., that the value of ψ depends only on the chosen equivalence class and not on the particular representative of that equivalence class:

$$[a] = [b] \Rightarrow a E_f b \qquad \text{Lemma 1 of Sec. 3.2}$$
$$\Rightarrow f(a) = f(b) \qquad \text{defn. of } E_f$$
$$\Rightarrow \psi([a]) = \psi([b]) \qquad \text{defn. of } \psi,$$

so ψ is indeed single-valued.

Moreover, each "$\Rightarrow$" in the above can be replaced by "$\Leftrightarrow$". Thus ψ is injective.

(iii) For any $a \in A$, we have

$$(\psi\nu)(a) = \psi(\nu(a))$$
$$= \psi([a]) \qquad \text{defn. of } \nu$$
$$= f(a) \qquad \text{defn. of } \psi,$$

giving $f = \psi\nu$, as claimed. $\square$

Corollary. If f is surjective, then ψ is bijective.

EXERCISES

Section 1

1. Test A for (1) inclusion in B, (2) equality with B, and (3) membership in B.
   (a) $A = \{\varnothing\}$, $B = \{\{\varnothing\}\}$;
   (b) $A = \{\varnothing, \{\varnothing\}\}$, $B = \{\varnothing, \{\varnothing, \{\varnothing\}\}\}$;
   (c) $A = \{\{\varnothing\}, \{\varnothing, \varnothing\}\}$, $B = \{\{\varnothing\}\}$.

Section 2

1. What are (a) $P(\varnothing)$, (b) $P(\{\varnothing\})$?

2. Let $A_1 = \{x \in \mathbf{R} | 0 \leq x < 2\}$, $A_2 = \{x \in \mathbf{R} | 0 < x \leq 2\}$, $A_3 = \{x \in \mathbf{R} | -1 \leq x < 1\}$. What are the sets:
   (a) $A_1 \cup A_2$,   (b) $A_1 \cup A_3$,   (c) $\bigcup_{i=1}^{3} A_i$,
   (d) $A_1 \cap A_2$,   (e) $A_1 \cap A_3$,   (f) $\bigcap_{i=1}^{3} A_i$,
   (g) $A_1 - A_2$,   (h) $A_2 - A_1$,   (i) $A_1 - A_3$?

3. Prove the distributive and consistency laws of Theorem 1, Section 2.2.

4. Prove: If A, B, and C are sets such that
$$A \cup B = A \cup C \quad \text{and} \quad A \cap B = A \cap C,$$
   then $B = C$. [*Hint:* Consider the absorption laws.]

5. (*Uniqueness of complements.*) Prove: If $A, B \subseteq I$ satisfy $A \cup B = I$ and $A \cap B = \varnothing$, then $B = A'$. Conclude the *involution* property of complementation: $(A')' = A$.

6. Prove *de Morgan's Laws:*
$$(A \cup B)' = A' \cap B',$$
$$(A \cap B)' = A' \cup B'.$$
   [*Hint:* Use Exercise 5.]

7. (*Ordered pairs in terms of unordered pairs.*)
   (a) Show that if we *define* (a, b) by $\{\{a\}, \{a, b\}\}$, then (a, b) satisfies the definitive property of ordered pairs: $(a, b) = (c, d) \Leftrightarrow a = c$ and $b = d$.
   (b) Show that $\{a, \{b\}\}$ fails to have this definitive property.

8. Let $A = \{x \in \mathbf{R} | 0 \leq x \leq 1\}$, $B = \{x \in \mathbf{R} | 0 \leq x \leq 2\}$. Describe geometrically the Cartesian products $A \times B$ and $B \times A$. Is $A \times B = B \times A$?

Section 3.1

1. Let ρ be the relation "is the father of," σ the relation "is the brother of," and τ the relation "is the husband of." What is the family relationship between a and b if:
   (i) $a \rho b$; (ii) $a \sigma b$; (iii) $a \tau b$; (iv) $a \rho \sigma b$; (v) $a \sigma \rho b$; (vi) $a \tau \rho b$; (vii) $a \tau \sigma b$; (viii) $a \rho \tau b$.

2. Let ρ be a relation on a set A. Prove:
   (a) ρ reflexive $\Rightarrow \rho \subseteq \rho^2$ ($\rho^2 = \rho\rho$);
   (b) ρ transitive $\Leftrightarrow \rho^2 \subseteq \rho$.

Section 3.2

1. Is the relation ρ on $\mathbf{Z}$ defined by
$$x \rho y \Leftrightarrow xy \neq 0$$
   an equivalence relation? Why (not)?

2. Show that the following relations on $\mathbf{R} \times \mathbf{R}$ are equivalence relations, and describe geometrically the equivalence classes.
   (a) $(a,b)\rho(c,d) \Leftrightarrow a^2 + b^2 = c^2 + d^2$;
   (b) $(a,b)\sigma(c,d) \Leftrightarrow ab = cd$.

3. Call a relation ρ on a set A *circular* if

   $$x \rho y \text{ and } y \rho z \Rightarrow z \rho x.$$

   Prove that a reflexive, circular relation is an equivalence relation.

4. (*Independence of the reflexive, symmetric, and transitive properties of an equivalence relation.*) Find three relations ρ, σ, τ on $\mathbf{Z}$ such that:
   (i) ρ is reflexive, symmetric, but not transitive;
   (ii) σ is reflexive, transitive, but not symmetric;
   (iii) τ is symmetric, transitive, but not reflexive.

5. What is wrong with the following "proof" that symmetry and transitivity imply reflexivity. (By the previous exercise something must be wrong.)
   Let ρ be a symmetric and transitive relation defined on a set A. For any $x, y \in A$,

   $$x \rho y \Rightarrow x \rho y \text{ and } y \rho x \qquad \rho \text{ is symmetric}$$

   $$\Rightarrow x \rho x \qquad \rho \text{ is transitive,}$$

   whence ρ is reflexive.

6. Re Theorem 2: Prove carefully that E_π is an equivalence relation.

7. Let A be a set, E an equivalence relation on A, and π a partition of A. In Theorem 2 we showed that $\pi = \pi_{E_\pi}$. Establish the analogous $E = E_{\pi_E}$.

Section 3.3

1. True or false: The set $\mathbf{Z}$ of all integers is a poset under divisibility ($a \mid b$ means $b = ka$ for some $k \in \mathbf{Z}$).

2. (a) Show that the set $\Pi(A)$ of all partitions of a set A under refinement is a poset.
   (b) Identify the greatest and least elements of this poset.
   (c) Draw the poset diagram for $\Pi(3)$; for $\Pi(4)$.

3. Draw a poset diagram for each of the following subsets of $\mathbf{P}$ partially ordered by divisibility ($\mid$), and specify which are lattices.
   (a) $\{1, 2, 3, 6, 12\}$;
   (b) $\{1, 2, 3, 12, 18, 36\}$;
   (c) $\{1, 2, 3, 5, 12, 60\}$;
   (d) $\{1, 2, 3, 6, 8, 9, 72\}$.

4. (a) Show that any finite poset has minimal and maximal elements. What about the infinite case?
   (b) Show that any finite lattice has a least and greatest element. What about the infinite case?

5. Prove carefully that if $[P; \leq]$ is a poset, then its dual, $[P; \geq]$, is also a poset.

6. Re Theorem 2: Prove the associative and consistency laws, using duality where applicable.

7. True or false: A poset in which every pair of elements has a g.l.b. is a lattice. Why does the existence of l.u.b.'s not follow by duality?

8. Show that the poset $\Pi(3)$ (Exercise 2) is a lattice in which neither distributive law (meet over join or join over meet) holds.

Section 4.1

1. Which of the following relations are functions? $x \rho y \Leftrightarrow x^2 + y^2 = 1$ (x, y real), where
   (a) $0 \le x \le 1, 0 \le y \le 1$;
   (b) $-1 \le x \le 1, -1 \le y \le 1$;
   (c) $-1 \le x \le 1, 0 \le y \le 1$;
   (d) x arbitrary, $0 \le y \le 1$.

2. Which of the following functions are injective? Which are surjective?
   (a) $x \mapsto x + 1$ on $\mathbf{N}$, on $\mathbf{Z}$;
   (b) $x \mapsto x^2$ on $\mathbf{N}$, on $\mathbf{Z}$, on $\mathbf{R}$;
   (c) $x \mapsto x^3$ on $\mathbf{N}$, on $\mathbf{Z}$, on $\mathbf{R}$.

Section 4.2

1. Write down all the left and right inverses of the functions $f: A \to B$ of Figs. 9(a) and (b).

2. Prove part 2 of Theorem 2: A function $f: A \to B$ is right invertible $\Leftrightarrow f$ is surjective.

3. For functions $f: A \to B$, $g: B \to C$, prove:
   (a) If f and g are injective (surjective), then gf is injective (surjective).
   (b) If f and g are invertible, then gf is invertible, with
   $$(gf)^{-1} = f^{-1}g^{-1}.$$

4. Let f be a function $A \to A$. Suppose that $f^n = 1_A$ for some positive integer n (such a function is called *periodic*). Prove that f is bijective.

5. (*Characterization of injective and surjective functions in terms of composition.*) Prove:
   (a) A function $f: A \to B$ is injective $\Leftrightarrow$ [for any functions $g: X \to A$ and $h: X \to A$, $fg = fh \Rightarrow g = h$ (*left-cancellability*)].
   (b) A function $f: A \to B$ is surjective $\Leftrightarrow$ [for any functions $g: B \to X$ and $h: B \to X$, $gf = hf \Rightarrow g = h$ (*right-cancellability*)].
   [*Hint:* In both (a) and (b), one direction ($\Rightarrow$) is straightforward. As for the other direction ($\Leftarrow$), the functions g and h are up for grabs—so choose them wisely.]

6. Let f be a function $A \to B$. Prove:
   (a) For $S \subseteq A$, $S \subseteq f^{-1}f(S)$.
   (b) For $T \subseteq B$, $ff^{-1}(T) \subseteq T$.
   When does equality hold in (a) and (b)?

Section 4.3

1. Prove Proposition 2.

2. Let ρ be a relation on a set $A = \{a_1, a_2, \ldots, a_n\}$. We see that the relation matrix $R_\rho = (r_{ij}^\rho)$,
$$r_{ij}^\rho = 1 \quad \text{if } a_i \rho a_j,$$
$$= 0 \quad \text{otherwise,}$$
is essentially the characteristic function of $\rho \subseteq A \times A$, with $c_\rho((a_i, a_j)) = r_{ij}^\rho$.

(a) Define the operations meet ($\wedge$), join ($\vee$), and complement ($'$) on binary (zero-one) matrices: If $R = (r_{ij})$, $S = (s_{ij})$, $T = (t_{ij})$ are $n \times n$ binary matrices, then

$$T = R \wedge S \text{ means } t_{ij} = r_{ij} \wedge s_{ij},$$

$$T = R \vee S \text{ means } t_{ij} = r_{ij} \vee s_{ij},$$

$$T = R' \qquad \text{means } t_{ij} = r'_{ij},$$

where the right-hand side operations are the meet, join, and complement defined on $\{0, 1\}$. Prove:

(i) $R_{\rho \cup \sigma} = R_\rho \vee R_\sigma$;
(ii) $R_{\rho \cap \sigma} = R_\rho \wedge R_\sigma$;
(iii) $R_{\rho'} = R_\rho{}'$.
(Cf. Proposition 2.)

(b) Now define the *Boolean matrix product* of R and S:

$$T = R * S \text{ means } t_{ij} = \bigvee_{k=1}^{n} (r_{ik} \wedge s_{kj})$$

(cf. the formula $t_{ij} = \Sigma_{k=1}^{n} r_{ik} \cdot s_{kj}$ for the numerical matrix product). [Note: $\bigvee_{k=1}^{n} a_k = 1$ means $a_k = 1$ for some $1 \le k \le n$.] Prove:

$$R_{\rho\sigma} = R_\rho * R_\sigma.$$

Section 4.4

1. Determine the kernel relation of the following functions $\mathbf{R} \to \mathbf{R}$:
   (a) $f(x) = x^3$;
   (b) $f(x) = \sin x$;
   (c) $f(x) = |\sin x|$.

2. Let $f: A \to B$ and $h: A \to C$ be functions defined on the same domain A. If $E_f \subseteq E_h$, show that there exists a function $g: B \to C$ such that $h = gf$. The situation is illustrated by the following mapping diagram.

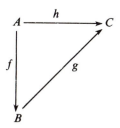

NOTES

Section 2

1. (*Russell's Paradox*.) We have already mentioned Georg Cantor as the founder of set theory. Cantor's set theory was founded on an intuitive notion of set which assumed (in our notation) that *any* meaningful assertion $p(x)$ specified a set, namely

$\{x\,|\,p(x)\}$. In Cantor's words, "[a set is] any collection into a whole of definite and separate objects of our intuition and thought." [G. Cantor, *Contributions to the Founding of the Theory of Transfinite Numbers*, translated, with notes and introduction, by P. Jourdain (New York: Dover, 1915), p. 85.]

It was precisely to this intuitive and seemingly innocuous notion of set that the British philosopher-mathematician Bertrand Russell (1877–1970) addressed his celebrated bombshell which revealed that intuitive set theory is inherently contradictory.

For the assertion $p(x)$ Russell took "$x \notin x$," thus obtaining $R = \{x\,|\,x \notin x\}$, the set of all sets that do not belong to themselves (R for Russell). Is this meaningful? Yes: A set either belongs to itself or it doesn't. To be sure, most sets that come to mind do not belong to themselves: the set of all integers is not an integer; the set of all countries in the world is not a country. But consider the set of all sets having more than five elements. This set clearly has more than five elements (because there are clearly more than five sets around having more than five elements) and so by its prescription must belong to itself. We have thus identified two members of R and one nonmember, and we could proceed likewise to test any other candidate for membership. In any case, intuition compels us to accept R as a bona fide (albeit awfully large) set.

The finale of Russell's intellectual symphony was to consider a very special set for membership in R, namely R itself. Again, either R belongs to itself or it doesn't. If R belongs to itself, then by the definition of R (R contains *only* sets not belonging to themselves) R does not belong to itself. Contradiction. On the other hand, if R does not belong to itself, then by the definition of R (R contains *all* sets not belonging to themselves) R *does* belong to itself. Another contradiction. There is no way out; we have arrived at

$$R \in R \text{ if and only if } R \notin R.$$

This, then, is *Russell's Paradox*, communicated in 1903. It led to the inescapable conclusion that the intuitive foundations of Cantorian set theory were logically flawed.

The reaction to this state of affairs was the development of an axiomatic approach to set theory, first undertaken by Ernst Zermelo (1871–1953). The idea here was to resolve the logical difficulties uncovered by Russell by greatly restricting Cantor's free-wheeling notion of a set. Precise rules—axioms—were laid down for producing new sets from old ones with the goal of admitting only "safe" sets into set theory which would be free of contradiction.

Now Russell's paradox might lead one to suspect that any treatment of sets short of a full blown axiomatic one is irresponsible. To be sure, Russell's paradox constitutes a grave indictment against the unbridled use of intuition (such as the contemplation of the set of all sets) in mathematics. But experience has clearly shown that the sets arising in mathematics and its applications are free of contradictions, that the discovery of paradoxical sets requires ingenuity, not bad luck.

For an eminently readable treatment of axiomatic set theory from an intuitive point of view, we refer the reader to P. Halmos's *Naive Set Theory* (Princeton, N.J.: Van Nostrand, 1960).

2. The algebra of sets is called *Boolean algebra* in honor of the British mathematician George Boole (1815–1864), who first investigated the algebraic laws that apply to logical assertions (under the logical connectives "or," "and," and "not") and sets (under the operations $\cup$, $\cap$, and $'$). Boole's original ideas are to be found in his *An Investigation of the Laws of Thought*, first published in 1854.

Section 3

1. We are restricting our attention to *binary* relations: $\rho \subseteq A_1 \times A_2$. The more general case is that of *n*-ary relations: $\rho \subseteq A_1 \times A_2 \times \cdots \times A_n$. For example, betweenness (β) on $\mathbf{Z}$ can be defined as the ternary relation

$$(x, y, z) \in \beta \Leftrightarrow x \le y \le z.$$

CHAPTER

THE INTEGERS

Of all the mathematical structures that have been devised over the centuries, none is more useful or important than the familiar system of integers. A great deal of applied algebra and computing is concerned with manipulating integers or reasoning about integers, and much of abstract algebra is inspired by properties of the integers (as we shall see in later chapters—especially in Chapter IV).

In the first section, we present as axioms a few self-evident properties of the integers; then in the following two sections we deduce some further, perhaps not quite so obvious properties of the integers that we shall need later.

Besides providing the reader with some useful facts about the integers, this chapter should serve as an introduction to the deductive method that lies at the heart of modern algebra.

1. BASIC PROPERTIES

We shall take for granted the following basic properties of the integers which have been ingrained in the consciousness of every educated person since childhood (a, b, and c below represent arbitrary integers).

1. Associativity and commutativity of $+$ and $\cdot$:
   (i) $a + (b + c) = (a + b) + c,$ $\quad a(bc) = (ab)c;$
   (ii) $a + b = b + a,$ $\qquad\qquad ab = ba.$

John D. Lipson, Elements of Algebra and Algebraic Computing. ISBN 0-201-04115-4.

2. Distributivity of $\cdot$ over $+$:
$$a(b + c) = ab + ac.$$

3. Properties of 0 and 1:
    (i) $0 + a = a$, $1a = a$;
    (ii) $0a = 0$.

4. Properties of $-$:
    (i) $(-a) + a = 0$;
    (ii) $(-a)b = -ab$;
    (iii) $(-a)(-b) = ab$.

5. Cancellation laws:
    (i) $a + b = a + c \Rightarrow b = c$;
    (ii) $ab = ac \Rightarrow b = c$ if $a \neq 0$.

Recall: $a \leq b$ means that $b - a$ is nonnegative; $a < b$ means that $b - a$ is positive.

6. Properties of order:
    (i) $a, b > 0 \Rightarrow a + b, ab > 0$;
    (ii) for each integer a, exactly one of the following holds:
$$a < 0, \quad a = 0, \quad \text{or } a > 0 \qquad (trichotomy).$$

These properties are not independent. For example, we prove the cancellation law for addition, property 5(i), from properties 1–4:

$$a + b = a + c$$

$\Rightarrow (-a) + (a + b) = (-a) + (a + c)$

$\Rightarrow ((-a) + a) + b = ((-a) + a) + c$ property 1(i)

$\Rightarrow 0 + b = 0 + c$ property 4(i)

$\Rightarrow b = c$ property 3(i). $\square$

Other dependencies also exist, some quite subtle. These we shall investigate in our study of algebraic systems inspired by the integers (Chapter IV in particular).

We observe that the above properties of the integers are shared by the two other number systems of "ordinary" arithmetic, namely the rationals and the reals. So we ask, *what makes the integers special?* The answer lies in the following

Definition. A totally ordered set $[P; \leq]$ is *well-ordered* (by $\leq$) if every nonempty subset of P has a least element.

The distinctive property of the integers that we seek is simply this:

$$\boxed{\textbf{N is well-ordered by } \leq}.$$

Stated otherwise: Any nonempty set of nonnegative integers contains a least element.

Like the other properties we have postulated in the section, we regard the well-orderedness of the natural numbers as an axiomatic property—one to be understood rather than proved.

Example 1.

(a) The set $S \subseteq \mathbf{N}$ of all even positive numbers has a least element, namely 2.
(b) The set $S \subseteq \mathbf{N}$ of all prime numbers > 20 has a least element, namely 23.
(c) The set $S \subseteq \mathbf{N}$ of all *even* prime numbers > 20 does not have a least element —because it is empty. $\square$

The well-orderedness property of the nonnegative integers really does distinguish the integers from the rationals and reals. For consider the set $\mathbf{Q}^+$ of all positive rationals. This subset of the nonnegative rationals has no least element.

Proof. If r is claimed to be the least element of $\mathbf{Q}^+$, then $r/2$ satisfies $0 < r/2 < r$, thus contradicting the claim and forcing the desired conclusion that $\mathbf{Q}^+$ has no least element. The same argument also shows that the nonnegative reals are not well-ordered by $\leq$.

An important consequence of the well-orderedness of $\mathbf{N}$ is

Theorem 1 (*Induction Property of* $\mathbf{N}$). If $S \subseteq \mathbf{N}$ satisfies

1. $0 \in S$,
2. $n \in S \Rightarrow n + 1 \in S$,

then $S = \mathbf{N}$. (In words: Any subset of $\mathbf{N}$ that contains 0, and contains $n + 1$ whenever it contains n, must be all of $\mathbf{N}$.)

Proof. Let S satisfy the hypotheses of the theorem. Let $T = \mathbf{N} - S$. If T is empty, then the theorem is proved. Assume, to the contrary, that T is nonempty. By well-orderedness, T contains a least element, call it l. Now $0 \notin T$ because of hypothesis 1; hence $l > 0$. It then follows that $l - 1$ is *not* in T (because l is the *least* element of T) while l *is* in T, which is to say that $l - 1$ is in S while l is not in S. But this contradicts hypothesis 2. Thus the assumption that T is nonempty is untenable, giving $S = \mathbf{N}$ as required. $\square$

2. INDUCTION AND RECURSION

The Induction Property just proved yields a powerful method for proving facts about the natural numbers.

Proof by induction. Let $p(n)$ be an assertion about the natural number n. Then to prove $p(n)$ for all $n \geq b$ (typically $b = 0$ or 1), it is sufficient to carry out the following two steps:

1. *Basis.* Prove $p(b)$.
2. *Induction.* For any $n \geq b$, prove $p(n) \Rightarrow p(n + 1)$.

Discussion. Both steps of proof by induction are vital. The basis step shows that the assertion $p(n)$ holds for the initial case, $n = b$. The induction step serves to permit unlimited broadening of the basis step to any $n \geq b$.

We emphasize with force that the assumption of $p(n)$ in the induction step, the so-called *induction hypothesis*, is not tantamount to assuming what is to be proved. $p(n)$ is assumed for a *fixed* (but arbitrary) natural number $\geq b$, and from this assumption one must establish $p(n + 1)$. The quantification "for any $n \geq b$" in the induction step has the force of requiring that the proof of "$p(n) \Rightarrow p(n + 1)$" be valid for any $n \geq b$. (In Exercise 2 we explore the possible dire consequences of a careless attitude towards this quantification.)

Theorem 1 (*Proof by induction works*).   If

1. $p(b)$, and
2. for any $n \geq b$, $p(n) \Rightarrow p(n + 1)$,

then $p(n)$ for all $n \geq b$.

Proof. Let $q(n)$ be the assertion $p(n + b)$, and let $S = \{n \mid q(n)\}$. Then S contains 0 by hypothesis 1 and contains $n + 1$ whenever it contains n by hypothesis 2. Hence $S = \mathbf{N}$ by the Induction Property of $\mathbf{N}$. We thus have $q(n)$ for all $n \geq 0$, which means $p(n + b)$ for all $n \geq 0$, or $p(n)$ for all $n \geq b$. $\square$

Example 1.   Prove that the sum of the first n odd positive integers is n^2.

The proof is by induction on n. Let $p(n)$ be the assertion

$$\sum_{k=1}^{n} (2k - 1) = [1 + 3 + 5 + \cdots + (2n - 1)] = n^2.$$

Basis ($n = 1$). Since $\sum_{k=1}^{1} (2k - 1) = 1 = 1^2$, we have $p(1)$.
Induction. Assume $p(n)$ for some arbitrary $n \geq 1$. Then

$$\sum_{k=1}^{n+1} (2k - 1) = \sum_{k=1}^{n} (2k - 1) + (2n + 1)$$

$$= n^2 + (2n + 1) \qquad \text{induction hypothesis}$$

$$= (n + 1)^2,$$

which establishes $p(n + 1)$. Hence by induction we have $p(n)$ for all $n \geq 1$. $\square$

A scheme analogous to proof by induction can be used to define concepts involving the natural numbers.

Definition by recursion (*recursive definition*).   To define a concept for all natural numbers $\geq b$, it is sufficient to carry out the following two steps:

1. *Basis.* Define the concept for $n = b$.
2. *Recursion.* Define the concept for $n + 1$, assuming that the concept has been defined for n, where $n \geq b$.

Successful completion of the above recursive definition schema means that the concept has been defined for all natural numbers $\geq b$. (Proof: Let $p(n)$ be the assertion that the concept is defined for the natural number n.)

In our next example we follow in the footsteps of the Italian mathematician Guiseppe Peano (1858–1932), who provided an axiomatic development of the natural numbers in terms of the *successor function* $'$: $n' = n + 1 =$ the "successor" of n.

Example 2.   In terms of the successor function, we can define addition, $m + n$, and multiplication, mn, of natural numbers m and n by recursion on n.

(a) *Addition*
      Basis. $m + 0 = m$.
      Recursion. $m + n' = (m + n)'$.
(b) *Multiplication* (assuming addition)
      Basis. $m0 = 0$.
      Recursion. $mn' = mn + m$.

Let us examine in detail how our definitions compute $4 + 2$ and $4 \cdot 2$.

$$4 + 2 = 4 + 1' \qquad \text{defn. of } '$$
$$= (4 + 1)' \qquad \text{recursion}$$
$$= (4 + 0')' \qquad \text{defn. of } '$$
$$= ((4 + 0)')' \qquad \text{recursion}$$
$$= ((4)')' \qquad \text{basis}$$
$$= (5)' = 6 \qquad \text{defn. of } '.$$
$$4 \cdot 2 = 4 \cdot 1' \qquad \text{defn. of } '$$
$$= 4 \cdot 1 + 4 \qquad \text{recursion}$$
$$= 4 \cdot 0' + 4 \qquad \text{defn. of } '$$
$$= (4 \cdot 0 + 4) + 4 \qquad \text{recursion}$$
$$= (0 + 4) + 4 \qquad \text{basis}$$
$$= 4 + 4 = 8 \qquad \text{defn. of addition. } \square$$

With the above recursive definitions in hand we can now prove the familiar properties of addition and multiplication of natural numbers without having to appeal to experience or empirical evidence. We prove below the associative law of addition and the distributive law of multiplication over addition, leaving the proofs of other well-known properties to the exercises (Exercise 5).

Example 3.   (a) *Associative law of addition.*   Let $p(n)$ be the assertion "$k + (m + n) = (k + m) + n$ for arbitrary $k, m \in \mathbf{N}$." The proof of $p(n)$ for all $n \in \mathbf{N}$ is by induction on n:

Basis. $k + (m + 0)$
$$= k + m \qquad \text{defn. of addition}$$
$$= (k + m) + 0 \qquad \text{defn. of addition,}$$

which establishes $p(0)$.

Induction. Assume $p(n)$ for some arbitrary $n \in \mathbf{N}$. Then

$k + (m + n')$
$$= k + (m + n)' \qquad \text{defn. of addition}$$
$$= (k + (m + n))' \qquad \text{defn. of addition}$$
$$= ((k + m) + n)' \qquad \text{induction hypothesis}$$
$$= (k + m) + n' \qquad \text{defn. of addition,}$$

which establishes $p(n + 1)$ and completes the proof by induction of $p(n)$ for all $n \in \mathbf{N}$.

(b) *Distributive law of multiplication over addition.*   Let $p(n)$ be the assertion "$k(m + n) = km + kn$ for arbitrary $k, m \in \mathbf{N}$." The proof of $p(n)$ for all $n \in \mathbf{N}$ is by induction on n.

Basis. $k(m + 0)$
$$= km \qquad \text{defn. of addition}$$
$$= km + 0 \qquad \text{defn. of addition}$$
$$= km + k0 \qquad \text{defn. of multiplication,}$$

which establishes $p(0)$.

Induction.   Assume $p(n)$ for some arbitrary $n \in \mathbf{N}$. Then

$k(m + n')$
$$= k[(m + n)'] \qquad \text{defn. of addition}$$
$$= k(m + n) + k \qquad \text{defn. of multiplication}$$
$$= (km + kn) + k \qquad \text{induction hypothesis}$$
$$= km + (kn + k) \qquad \text{associativity of addition}$$
$$= km + kn' \qquad \text{defn. of multiplication,}$$

which establishes $p(n + 1)$ and completes the proof by induction of $p(n)$ for all $n \in \mathbf{N}$. $\square$

Example 4 (*Exponentiation by positive integral powers*).   Let $*$ be an associative operation, call it multiplication, defined on some set S (e.g., $*$ could be ordinary multiplication of integers or real numbers, or composition of functions from a set to itself). The *powers* x^n of $x \in S$ by the positive integral exponent n are defined recursively by

Basis. $x^1 = x$;
Recursion. $x^{n+1} = x^n * x$.

Thus $x^n = x * \cdots * x$ (n factors of x); the recursive definition renders precise mathematical meaning to the dots, "$\cdots$".

With the above recursive definition in hand, we can prove the familiar laws of exponents:

(a) $x^m * x^n = x^{m+n}$;
(b) $(x^m)^n = x^{mn}$;
(c) $(x^n) * (y^n) = (x * y)^n$ if $*$ is commutative.

We prove (a) below, leaving (b) and (c) for an exercise (Exercise 6). Let $p(n)$ be the assertion "$x^m * x^n = x^{m+n}$ for all $m \in \mathbf{P}$." The proof of $p(n)$ for all $n \in \mathbf{P}$ is by induction on n.

Basis $(n = 1)$.

$$x^m * x^1 = x^m * x \qquad \text{basis}$$
$$= x^{m+1} \qquad \text{recursion,}$$

which establishes $p(1)$.

Induction.   Assume $p(n)$ for some arbitrary $n \in \mathbf{P}$. Then

$$x^m * x^{n+1} = x^m * (x^n * x) \qquad \text{recursion}$$
$$= (x^m * x^n) * x \qquad \text{associativity of } *$$
$$= x^{m+n} * x \qquad \text{induction hypothesis}$$
$$= x^{(m+n)+1} \qquad \text{recursion}$$
$$= x^{m+(n+1)} \qquad \text{associativity of } +,$$

which establishes $p(n + 1)$. $\square$

Course-of-values induction.   In the induction step of our proof by induction schema, we are allowed to assume $p(n)$ in order to establish $p(n + 1)$. In *course-of-values* induction, we assume not only $p(n)$, but all of its predecessors $p(n - 1), p(n - 2), \ldots$ as well, in establishing $p(n + 1)$.

Course-of- values proof by induction.   Let $p(n)$ be an assertion about the natural number n. Then to prove $p(n)$ for all $n \geq b$, it is sufficient to carry out the following two steps:

1. *Basis.* Prove $p(b)$.
2. *Induction.* For any $n \geq b$, prove
$$[\,p(m) \text{ for all } b \leq m \leq n\,] \Rightarrow p(n + 1).$$

Theorem 2 (*Course-of-values proof by induction works*).   If

1. $p(b)$, and
2. for any $n \geq b$, $[\,p(m) \text{ for all } b \leq m \leq n\,] \Rightarrow p(n + 1)$,

then $p(n)$ for all $n \geq b$.

Remark.   The point at issue here is the permissive induction hypothesis of course-of-values induction, which allows one to assume not only $p(n)$ as in ordinary induction, but also $p(n - 1), p(n - 2), \ldots, p(b)$. Can all this be assumed [in order to prove $p(n + 1)$] without invalidating the desired conclusion $p(n)$ for all $n \geq b$? Theorem 2 says yes.

Proof of Theorem 2.   Let $p(n)$ satisfy the two hypotheses in the statement of Theorem 2, and let $q(n)$ be the assertion "$p(m)$ for all $b \leq m \leq n$." We note first that

$$q(n) \text{ for all } n \geq b$$
$$\Rightarrow [\, p(m) \text{ for all } b \leq m \leq n] \text{ for all } n \geq b \qquad \text{defn. of } q$$
$$\Rightarrow p(m) \text{ for all } m \geq b \qquad\qquad\qquad\qquad \text{logic.}$$

So if we can prove $q(n)$ for all $n \geq b$, we are done. This we do by ordinary induction (i.e., assuming only $q(n)$ in the induction step).

Basis.   We have $p(b)$ by hypothesis 1, which trivially means $p(m)$ for $b \leq m \leq b$, which is $q(b)$.

Induction.   Assume $q(n)$ for some arbitrary $n \geq b$. Now,

$$q(n) \Rightarrow [\, p(m) \text{ for all } b \leq m \leq n] \qquad \text{defn. of } q$$
$$\Rightarrow p(n + 1) \qquad\qquad\qquad\qquad\qquad \text{hypothesis 2}$$
$$\Rightarrow [\, p(m) \text{ for all } b \leq m \leq n + 1] \qquad \text{combining}$$
$$\Rightarrow q(n + 1) \qquad\qquad\qquad\qquad\qquad \text{defn. of } q.$$

Thus by ordinary induction we conclude $q(n)$ for all $n \geq b$, which completes the proof of Theorem 2. $\square$

The moral of Theorem 2 is simply this: course-of-values induction, like ordinary induction, is a bona fide proof technique. Use it whenever it is convenient to do so.

Example 5 (*Division Property of the natural numbers*).   The reader is undoubtedly familiar with the Division Property of the natural numbers: given $a, b \in \mathbf{N}$ ($b \neq 0$), there exist (unique) q and r such that

(DP) $a = bq + r \qquad (0 \leq r < b)$.

q and r are the *quotient* and *remainder* when a (the dividend) is divided by b (the divisor). Example: If $a = 7$ and $b = 3$, then $q = 2$ and $r = 1$ satisfy (DP).

The Division Property is simply a statement that division works. How can we prove it?

Let $p(a)$ be the assertion "for any positive integer b, there exist q and r satisfying (DP)." (We shall worry about the uniqueness of q and r later.) The Division Property is then the statement "$p(a)$ for all $a \in \mathbf{N}$." This we prove by course-of-values induction on a.

Basis.   If $a = 0$, then $q = 0$, $r = 0$ satisfy (DP) for any $b > 0$. This establishes $p(0)$.

Induction.   Assume $p(m)$ for $0 \leq m \leq a$. (Our goal is to establish $p(a + 1)$.)

CASE 1. $b > a + 1$. Then $q = 0$, $r = a + 1$ satisfy (DP). (No need to use the induction hypothesis here.)

CASE 2. $b \leq a + 1$ (the interesting case). If $0 < b \leq a + 1$, then $0 \leq (a + 1) - b \leq a$. Applying our induction hypothesis to $(a + 1) - b$ yields

$$(a + 1) - b = bq + r \qquad (0 \leq r < b),$$

so that

$$a + 1 = b(q + 1) + r \qquad (0 \le r < b).$$

This establishes $p(a + 1)$ and completes the proof by course-of-values induction of the Division Property (at least as far as existence of q and r is concerned).

As for uniqueness, assume that we have

$$\begin{aligned} a &= bq + r \\ &= bq' + r'. \end{aligned} \qquad (0 \le r, r' < b)$$

Subtracting gives

$$(*) \qquad\qquad b(q - q') = r' - r.$$

Suppose $q \ne q'$, say $q > q'$. Then $b(q - q') \ge b$ (since $b > 0$), whereas $r' - r < b$ (since $0 \le r,\ r' < b$). But this contradicts $(*)$. Thus $q = q'$, which forces $r = r'$, proving uniqueness. $\square$

The Division Property just proved is readily extended from the nonnegative integers to all the integers (Exercise 9).

Analogous to course-of-values induction we also have course-of-values recursion (cf. "ordinary" recursion):

Course-of-values recursion.   To define a concept for all natural numbers $\ge b$ it is sufficient to carry out the following two steps:

1. *Basis.* Define the concept for $n = b$.
2. *Recursion.* Define the concept for $n + 1$, assuming that it has been defined for all m such that $b \le m \le n$.

Successful completion of the above recursive definition schema means that the concept has been defined for all natural numbers $\ge b$. (Proof: Let $p(n)$ be the assertion that the concept is defined for the natural number n; use course-of-values induction.)

3. DIVISION AND DIVISIBILITY

Let us take stock. The Division Property of the integers (Example 5 and Exercise 9 of Section 2) states: for $a, m \in \mathbf{Z}$, $m \ne 0$, there exist a unique quotient q and remainder r such that

$$\text{(DP)} \qquad\qquad a = mq + r \qquad (0 \le r < m).$$

We embellish the notation for r in (DP) to $r_m(a)$. (Examples: $r_3(7) = 1$, $r_3(1) = 1$, $r_3(-7) = 2$.) We also denote $r_m(a)$ by $a \bmod m$ and refer to r_m as the *remainder mod m function.*

If $r_m(a) = 0$, then we say the m *divides* a (m is a *factor*, or *divisor, of a; a is a multiple* of m) and write $m \mid a$. Thus

$$m \mid a \quad \Leftrightarrow \quad a = km \text{ for some } k \in \mathbf{Z},$$

which extends the divisibility relation that we previously defined on **P** in Example 3 of Section I.3.3. (Examples: $3 \mid 6$, $3 \mid -3$, $3 \nmid 7$.)

In terms of divisibility our equivalence mod m relation (Example 3 of Section I.3.2) becomes

$$a \equiv_m b \iff m \mid (a - b) \iff r_m(a - b) = 0.$$

3.1 Equivalence and Remainders mod *m*

In the following lemma, we catalog some useful relationships between equivalence mod m and remainders mod m—ones that we will make frequent use of in later developments.

Lemma 1. Let m be a positive integer. Then

(a) $a \equiv_m r_m(a)$;
(b) $a = r_m(a) \iff 0 \le a < m$;
(c) $a \equiv_m b \iff r_m(a) = r_m(b)$.

Proof. (a) By the Division Property, $a = mq + r_m(a)$, so that $m \mid (a - r_m(a))$.

(b) One direction ($\Rightarrow$) is trivial. As for the other direction, if $0 \le a < m$, then $a = m0 + a$ satisfies the Division Property $a = mq + r$ $(0 \le r < m)$. By the uniqueness of the remainder $r = r_m(a)$, it follows that $r_m(a) = a$.

(c) By the Division Property,

$$a = mq + r_m(a)$$
$$b = mq' + r_m(b) \qquad (0 \le r_m(a), r_m(b) < m)$$

giving

$$(*) \qquad\qquad a - b = m(q - q') + (r_m(a) - r_m(b)).$$

($\Leftarrow$) If $r_m(a) = r_m(b)$, then from $(*)$, $m \mid (a - b)$.

($\Rightarrow$) We first ask what $r_m(a - b)$ is:

CASE 1. $r_m(a) \ge r_m(b)$. Then $0 \le r_m(a) - r_m(b) < m$, so that $(*)$ satisfies the Division Property when $a - b$ is divided by m; hence $r_m(a - b) = r_m(a) - r_m(b)$ by uniqueness of $r_m(a - b)$. Now if $a \equiv_m b$, then $r_m(a - b) = 0$, so that $r_m(a) = r_m(b)$ as required.

CASE 2. $r_m(a) \le r_m(b)$. Then $r_m(a - b) = r_m(b) - r_m(a)$, with the same consequence as Case 1: $r_m(a) = r_m(b)$. $\square$

We define the set $\mathbf{Z}_m$ of "integers mod m"

$$\mathbf{Z}_m = \{0, 1, \ldots, m - 1\}.$$

Theorem 1.

(a) For any $[a] \in \mathbf{Z}/\equiv_m$, $[a] = [r_m(a)]$.
(b) For any $a, b \in \mathbf{Z}_m$, $a \ne b \Rightarrow [a] \ne [b]$.

Proof.   (a) Lemma 1(a).

(b) If $a, b \in \mathbf{Z}_m$ with $a \neq b$, then by Lemma 1(b), $r_m(a) = a \neq b = r_m(b)$. Hence by Lemma 1(c), $a \not\equiv_m b$, so that $[a] \neq [b]$. $\square$

Stating Theorem 1 in other terms: each equivalence class $[a]$ has a *unique* representative in $\mathbf{Z}_m$, namely $r_m(a)$.

3.2   Greatest Common Divisors

Note.   In this and Section 3.3 we investigate two important concepts about the integers that revolve around the relation of divisibility—greatest common divisors and primes. Since the sign of an integer plays no essential role in the divisibility relation (if $a|b$ then $\pm a| \pm b$), we shall restrict our attention in these sections to nonnegative integers.

Consider the integers 12 and 18. 12 has divisors 1, 2, 3, 4, 6, and 12 (remember, we are restricting our attention to nonnegative integers); 18 has divisors 1, 2, 3, 6, 9, and 18. Thus 12 and 18 have *common* divisors 1, 2, 3, and 6, hence *greatest* common divisor 6.

In general, the greatest common divisor (g.c.d.) g of integers a and b (a and b not both zero) is characterized by the properties:

1. $g|a$ and $g|b$ (g is a *common* divisor of a and b);
2. $c|a$ and $c|b \Rightarrow c|g$ (g is the *greatest* common divisor of a and b).

Our reference to *the* g.c.d. (should it exist) is justified by

Proposition 1.   The (positive) g.c.d. of integers a and b is unique.

Proof.   If g and g' are both g.c.d.'s of a and b, then $g|g'$, since g' is a *greatest* common divisor and g is a common divisor of a and b. Symmetrically, $g'|g$. Hence $g' = \pm g$, but since g and g' are both positive, $g' = g$ as required. $\square$

We shall henceforth denote the unique positive g.c.d. of integers a and b by $\gcd(a, b)$ or simply by (a, b) (the latter not to be confused with the ordered pair containing a and b). Our goal now is to show the existence of integer g.c.d.'s, moreover the existence in a special form. (This special form plays a key role in a number of subsequent developments.)

Theorem 2.   Let a and b be nonnegative integers, not both zero. Then

$$\gcd(a, b) = sa + tb$$

for some $s, t \in \mathbf{Z}$.

The proof of Theorem 2 which follows makes interesting use of the well-orderedness of $\mathbf{N}$ and the division property of $\mathbf{Z}$.

Proof.   Let $S(a, b) = \{sa + tb | s, t \in \mathbf{Z}\}$. (Our goal is to show that $\gcd(a, b)$ is in $S(a, b)$.) Now $a = 1a + 0b$ and $b = 0a + 1b$ are both in $S(a, b)$, so that $S(a, b)$

contains positive integers. By well-orderedness, $S(a,b)$ contains a smallest positive integer $g = s_g a + t_g b$. We claim that $g = \gcd(a,b)$. The following result is the key to proving the claim.

Lemma. $S(a,b) = (g)$, where (g) denotes the set of all integral multiples of g, i.e., $(g) = \{kg \mid k \in \mathbf{Z}\}$.

Proof. (i) $(g) \subseteq S(a,b)$: For any $kg \in (g)$, we have $kg = k(s_g a + t_g b)$ $= (ks_g)a + (kt_g)b \in S(a,b)$.

(ii) $S(a,b) \subseteq (g)$: For any $x = s_x a + t_x b \in S(a,b)$, we can divide x by g to obtain

$$x = gq + r \qquad (0 \leq r < g),$$

whence

$$r = x - gq$$
$$= (s_x a + t_x b) - (s_g a + t_g b)q$$
$$= (s_x - s_g q)a + (t_x - t_g q)b.$$

Thus r is in $S(a,b)$. But then $r = 0$, otherwise we would have a contradiction to the fact that g is the *smallest* positive integer in $S(a,b)$. Thus $x = gq \in (g)$, so that $S(a,b) \subseteq (g)$. $\square$

Returning to the proof of Theorem 2, we have already noted that $a, b \in S(a,b)$. Hence, $a, b \in (g)$ by the above lemma, so that $g \mid a$ and $g \mid b$; i.e., g is a common divisor of a and b. But g is also a *greatest* common divisor. For if $c \mid a$ and $c \mid b$, then $c \mid (s_g a + t_g b)$ trivially. $\square$

Remark. Theorem 2 assures the existence of $\gcd(a,b)$ in the form $sa + tb$ (Example: $\gcd(12, 18) = 6 = 2 \times 12 + (-1) \times 18$), but it does not tell us how to compute $\gcd(a,b)$, s, or t. Early in Part Three we shall investigate Euclid's celebrated algorithm for carrying out this computation.

3.3 Factorization into Primes

By the *factorization* of a (positive) integer, we mean a decomposition of the integer in question into constituent factors. For example, $4 \times 21, 6 \times 14, 3 \times 4 \times 7$ are all factorizations of 84.

A particularly important case of integer factorization is factorization into *primes*. First recall the following familiar

Definition. An integer $n \geq 2$ is said to be *prime* if n has no proper divisors, i.e., if its only (positive) divisors are 1 and n. An integer $n \geq 2$ that is not prime is called *composite*. $\square$

Thus 7 is a prime, because $7 = ab$ for no $1 < a, b < 7$. On the other hand, $8 = 4 \times 2$ is composite. The first several primes are undoubtedly well-known to all: $2, 3, 5, 7, 11, 13, 17, 19, 23, 29, 31, \ldots$.

Returning to the factorization of 84, we have

$$84 = 4 \times 21 = 2 \times 2 \times 3 \times 7.$$

$2 \times 2 \times 3 \times 7$ is the factorization of 84 into primes; no further factorization is possible (by the definition of prime).

Our goal now is to establish two key properties of factorization into primes:

1. Every integer ≥ 2 can be factored into a product of prime factors (we permit a product of prime factors to consist of a single prime).

2. This factorization into primes is essentially unique: any two factorizations of the same integer will differ only in the order in which constituent primes occur.

We have already illustrated the first property in our factorization of 84 into primes: $2 \times 2 \times 3 \times 7$. As an illustration of the second property, we also have

$$84 = 6 \times 14 = 2 \times 3 \times 2 \times 7;$$

we have obtained essentially the same factorization as before, differing only in the order in which the primes occurs.

These two properties of prime factorization establish the primes as the "building blocks" of the integers. Every positive integer > 1 can be composed from prime factors (property 1) in essentially one way (property 2).

Theorems 1A and 1B, which establish the existence and uniqueness of prime factorization, are together regarded as "the fundamental theorem of arithmetic." Moreover, the proofs of these theorems provide good case studies of the power of induction for establishing general facts about the integers.

Theorem 1A (*Existence of a prime factorization*).   Any integer $n \geq 2$ can be expressed as a product of primes.

Proof.   The proof is by course-of-values induction on n.

Basis ($n = 2$).   The statement of the theorem holds for $n = 2$ trivially, since 2 is a prime.

Induction.   Assume the statement of the theorem for any integer m such that $2 \leq m \leq n$, and consider $n + 1$. If $n + 1$ is prime, then we are done. If $n + 1$ is not prime, then it has a factorization of the form

$$n + 1 = ab$$

where $1 < a,\ b < n + 1$. Our induction hypothesis then applies to a and b, allowing us to express a and b as a product of primes: $a = p_1 p_2 \cdots p_s$, $b = q_1 q_2 \cdots q_t$ $(s, t \geq 1)$. We then obtain $n + 1$ as

$$n + 1 = ab = p_1 p_2 \cdots p_s q_1 q_2 \cdots q_t,$$

i.e., as a product of primes. We have thus established the statement of the theorem for the integer $n + 1$, which completes the proof of Theorem 1A by course-of-values induction. $\square$

The uniqueness of prime factorization depends heavily on the following important

Lemma 1. Let p be a prime. Then

$$p \mid ab \Rightarrow p \mid a \text{ or } p \mid b.$$

(In words: If a prime divides a product, then it must divide one of the factors.)

Before proving Lemma 1, we note that it is false in general for composite numbers. For example $12 \mid (4 \times 6)$, whereas $12 \nmid 4$ and $12 \nmid 6$.

Proof of Lemma 1. Assume that $p \mid ab$ and $p \nmid a$. (We must show that $p \mid b$.) Since $p \nmid a$, it follows that a and p have only 1 as a common divisor. Hence $\gcd(a,p) = 1$. By Theorem 2 of the previous section, we have

$$1 = \gcd(a,p) = sa + tp$$

for some $s, t \in \mathbf{Z}$. Multiplying both sides by b gives

$$b = sab + tpb.$$

By assumption, $p \mid ab$. Hence $p \mid (sab + tpb)$; i.e., $p \mid b$, as required. $\square$

Terminology. When $\gcd(a,b) = 1$, we say that a and b are *relatively prime*.

Corollary. Let p be a prime. Then

$$p \mid a_1 a_2 \cdots a_n \Rightarrow p \mid a_i \text{ for some } 1 \le i \le n.$$

Proof. Lemma 1 + induction. $\square$

Theorem 1B (*Uniqueness of the prime factorization*). Every integer $n \ge 2$ can be expressed as a product of primes in essentially one way: if

$$n = p_1 p_2 \cdots p_s = q_1 q_2 \cdots q_t$$

(p_i, q_j primes; $s, t \ge 1$), then $s = t$ and there exists a reordering of the q_j's such that $p_1 = q_1, p_2 = q_2, \ldots, p_s = q_s$.

Proof. The proof is by course-of-values induction on n.

Basis ($n = 2$). The statement of the theorem holds trivially for $n = 2$, since 2 is a prime.

Induction. Assume the statement of the theorem for any integer m such that $2 \le m \le n$, and consider $n + 1$. If $n + 1$ is prime, then we are done. If $n + 1$ is not prime, then by Theorem 1A we can express $n + 1$ as a product of primes, say in two ways:

$$(*) \qquad n + 1 = p_1 p_2 \cdots p_s = q_1 q_2 \cdots q_t \qquad (s, t \ge 2).$$

Now $p_1 \mid n + 1$, whence $p_1 \mid q_1 q_2 \cdots q_t$, so that $p_1 \mid q_i$ for some $1 \le i \le t$ by the corollary to Lemma 1. By reordering the q_j's if necessary, we can assume that $p_1 \mid q_1$. Since q_1 is a prime, it has only itself as a divisor > 1. Hence $p_1 = q_1$.

Dividing through (∗) by p_1 $(= q_1)$ gives
$$(n+1)/p_1 = p_2 \cdots p_s = q_2 \cdots q_t.$$
Since $2 \le (n+1)/p_1 \le n$, our course-of-values induction hypothesis applies: $s - 1 = t - 1$ (hence $s = t$), and by reordering the q_j's we have $p_2 = q_2, \ldots, p_s = q_s$. This completes the proof of Theorem 1B by course-of-values induction. ◻

As an immediate consequence of Theorems 1A and 1B, we have the

Corollary. Any integer $n > 1$ has a unique "prime decomposition" of the form
$$n = p_1^{e_1} p_2^{e_2} \cdots p_r^{e_r} \quad (e_i \ge 1)$$
where the p_i's are the distinct prime factors of n. (Example: $360 = 2^3 \times 3^2 \times 5^1$.)

EXERCISES

Section 1

1. Appealing only to the postulated properties of the integers, prove:
   (a) There is no x such that $0 < x < 1$ (i.e., 1 is the least positive integer). [*Hint:* Use well-ordering to show that $\{x \mid 0 < x < 1\}$ must be empty.]
   (b) For any integer n, there is no x such that $n < x < n + 1$.

2. Let $[S; \le]$ be a totally ordered set. Show that $[S; \le]$ is well-ordered if and only if there is no infinite sequence $x_1, x_2, x_3, \ldots \in S$ such that $x_i > x_{i+1}$ for all $i \ge 1$.

Section 2

1. Prove:
   (a) $\sum_{i=1}^{n} i = n(n+1)/2$;
   (b) $\sum_{i=1}^{n} i^2 = n(n+1)(2n+1)/6$.

2. What is wrong with the proof of the following "theorem"?
   "Theorem". Any set $\{a_1, a_2, \ldots, a_n\}$ has the property that all the elements are the same. (!)
   "Proof". The proof is by induction on n.
   Basis. Trivially any set $\{a_1\}$ has the property that all elements are the same.
   Induction. Assuming the statement of the theorem for any set $\{a_1, a_2, \ldots, a_n\}$, we consider a set $\{a_1, a_2, \ldots, a_{n+1}\}$. Applying the induction hypothesis to the subset $\{a_1, a_2, \ldots, a_n\}$ gives $a_1 = a_2 = \cdots = a_n$; applying the induction hypothesis to the subset $\{a_2, a_3, \ldots, a_{n+1}\}$ gives $a_2 = a_3 = \cdots = a_{n+1}$. Since a_2 is common to both equality chains, we obtain $a_1 = a_2 = \cdots = a_n = a_{n+1}$, which completes the proof by induction of our "theorem".

3. (*Binomial coefficients.*) Let $\binom{n}{k}$ denote the number of k element subsets of an n element set $(0 \le k \le n)$.
   (a) Show that $\binom{n}{k} = \binom{n-1}{k-1} + \binom{n-1}{k}$.
   [*Hint:* Let S_n denote an n element set, and let $S_n = \{x\} \cup S_{n-1}$. Now consider

two classes of k element subsets of S_n: those containing x and those not containing x.]

(b) Now show that

$$\binom{n}{k} = \frac{n!}{k!(n-k)!} .$$

4. For which positive integers n do we have $2^n > n$? Prove your answer.

5. With addition and multiplication in $\mathbf{N}$ as defined in Example 2, prove:
   (a) the associative law of multiplication;
   (b) the commutative law of addition;
   (c) the commutative law of multiplication.

6. Prove the laws of exponents (b) and (c) of Example 4.

7. (*Powers of a relation; transitive closure.*) Let ρ be a relation on a set A, and let $G(\rho)$ be the graph of ρ as discussed in Section I.3.1. There is an edge from node x to node y whenever $x\rho y$.

$$\Leftrightarrow \quad x\rho y \quad (x,y \in A).$$

The powers ρ^n ($n = 1,2,3,\ldots$) with respect to relational composition have a natural interpretation in terms of "walks" along paths in the graph $G(\rho)$ (path = sequence of edges). $x\rho y$ means node y is reachable from node x by a path of length 1, i.e., by an edge; $x\rho^2 y$, which is to say $x\rho z$ and $z\rho y$ for some $z \in A$, means that node y is reachable from node x by a path of length 2 (length = number of edges); and, inductively, $x\rho^n y$, which is to say $x\rho z_1, z_1\rho z_2, \ldots, z_{n-1}\rho y$, means that node y is reachable from node x by a path of length n.

Definition. The *transitive closure* ρ_t of a relation ρ is the least transitive relation that includes ρ; i.e., ρ_t satisfies
    1. ρ_t is transitive;
    2. $[\sigma$ transitive and $\rho \subseteq \sigma] \Rightarrow \rho \subseteq \rho_t \subseteq \sigma$.

(a) (*Preliminary result.*) Let $\rho \subseteq \sigma$, with σ transitive. Prove that $\rho^n \subseteq \sigma$ for all $n \in \mathbf{P}$.
(b) Show that ρ_t, if it exists, is unique.
(c) (*Main result.*) Prove that $\rho_t = \bigcup_{i=1}^{\infty} \rho^i$. (*Note:* $x(\bigcup_{i=1}^{\infty} \rho^i)y$ means $x\rho^i y$ for some $i \in \mathbf{P}$.)
(d) In terms of $G(\rho)$, what does it mean to say that $x\rho_t y$?
(e) Let ρ be defined on an n element set A. Prove that $\rho_t = \bigcup_{i=1}^{n} \rho^i$.

8. (*Generalized associative law.*) Let $*$ be an associative operation, call it multiplication, on a set S. Prove that any two ways of multiplying the elements $a_1, a_2, \ldots, a_n$ of S in this order yield the same element of S. [E.g., for $n = 4$, the elements a_1, a_2, a_3, a_4 can be multiplied according to $a_1*(a_2*(a_3*a_4))$, $a_1*((a_2*a_3)*a_4)$, $(a_1*(a_2*a_3))*a_4$, $((a_1*a_2)*a_3)*a_4$, or $(a_1*a_2)*(a_3*a_4)$.]
   [*Hint:* Use course-of-values induction on n, where $n \geq 3$. Since $*$ is a *binary* operation (i.e., taking two operands), any product $C = a_2*a_2* \cdots *a_n$ of n a_i's (with a specified association) must have a topmost $*$, hence must be of the form $A * B$. Now observe that A and B are products involving fewer factors than C.]

9. Extend the proof of the Division Property (Example 5) from $\mathbf{N}$ to $\mathbf{Z}$: for any
    $a, b \in \mathbf{Z}$ ($b \neq 0$), show that there exist (unique) q and r such that

$$a = bq + r \qquad (0 \leq r < |b|).$$

[*Hint:* Consider in turn the cases (i) $a \geq 0$, $b > 0$, (ii) $a \geq 0$, $b < 0$, (iii) $a \leq 0$, $b < 0$,
(iv) $a \leq 0$, $b > 0$; in each case appeal to the Division Property of $\mathbf{N}$.]

10. (*Induction on well-ordered sets.*)
    (a) Let $[W; \leq]$ be a well-ordered set. Let $p(x)$ be an assertion about $x \in W$. Prove:
        If
        1. $p(x_0)$ (basis),
        2. for any $x \geq x_0$,

$$[p(y) \text{ for } x_0 \leq y < x] \Rightarrow p(x) \quad \text{(induction)},$$

then $p(x)$ for all $x \geq x_0$. [*Hint:* Let S be the set of all $x \geq x_0$ for which $p(x)$ is
false.]
    (b) Let $[W; \leq]$ be a well-ordered set. Show that $W \times W$ is well-ordered under
        lexicographic order:

$$(a, b) \leq (c, d) \text{ if and only if}$$

$$\text{(i) } a < c \quad \text{or} \quad \text{(ii) } a = c \text{ and } b \leq d.$$

(c) (*An application.*) For any $m, n \in \mathbf{P}$, define $(m)_n = m(m+1) \cdots (m+n-1)$.
    Show that $n! \mid (m)_n$. (Thus the product of n consecutive positive integers is
    divisible by $n!$.)

Section 3

1. Prove that there exist infinitely many primes. [*Hint:* Suppose, to the contrary, that
    the nth prime, p_n, is the last one. Now consider the product $N = p_1 p_2 \cdots p_n + 1$.]

2. Prove: For $a, b \in \mathbf{Z}$, if $sa + tb = 1$ for some $s, t \in \mathbf{Z}$, then $(a, b) = 1$.

3. (*Least common multiples.*) The *least common multiple* (l.c.m.) l of two (say positive)
    integers a and b is characterized by the properties
        1. $a \mid l$ and $b \mid l$ (l is a *common* multiple of a and b);
        2. $a \mid c$ and $b \mid c \Rightarrow l \mid c$ (l is the *least* common multiple of a and b).
    (Cf. the *dual* properties of the g.c.d.)
    (a) Show that the l.c.m. of a and b is unique up to sign.

    Notation: We denote the (unique) positive l.c.m. of a and b by $\mathrm{lcm}(a, b)$ or by
    $[a, b]$.

    (b) Prove that if $(a, b) = 1$, then $[a, b] = ab$.
    (c) Prove that $[ka, kb] = k[a, b]$ ($k > 0$).
    (d) (*A useful relationship between g.c.d.'s and l.c.m.'s.*) Conclude that $[a, b] = ab/(a, b)$.

4. (*G.c.d.'s and l.c.m.'s by prime decomposition.*) Let the positive integers a and b have prime decompositions

$$a = p_1^{e_1} p_2^{e_2} \cdots p_n^{e_n} \qquad (e_i \geq 0)$$

$$b = p_1^{f_1} p_2^{f_2} \cdots p_n^{f_n} \qquad (f_i \geq 0)$$

where $p_1, \ldots, p_n$ are distinct prime factors of either a or b. Define

$$g = p_1^{g_1} p_2^{g_2} \cdots p_n^{g_n} \quad \text{where } g_i = \min(e_i, f_i),$$

$$l = p_1^{l_1} p_2^{l_2} \cdots p_n^{l_n} \quad \text{where } l_i = \max(e_i, f_i).$$

Show (i) that $g = (a, b)$ and (ii) that $l = [a, b]$.

PART
TWO
ALGEBRAIC SYSTEMS

PROLOGUE TO ALGEBRA: ALGEBRAIC SYSTEMS AND ABSTRACTION

In simplest terms an *algebraic system* is a set with operations, e.g., the set of integers under the operations of addition and multiplication. We embellish this idea with a formal

Definition. An *algebraic system* (*algebra*, Ω-*algebra*) is a pair $[A; \Omega]$ where
1. A is a set (the "carrier" of the algebraic system).
2. Ω is a collection of operations defined on A: each operation $\omega \in \Omega$ is a function $A^n \to A$, taking operands $a_1, \ldots, a_n \in A$ into $\omega(a_1, \ldots, a_n) \in A$, where the "arity" n of ω is a nonnegative integer that depends on ω, i.e., $n = n(\omega)$.

We often denote an algebraic system $[A; \Omega]$ simply by "A" if the set Ω of operations is fixed and understood, letting context decide whether we mean the algebra or just the carrier of the algebra. As a minor notational point, we write $[A; \omega, \sigma, \ldots]$ for $[A; \{\omega, \sigma, \ldots\}]$; e.g., $[\mathbf{Z}; +, \cdot]$ denotes the set of integers under addition and multiplication.

1. LEVELS OF ABSTRACTION

While the idea of an algebraic system captures the substance of algebra, it is *abstraction*—generality, suppression of detail—that captures the spirit. Three

John D. Lipson, Elements of Algebra and Algebraic Computing. ISBN 0-201-04115-4.

57

quite distinct levels of abstraction have evolved over the long course of algebra's development.

A. *Concrete level.* This is the level of the empirical (real) world of mathematics and its applications, where one deals with specific sets of elements and operations on these sets that are defined by rules or algorithms of combination. In general one "knows" an algebraic system $[A; \Omega]$ at the concrete level when one knows what the elements of the carrier A are and how to evaluate the operations of Ω over A.

Several miscellaneous examples of algebraic systems follow.

Example 1. $[S; +, \cdot]$ where for S we take any of the familiar number sets **P, N, Q, R,** or **C.** $\square$

Example 2. $[\mathbf{P}; \gcd, \mathrm{lcm}]$, the set of positive integers under the binary (arity $= 2$) operations of taking the greatest common divisor and least common multiple. (E.g., $\gcd(4, 6) = 2$, $\mathrm{lcm}(4, 6) = 12$.) $\square$

For the number of elements of a set A we use the notation $|A|$. For the case of an algebraic system $[A : \Omega]$ we refer to $|A|$ as its *order*. Further, we call an algebraic system *finite* or *infinite* according to whether its order is finite or infinite.

The algebraic systems of Examples 1 and 2 are infinite. An example of a finite algebraic system is given by

Example 3. $[\mathbf{Z}_m; \oplus, \odot]$, the set $\mathbf{Z}_m = \{0, 1, \ldots, m - 1\}$ under *mod m addition* and *mod m multiplication:*

$$a \oplus b = r_m(a + b), \qquad a \odot b = r_m(a \cdot b),$$

where $r_m(x)$ is the least positive remainder ($0 \le r_m(x) < m$) on dividing x by m. (Thus, if $m = 6$ then $2 \oplus 3 = 5$, $2 \oplus 5 = 1$, $2 \odot 3 = 0$, $2 \odot 5 = 4$.) $\square$

Example 4. $[P(I); \cup, \cap, ']$, the power set of some set I under the binary operations of union and intersection, and the unary (arity $= 1$) operation of complementation. $\square$

Example 5. $[A^A; \circ]$, the set of all functions $A \to A$ under composition. (Recall: $h = f \circ g$ means $h(x) = f(g(x))$ for all $x \in A$.) $\square$

Example 6. $[\mathbf{R}^n; +, \{\lambda_r\}_{r \in \mathbf{R}}]$, the set $\mathbf{R}^n$ of all *n*-tuples (vectors) $\mathbf{a} = (a_1, \ldots, a_n)$ of real numbers under (i) *n*-tuple (component-wise) addition,

$$\mathbf{a} + \mathbf{b} = (a_1 + b_1, \ldots, a_n + b_n),$$

and (ii) a family of unary operations λ_r, one for each real number r, defined by

$$\lambda_r(\mathbf{a}) = (ra_1, \ldots, ra_n) \quad (scalar\ multiplication).$$

Thus the algebra of vectors under vector addition and scalar multiplication has been captured as an algebraic system according to our general definition by the expedient of defining scalar multiplication (ordinarily regarded as a map $\mathbf{R} \times \mathbf{R}^n$ $\rightarrow \mathbf{R}^n$) as a *family* of unary operations $\mathbf{R}^n \rightarrow \mathbf{R}^n$, one such operation for each element ("scalar") $r \in \mathbf{R}$.

Total vs. partial algebras. The statement that ω is an n-ary operation $A^n \rightarrow A$ tacitly carries with it the force that ω is a *total* operation, i.e., defined on all of A^n. If we wish to emphasize that ω is a total operation, we say that A is *closed* under ω.

The vast majority of the operations that we shall deal with will indeed be closed on their carriers, but on occasion we shall wish to permit some of the operations of an algebraic system to be only partially defined. This leads to the notion of a *partial algebra*, which differs from an ordinary (total) algebra in that some of the operations may only be partial.

Familiar examples of partial algebras include:

Example 7. $[\mathbf{P}; -]$, the positive integers under binary subtraction. This is a partial algebra because $m - n$ is undefined if $m \le n$. $\square$

Example 8. $[\mathbf{R}; +, -, \times, /]$, the real numbers under addition, subtraction, multiplication, and division. This is a partial algebra because of division: one cannot divide by zero. $\square$

Remark. The partial algebra of Example 7 can readily be converted to a total algebra by extending the carrier, $\mathbf{P}$, to the set $\mathbf{Z}$ of *all* integers. No such expedient works for Example 8, however. One might exclude zero from $\mathbf{R}$; this would make division a total operation. But addition would then become a partial operation: $a + (-a) = 0$ for any real number a. $[\mathbf{R}; +, -, \times, /]$ is irrevocably partial.

We now pass to our next, and for us the most important, level of abstraction.

B. *Postulational level.* At the *postulational*, or *axiomatic*, level of abstraction, an algebraic system is defined by *postulates* (*axioms*, *laws*) which specify general properties, rather than values, of operations. The axiomatic approach to the study of algebraic systems forms the cornerstone of so-called modern algebra. (Note 1.)

Example 9. A *monoid* $[M; \cdot, 1]$ is a postulationally defined algebraic system in which the binary operation $\cdot$ (call it multiplication) and the nullary (arity $= 0$) operation 1 (select the *identity element* 1 of M) satisfy the laws

    1. $x \cdot (y \cdot z) = (x \cdot y) \cdot z$ (associative law);
    2. $1 \cdot x = x \cdot 1 = x$ (identity law). $\square$

Remarks. 1. A *nullary* (arity $= 0$) operation is an operation which takes no operands. One can regard a nullary operation as selecting a fixed element of the carrier; this fixed element is the constant value of the nullary operation. We usually denote a nullary operation by the element it selects; e.g., in **Z**, the nullary operations 0 and 1 select the elements 0 and 1 of **Z**.

2. To say that the above equations are laws (axioms, postulates) means that they hold for *all* $x, y, z \in M$, i.e., that they are *identities* over *M*.

The above notion of an *abstract* monoid (abstract in the sense that the set-theoretic nature of *M*, ·, and 1 are left unspecified) is seen to generalize many familiar *concrete* monoids (the reader is invited to fill in the missing identity elements):

$$[\mathbf{N}; +, 0]$$

$$[\mathbf{P}; \cdot, 1]$$

$$[P(A); \cup, ?]$$

$$[P(A); \cap, ?]$$

$$[A^A; \circ, ?]$$

There are a number of interrelated factors that contribute to the tremendous success and popularity of the postulational approach to algebra:

(a) *Simplicity*. In order to understand the mass of detail that is to be found in the tremendous variety of concrete algebraic systems, it is necessary to simplify and distinguish the significant features of an algebraic system from the merely incidental. The significant facts about algebraic systems, facts from which nontrivial consequences flow, have invariably turned out to involve laws governing the general behavior of operations. (Thus knowledge that integer addition and multiplication is associative and commutative undoubtedly yields a more penetrating insight into the structure of integers—into what makes them tick—than any sampling of "concrete" facts, such as $3 + 5 = 8$, or $2^{31} - 1$ is a prime.) The postulational approach focuses attention on these significant aspects of algebraic systems—laws of general behavior of operations—by the sensible expedient of ignoring inessential (concrete) details. (Note 2.)

(b) *Economy of effort*. The postulational approach offers the possibility of a tremendous return on invested effort, because any concept or result that is established for an abstract system applies with full force to any of its concrete instances.

(c) *Structuring of complexity*. The postulational approach permits one to investigate ever more complex algebraic systems by progressively adding postulates and operations to already existing abstractions—bricks to walls to houses.

Example 10 (*Monoids to groups to rings*). With the definition of monoid in hand, we arrive at the important concept of a *group* by focussing attention on invertibility with respect to the postulated binary operation.

A *group* $[G; \cdot, {}^{-1}, 1]$ is an algebraic system that satisfies

1. $[G; \cdot, 1]$ is a monoid;
2. $x \cdot x^{-1} = x^{-1} \cdot x = 1$ (inverse law).

By postulate 1, group theory is a specialization of monoid theory. Hence any concept or result about monoids applies *a fortiori* to groups.

Familiar examples of numerical groups include the additive group of integers, $[\mathbf{Z}; +, -, 0]$, in which $-m$ is the (additive) inverse of m, and the multiplicative group of nonzero numbers, $[\mathbf{R} - \{0\}; \cdot, {}^{-1}, 1]$, in which the reciprocal $1/x$ is the (multiplicative) inverse of x. (We use the modifiers "additive" or "multiplicative" according to whether the binary operation of the group is denoted by an addition or a multiplication symbol.)

From groups, everyday arithmetic leads naturally to the concept of a *ring*, where addition and multiplication are dealt with simultaneously.

A *ring* $[R; +, -, 0, \cdot, 1]$ is an algebraic system that satisfies

1. $[R; +, -, 0]$ is a group   (the *additive group* of R);
2. $[R; \cdot, 1]$ is a monoid   (the *multiplicative monoid* of R);
3. $a(b + c) = ab + ac, (b + c)a = ba + ca$ (distributive laws).

The concept of an abstract ring constitutes one of the most fruitful and beautiful abstractions of all algebra. It models the structure of a host of important concrete algebraic systems, including the familiar number systems (integers, reals, rationals, complex numbers), polynomials, matrices, and real and complex functions. And it does so in a structured way. The postulates decompose a ring into a purely additive part, having the structure of a group; a purely multiplicative part, having the structure of a monoid (noncommutativity accommodates the matrix example); and the familiar distributive laws which relate the additive and multiplicative parts. (If multiplication is commutative, then either one of the distributive laws is of course sufficient.)

The relevance of rings to computing is obvious. But most significant for us in this regard is the combination of ideas from modern algebra and computer science that have injected a vitality and depth into the algorithmic side of algebra not to be found in pre-computer treatments. (These ideas shall be our preoccupation in Part Three of this book.) □

C. *Universal level.* The universal level of abstraction, the viewpoint of *universal algebra*, represents for us the high water mark of abstraction in the treatment of algebraic systems. The new idea here in the direction of increased generality is to leave the postulates unspecified and to regard a (universal) algebraic system as a set with operations—period.

Now, accepting the truism in mathematics that the more abstract the viewpoint the more superficial the attainable results, one might wonder whether *anything* of interest can be said about something as general as a universal algebraic system. Remarkably enough, it turns out that a great deal can be said (as we shall see, in particular, in Chapters III and V). (Note 3.)

2. HETEROGENEOUS ALGEBRAS

Preliminary remark.   We include this brief section on heterogeneous algebras to give the reader a sense of the wide applications of algebraic ideas and techniques to computer science related topics. (Note 1.)

The main idea behind the heterogeneous algebra concept is to permit one to deal simultaneously with more than one carrier.

Definition.   A *heterogeneous algebra* $[\{A_i\}_{i \in I}; \Omega]$ consists of a family of different "types" A_i of elements together with a collection Ω of operations defined on these types in the following way: Associated with each n-ary operation $\omega \in \Omega$ $(n = n(\omega))$ is an $n + 1$-tuple $(i_1, \ldots, i_n, i_{n+1}) \in I^{n+1}$; ω is then a map (a *heterogeneous* operation)

$$\omega \colon A_{i_1} \times \cdots \times A_{i_n} \to A_{i_{n+1}}.$$

Thus the kth operand of ω is taken from A_{i_k} $(k = 1, \ldots, n)$, and the value of ω lies in $A_{i_{n+1}}$.

Example 1 (*Automata*).   An *automaton* (*sequential machine*) is a heterogeneous algebra $[S, X, Z; \delta, \lambda]$ with three carriers:

S is the set of *states*,
X is the *input alphabet*,
Z is the *output alphabet*;

and two operations:

$\delta \colon S \times X \to S$ is the *next-state* function: $\delta(s, x)$ is the next state that results when the machine is in state s and receives input x;
$\lambda \colon S \times X \to Z$ is the *output* function: $\lambda(s, x)$ is the output produced by the machine when it is in state s and receives input x.

This purely mathematical concept of a finite automaton models the salient features of many information-processing, or computer-like, devices. We think of an automaton as starting in some initial state s_0 (state = configuration of memory) and then processing an "input tape" $x_1 x_2 \cdots x_n$ consisting of a sequence of symbols from the input alphabet by (1) changing states $s_0 \to s_1 \to \cdots \to s_n$ according to the next-state function δ and (2) producing an "output tape" $z_1 z_2 \cdots z_n$ according to the output function λ:

$$s_i = \delta(s_{i-1}, x_i),$$
$$z_i = \lambda(s_{i-1}, x_i) \qquad (i = 1, 2, \ldots, n).$$

(Note 2.)

As an illustration of the setup, an automaton for performing binary addition is given below. (In a graphical description of an automaton,

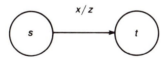

means $\delta(s, x) = t$, $\lambda(s, x) = z$.)

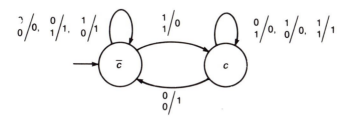

Here we have

$$S = \{c \text{ (carry state)}, \bar{c} \text{ (no carry state)}\} \qquad (\text{initial state} = \bar{c}),$$

$$X = \left\{ \frac{0}{0}, \frac{0}{1}, \frac{1}{0}, \frac{1}{1} \right\} \qquad (\text{each input is a pair of bits}),$$

$$Z = \{0, 1\}.$$

Sample execution.

						←(pairs of bits read from right to left)

Input tape: 0 0 1 1   (= 3)
               1 0 0 1   (= 9)

States: $\bar{c}$ $\bar{c}$ c c $\bar{c}$
                                        ↑
                                  (initial state)

Output: 1 1 0 0   (= 12 ✓) □

Example 2 (*Structured data types*). The structured data types of computer programming—arrays, stacks, queues, trees, and the like—can be conveniently formulated as heterogeneous algebras. We illustrate this for the data type stack.

Informally, a stack is a list, or vector, with a special accessing discipline: last-in first-out, with only the first, or top, element of the stack being accessible at any instant (this in contrast with an ordinary vector, which permits *random* access, i.e., access to any component). Now for the algebraic definition.

A *stack of* (say) *integers* is a heterogeneous algebra

[INTEGERSTACKS, INTEGERS, ERRORS; PUSH, POP, TOP, EMPTY]

(we are abiding by the convention in programming languages of spelling out the names of carriers and operations) with carriers

INTEGERSTACKS = the set of all stacks with integer components, includ-
ing the empty stack, EMPTY,
INTEGERS = the set of integers,
ERRORS = {STACKERROR, INTEGERERROR};

and operations

PUSH: INTEGERSTACKS × INTEGERS → INTEGERSTACKS,
POP: INTEGERSTACKS → INTEGERSTACKS ∪ ERRORS,
TOP: INTEGERSTACKS → INTEGERS ∪ ERRORS,
EMPTY: a nullary operation returning EMPTY (the empty stack).

The *semantics* of the data type stack, i.e., the intended behavior of the stack operations, is at least suggested by the names of the operations. Examples:

if stack = (4, 9, 3), then PUSH(stack, 7) = (7, 4, 9, 3);
if stack = EMPTY, then PUSH(stack, 7) = (7);
if stack = (4, 9, 3), then POP(stack) = (9, 3);
if stack = EMPTY, then POP(stack) = STACKERROR;
if stack = (4, 9, 3), then TOP(stack) = 4;
if stack = EMPTY, then TOP(stack) = INTEGERERROR.

Examples are fine, but of course it behooves us to find a formal, more comprehensive method for specifying the semantics of a data type. Interestingly enough, the postulational approach of modern algebra provides just such a method, which we now illustrate for the case of stacks.

A stack of integers is now defined as a heterogeneous algebra as above (this is the *syntactic* part of the definition) that satisfies the following postulates (this is the *semantic* part of the definition):

1. TOP(PUSH(stack, integer)) = integer,
2. TOP(EMPTY) = INTEGERERROR,
3. POP(PUSH(stack, integer)) = stack,
4. POP(EMPTY) = STACKERROR.

These "laws of stackhood" are to operations on stacks as the associative and commutative laws are to addition and multiplication of numbers. Most significant from a programming methodology viewpoint is that the postulational approach to data type definition yields a specification that is free of implementation (largely incidental) details.

For more on this relatively new subject of applying algebraic methods to data type specification, we refer the reader to the literature. (Note 3.) □

PROLOGUE NOTES

Section 1

1. The adjective "modern" persists even though the nature of modern algebra was incisively stated in an 1840 paper of D. F. Gregory entitled "On the Real Nature of Symbolical Algebra" [*Trans. Roy. Soc. Edinburgh* **14**, (1840), 208–216]:

> The light, then, in which I would consider symbolical algebra, is, that it is the science which treats of the combination of operations defined not by their nature, that is, by what they are or what they do, but by the laws of combination to which they are subject. And as many different kinds of operations may be included in a class defined in the manner I have mentioned, whatever can be proved of the class generally, is necessarily true of all the operations included under it. This, it may be remarked, does not arise from any analogy existing in the nature of the operations, which may be totally dissimilar, but merely from the fact that they are all subject to the same laws of combination.

Gregory (1813–1841) was a member of the British algebraic school, which included such notable members as G. Boole (1815–1864), W. R. Hamilton (1805–1865), A. De Morgan (1806–1871), and A. Cayley (1821–1895). This school vigorously developed the postulational approach to algebra during the mid 1800's, and was largely responsible for liberating algebra from its previously exclusive preoccupation with numbers.

2. But what are the inessential details? The history of algebra gives evidence that the answer is far from obvious, that abstraction is a delicate art indeed. Modern algebra itself is founded on the remarkable premise that the property of an operation to be discarded as "inessential detail" is its value: it no longer matters that $2 + 3 = 5$, only that '+' is associative, commutative, etc. The counter-intuitive nature of this abstraction—that discarding something as fundamental as the value of an operation should result in a useful abstraction—probably accounts for the fact that the axiomatic method was relatively late entering algebra, some two thousand years after it had exerted a profound influence on geometry. (It probably also explains why students typically find the first step from manipulative to abstract algebra a rather difficult one.)

3. Universal algebra was conceived, albeit in the vaguest of form, by A. N. Whitehead in his 1898 treatise *Universal Algebra*, but no real technical progress was achieved until the 1930's and the publication of G. Birkhoff's important papers on the subject [in particular, *Proc. Cambridge Phil. Soc.* **31** (1935), 433–454]. G. Grätzer, in the introduction to his book *Universal Algebra* (Princeton, N. J.: Van Nostrand, 1968), points out the vital role played by experience in the evolution of abstract ideas: "Some thirty years elapsed between Whitehead's book and Birkhoff's first paper, despite the fact that the goal of research was so beautifully stated in Whitehead's book. However, to generalize, one needs experience, and before the thirties most of the branches of modern algebra were not developed sufficiently to give impetus to the development of universal algebras."

Section 2

1. An early paper on heterogeneous algebras is P. J. Higgins [*Math. Nachr.* **27** (1963), 115–132]. For a development of the general theory of heterogeneous algebras, with applications to automata, see G. Birkhoff and J. D. Lipson [*J. Combinatorial Theory* **8** (1970), 115–133; *Proc. Symp. in Pure Math.*, Vol. XXV, (Providence, R.I.: AMS, 1974), 41–51].

2. For a systematic development of (finite) automata theory, from both the engineering and mathematical viewpoints, see Z. Kohavi, *Switching Theory and Finite Automata Theory* (New York: McGraw-Hill, 1970).

3. The algebraic approach to data type specification presented in this section was gleaned from the work of J. V. Guttag [*Comm. ACM* **20** (1977), 396–404; *IEEE Trans. Software Eng.* **SE-6** (1980), 13–23]. For further material on abstract approaches to the theory and practice of data types, see C. A. R Hoare [*Acta Informatica* **1** (1972), 271–281] and the collection *Current Trends in Programming Methodology*, Vol. IV: *Data Structuring*, R. T. Yeh (ed.) (Englewood Cliffs, N.J.: Prentice-Hall, 1978), in particular the chapters by C. A. R. Hoare (pp. 1–11); T. A. Standish (pp. 30–50); J. V. Guttag, E. Horowitz, and D. R. Musser (pp. 60–79); J. A. Goguen, J. W. Thatcher, and E. G. Wagner (pp. 80–149); also the chapter by B. Liskov and S. Zilles in Vol. I of the above *Programming Methodology* series (pp.1–32).

CHAPTER

SEMIGROUPS, MONOIDS, AND GROUPS

In this chapter we highlight some important *binary* algebraic systems. By a "binary" algebraic system we mean one that has a single binary operation. There may be other unary or nullary operations, but only one binary operation.

1. BASIC DEFINITIONS AND EXAMPLES

1.0 Groupoids

Definition. A *groupoid* $[G; \cdot]$ is an algebraic system with a binary operation $\cdot : (x,y) \mapsto x \cdot y$ ($x \cdot y$ is also denoted by xy). No postulates are specified.

We call a groupoid *multiplicative* or *additive* according to whether the binary operation is denoted by a multiplication symbol or an addition symbol.

The concept of a groupoid is too rudimentary to be interesting. We next bring some algebraic structure into the picture.

1.1 Semigroups

Definition. A *semigroup* $[S; \cdot]$ is a groupoid that satisfies the associative law: $x(yz) = (xy)z$.

John D. Lipson, Elements of Algebra and Algebraic Computing. ISBN 0-201-04115-4.

Some examples of semigroups:

Example 1 (*From ordinary arithmetic*).

(a) $[\mathbf{P}; +]$;   (b) $[\mathbf{P}; \cdot]$;
(c) $[\mathbf{R}; +]$;   (d) $[\mathbf{R}; +, \cdot]$;   (e) $[\{0, 1\}; \cdot]$. $\square$

Example 2 (*Integers mod m*).

(a) $[\mathbf{Z}_m; \oplus]$, where $a \oplus b = r_m(a + b)$ (addition mod m);
(b) $[\mathbf{Z}_m; \odot]$, where $a \odot b = r_m(ab)$ (multiplication mod m).

As a preliminary step to verifying the associativity of mod m addition and multiplication, we establish the useful

Proposition 1.

(a) $r_m(a + b) = r_m(r_m(a) + r_m(b))$;
(b) $r_m(ab) = r_m(r_m(a)r_m(b))$.

Proof.   It is trivially shown that

$$a + b \equiv_m r_m(a) + r_m(b),$$

$$ab \equiv_m r_m(a)r_m(b).$$

Proposition 1 then follows from Lemma 1(c) of Section II.3.1. $\square$

Now for the associativity of $\oplus$. For any $a, b, c \in \mathbf{Z}_m$,

$$
\begin{aligned}
a \oplus (b \oplus c) & \\
= r_m(a + r_m(b + c)) \qquad & \text{defn. of } \oplus \\
= r_m(r_m(a) + r_m(b + c)) \qquad & r_m(a) = a \\
= r_m(a + (b + c)) \qquad & \text{Proposition 1(a)} \\
= r_m((a + b) + c) \qquad & \text{associativity of } + \\
= r_m(r_m(a + b) + r_m(c)) \qquad & \text{Proposition 1(a)} \\
= (a \oplus b) \oplus c \qquad & \text{defn. of } \oplus
\end{aligned}
$$

The verification for $\odot$ is completely analogous. $\square$

We continue with our examples of semigroups.

Example 3 (*Set-related*).

(a) $[P(I); \cup]$; (b) $[P(I); \cap]$;
(c) $[\{0, 1\}; \vee]$; (d) $[\{0, 1\}; \wedge]$ (verify). $\square$

A semigroup (more generally a groupoid) is called *commutative* if its binary operation satisfies the commutative law: $xy = yx$. By convention, an additive semigroup is always taken to be commutative: $x + y = y + x$.

The above examples of semigroups are all commutative; now for a couple of noncommutative examples.

Example 4 (*Functions*). $[A^A; \circ]$, the set of all functions $A \to A$ under composition. (We noted the associativity of $\circ$ in Section I.4.2.) □

Example 5 (*Strings*). $[\Sigma(A); \circ]$, the set $\Sigma(A)$ of all finite strings $a = a_1 a_2 \cdots a_n$ over an alphabet A (each $a_i \in A$), including the empty ($n = 0$) string, e, under the operation of *concatenation:* If $a = a_1 \cdots a_n$ and $b = b_1 \cdots b_m$ are strings in $\Sigma(A)$, then

$$a \circ b = a_1 \cdots a_n b_1 \cdots b_m, \qquad a \circ e = e \circ a = a.$$

The associative law is immediate: If a and b are as above and $c = c_1 \cdots c_p$ is also a string in $\Sigma(A)$, then $a \circ (b \circ c)$ and $(a \circ b) \circ c$ both yield the string $a_1 \cdots a_n b_1 \cdots b_m c_1 \cdots c_p$. □

Example 6 (The *direct product*). Let $[S; \cdot]$ and $[T; \cdot]$ be two (multiplicative) semigroups. The algebraic system $[S \times T; \cdot]$ consisting of the Cartesian product of S and T under componentwise multiplication,

$$(s_1, t_1)(s_2, t_2) = (s_1 s_2, t_1 t_2),$$

is trivially verified to be a semigroup. This semigroup is called the *direct product* of the semigroups $[S; \cdot]$ and $[T; \cdot]$ (direct *sum* in the case of two additive semigroups). □

Two nonsemigroups:

Example 7. $[\mathbf{Z}; -]$, the integers under binary subtraction. $[\mathbf{Z}; -]$ is a perfectly good groupoid, but is not a semigroup, since the associative law obviously does not hold (e.g., $1 - (2 - 3) = 2$ whereas $(1 - 2) - 3 = -4$). □

Example 8. $[\{0, 1\}; +]$ (cf. Example 1(e)). Here we do not even have a groupoid. □

1.2 Monoids

The concept of a monoid focusses attention on the idea of an *identity*, or *neutral*, element.

> **Definition.** A *monoid* $[M; \cdot, 1]$ is an algebraic system with a binary operation $\cdot$ and a nullary operation 1 ("select the identity element"), that satisfies
>
> 1. $[M; \cdot]$ is a semigroup;
> 2. $x1 = 1x = x$ (identity law).

Recall from the Prologue (Example 9 of Section 1) that we denote a nullary (essentially a selector) operation by the element it selects. Thus, in a monoid $[M; \cdot, 1]$ the nullary operation 1 selects an identity element, 1, of M, i.e., an element satisfying the identity law of postulate 2.

For a multiplicative monoid $[M; \cdot, 1]$, we refer to the identity element as a "one" or "unit element." For an additive monoid the notation changes to $[M; +, 0]$, and we refer to the identity element 0 as a "zero."

Some examples of monoids (cf. our examples of semigroups—Examples 1–6 of Section 1.1):

Example 1.

(a) $[\mathbf{N}; +, 0]$;   (b) $[\mathbf{P}; \cdot, 1]$;
(c) $[\mathbf{R}; +, 0]$;   (d) $[\mathbf{R}^+; \cdot, 1]$;
(e) $[\{0, 1\}; \cdot, 1]$. □

Example 2.   (a) $[\mathbf{Z}_m; \oplus, 0]$;   (b) $[\mathbf{Z}_m; \odot, 1]$. □

Example 3.   (a) $[P(I); \cup, \varnothing]$;   (b) $[P(I); \cap, I]$;   (c) $[\{0, 1\}; \vee, 0]$;
(d) $[\{0, 1\}; \wedge, 1]$. □

Example 4. $[A^A; \circ, 1_A]$. (Recall: 1_A is the identity function on A: $1_A(a) = a$ for all $a \in A$.) □

Example 5. $[\Sigma(A); \circ, e]$. □

Example 6. $[M \times N; \cdot, (1_M, 1_N)]$, the direct product of monoids $[M; \cdot, 1_M]$ and $[N; \cdot, 1_N]$. ($M \times N$ under the indicated operations is trivially seen to be a monoid.) □

Three non-monoid semigroups:

Example 7. $[\mathbf{P}; +]$. □

Example 8. $[\{\text{even integers}\}; \cdot]$. □

Example 9. $[\mathbf{P}; \gcd]$ (Exercise 2). □

1.3   Groups

Our next algebraic system undoubtedly constitutes the most important and intensively studied of all binary algebraic systems.

Definition.   A *group* $[G; \cdot, {}^{-1}, 1]$ is an algebraic system with one binary operation $\cdot$, one unary operation ${}^{-1}$ ("take the inverse"), and one nullary operation 1, that satisfies

  1. $[G; \cdot, 1]$ is a monoid;
  2. $xx^{-1} = x^{-1}x = 1$ (inverse law).

The element x^{-1} is called the (*multiplicative*) *inverse* of x (whose uniqueness will be proved in Section 2.2). Using additive notation, a group would be denoted $[G; +, -, 0]$, with $-x$ denoting the (additive) inverse of x.

In a multiplicative group $[G; \cdot, ^{-1}, 1]$, we define *division* according to

$$a/b = ab^{-1};$$

and in an additive group $[G; +, -, 0]$, we define *subtraction* analogously according to

$$a - b = a + (-b).$$

Some examples of groups:

Example 1.

(a) $[\mathbf{N}; +, -, 0]$;   (b) $[\mathbf{Q}^+; \cdot, ^{-1}, 1]$;
(c) $[\mathbf{R}; +, -, 0]$;   (d) $[\mathbf{R}^*; \cdot, ^{-1}, 1]$;
(e) $[\{1\}; \cdot, ^{-1}, 1]$ (a trivial group but a group none the less).

We emphasize with force that the multiplicative monoids $\{0, 1\}$, $\mathbf{Q}$, $\mathbf{R}$, $\mathbf{C}$, and $\mathbf{Z}_m$ (under $\cdot$ and 1) are not groups: zero does not have a (multiplicative) inverse. $\square$

Example 2.   (a) $[\mathbf{Z}_m; \oplus, \ominus, 0]$, where

$$\ominus a = m - a \quad \text{if } a \neq 0,$$
$$= 0 \qquad \text{if } a = 0.$$

Since $a \oplus (\ominus a) = 0$ (trivially checked), $\ominus a$ is the additive inverse of a.

(b) We noted above that the multiplicative monoid $[\mathbf{Z}_m; \odot, 1]$ is not a group. It is tempting to discard zero and claim that $\mathbf{Z}_m^* = \mathbf{Z}_m - \{0\}$ is a monoid. But consider $\mathbf{Z}_4^* = \{1, 2, 3\}$. Here we have $2 \odot 2 = r_4(2 \times 2) = 0$, so we fail even to have closure. However, a certain class of $\mathbf{Z}_m^*$'s does yield multiplicative groups. This is the content of the important

Theorem 1.   Under mod m multiplication,

$$\mathbf{Z}_m^* \text{ is a multiplicative group} \Leftrightarrow m \text{ is prime.}$$

Proof.   ($\Rightarrow$) If m is composite, say $m = st$ with $1 < s, t < m$, then

$$s \odot t = r_m(st) = r_m(m) = 0$$

so that $\odot$ is not closed on $\mathbf{Z}_m^*$. In this case $\mathbf{Z}_m^*$ is not even a groupoid under $\odot$.

($\Leftarrow$) Let m be a prime p.

Closure under $\odot$.   Let $a, b \in \mathbf{Z}_p^*$. We must show that $a \odot b \in \mathbf{Z}_p^*$, i.e., that $a \odot b \neq 0$. Now

$$a \odot b = 0 \Rightarrow r_p(ab) = 0 \qquad \text{defn. of } \odot$$

$$\Rightarrow p \,|\, ab \qquad\qquad \text{defn. of } r_p$$

$$\Rightarrow p \,|\, a \text{ or } p \,|\, b \qquad \text{Lemma 1 of Sec. II.3.3,}$$

which is impossible, since $1 \leq a, b < p$. We are thus forced to conclude that $a \odot b \neq 0$, which proves closure.

Associativity of $\odot$.   This follows *a fortiori* from the associativity of $\odot$ in $\mathbf{Z}_p$ (Example 2 of Section 1.1). Hence $[\mathbf{Z}_p^*; \odot]$ is a semigroup.

Multiplicative identity element. $1 \in \mathbf{Z}_p^*$ is an identity element with respect to $\odot$. Hence $[\mathbf{Z}_p^*; \odot, 1]$ is a monoid.

Multiplicative inverses.   Let $a \in \mathbf{Z}_p^*$. Since p is prime and $1 \le a < p$, it follows that a and p have only 1 as a common (positive) divisor. Hence $\gcd(a,p) = 1$, and by Theorem 2 of Section II.3.2 we can write

$$(*) \qquad\qquad 1 = \gcd(a,p) = sa + tp$$

for integers s and t. The claim now is that $r_p(s)$ is a multiplicative inverse of a.

Proof of claim: We have

$$
\begin{aligned}
1 &= r_p(1) \\
&= r_p(sa + tp) && \text{from } (*) \\
&= r_p(sa) && sa \equiv_p sa + tp \\
&= r_p(r_p(s)r_p(a)) && \text{Proposition 1(b) of Sec. 1.1} \\
&= r_p(s) \odot r_p(a) && \text{defn. of } \odot \\
&= r_p(s) \odot a
\end{aligned}
$$

so that $a^{-1} = r_p(s) \in \mathbf{Z}_p^*$, as claimed. Hence $[\mathbf{Z}_p^*; \odot, {}^{-1}, 1]$ is a group, which completes the proof of Theorem 1. (See Fig. 1 for one such group.) $\square$

The groups that we have encountered thus far have all been commutative. (*Terminology.* Commutative groups are also called *abelian* in honor of the Norwegian mathematician N. H. Abel (1802–1829).) Now for an important noncommutative (nonabelian) example.

Example 3. $[S_X; \circ, {}^{-1}, 1_X]$, where X is a given set and S_X is the set of all bijections on X. This group is called the *symmetric group on the set X*. If X is finite having say n elements, then S_X is denoted by S_n and referred to as the *symmetric group on n letters* (Exercise 3). $\square$

$\odot$	1	2	3	4	5	6		x	x^{-1}
1	1	2	3	4	5	6		1	1
2	2	4	6	1	3	5		2	4
3	3	6	2	5	1	4		3	5
4	4	1	5	2	6	3		4	2
5	5	3	1	6	4	2		5	3
6	6	5	4	3	2	1		6	6

Fig. 1.   Multiplication and inverse tables for $\mathbf{Z}_7^*$.

Example 4. $[G \times H; \cdot, ^{-1}, (1_G, 1_H)]$, the direct product of groups $[G; \cdot, ^{-1}, 1_G]$ and $[H; \cdot, ^{-1}, 1_H]$, where inversion, like multiplication, is defined component-wise: $(g, h)^{-1} = (g^{-1}, h^{-1})$. (It is trivially verified that $G \times H$ under the indicated operations is a group—abelian if G and H are both abelian, nonabelian otherwise.) $\square$

2. BASIC PROPERTIES OF BINARY ALGEBRAIC SYSTEMS

2.1 Identity Elements in Semigroups

Can a semigroup have two distinct identity elements (in which case the same semigroup would represent two distinct monoids)? The answer, pleasantly enough, is no.

Proposition 1. If a semigroup $[S; \cdot]$ has an identity element, then it is unique.

Proof. Let 1 and 1' of S both be identity elements. Then

$$1 = 1' \cdot 1 \qquad 1' \text{ is an identity element,}$$

$$= 1' \qquad 1 \text{ is an identity element. } \square$$

It immediately follows that in any monoid $[M; \cdot, 1]$, the identity element is unique. (See Exercise 1.)

2.2 Inverses in Monoids and Groups

In our discussion of functions (Section I.4.2), we discussed invertibility and inverse functions. We now generalize these ideas to (abstract) monoids.

Definition. A *left* (*right*) *inverse* of an element a in a monoid $[M; \cdot, 1]$ is an element a_l (a_r) that satisfies $a_l a = 1$ $(a a_r = 1)$. If a_l (a_r) exists, then a is called *left* (*right*) *invertible*.

Proposition 1 (Cf. Theorem 1 of Section I.4.2). In a monoid $[M; \cdot, 1]$, if $a \in M$ has both a left inverse a_l and a right inverse a_r, then $a_l = a_r$.

Proof. $a_l = a_l 1$ identity law,

      $= a_l(a a_r)$ defn. of right inverse,

      $= (a_l a) a_r$ associative law,

      $= 1 a_r$ defn. of left inverse,

      $= a_r$ identity law. $\square$

Corollary 1. In a monoid, any element has at most one two-sided (left and right) inverse. (We usually omit the modifier "two-sided.")

The unique inverse of an element a is denoted, when it exists, by a^{-1} in the case of a multiplicative monoid $[M; \cdot, 1]$, $-a$ in the case of an additive monoid $[M; +, 0]$. This notation agrees with the notation that we have already introduced for inverses in groups. Of course, in groups every element has an inverse, and we have

Corollary 2.   In a group $[G; \cdot, ^{-1}, 1]$, a^{-1} is the unique element that satisfies the inverse law: $a^{-1}a = aa^{-1} = 1$.

Corollary 3.   In a group $[G; \cdot, ^{-1}, 1]$, if $ab = 1$ (or if $ba = 1$), then $b = a^{-1}$.

For a^{-1} is both a left and right inverse of a, hence by Proposition 1 must be equal to any other left or right inverse of a.

Consequence: In a group, to test whether b is the inverse of a, it is sufficient to test whether b is a left inverse (or alternatively a right inverse) of a.

Proposition 2.   Let $[G; \cdot, ^{-1}, 1]$ be a group, and let $a, b \in G$. Then

(a) $(a^{-1})^{-1} = a$;
(b) $(ab)^{-1} = b^{-1}a^{-1}$.

Proof.   (a) Since $aa^{-1} = 1$, a is a left inverse of a^{-1}, hence *the* inverse of a^{-1} by Corollary 3 to Proposition 1. Thus $a = (a^{-1})^{-1}$, as required.

(b) Appealing to Corollary 3 of Proposition 1, we need only check that $b^{-1}a^{-1}$ is a left inverse of ab:

$$(b^{-1}a^{-1})ab = b^{-1}(a^{-1}a)b = b^{-1}1b = b^{-1}b = 1. \quad \square$$

2.3   Solving Equations over Groups

Over the additive monoid $\mathbf{N}$ of nonnegative integers, the equation $a + x = b$ $(a, b \in \mathbf{N})$ may fail to have a solution $x \in \mathbf{N}$ (indeed, *will* fail if $a > b$). Over the additive group $\mathbf{Z}$ of integers, on the other hand, the equation $a + x = b$ $(a, b \in \mathbf{Z})$ always has a (unique) solution $x \in \mathbf{Z}$. We now show that the property of $\mathbf{Z}$ that allows for the solution of all equations of the form $a + x = b$ is essentially a group-theoretic one.

We first establish the important

Lemma 1 (*Cancellation law for groups*).   In a group $[G; \cdot, ^{-1}, 1]$, the following laws hold:

$$ax = ay \Rightarrow x = y \quad \text{(left cancellation)};$$

$$xa = ya \Rightarrow x = y \quad \text{(right cancellation)}.$$

Proof. $ax = ay$

$$\Rightarrow a^{-1}(ax) = a^{-1}(ay)$$
$$\Rightarrow (a^{-1}a)x = (a^{-1}a)y \qquad \text{associative law}$$
$$\Rightarrow 1x = 1y \qquad \text{inverse law}$$
$$\Rightarrow x = y \qquad \text{identity law.}$$

The right cancellation law follows similarly (by postmultiplying rather than premultiplying by a^{-1}). $\square$

Remark. Henceforth, we shall use the cancellation law (left or right) routinely when dealing with groups.

Now for the promised result.

Theorem 1. Let $[G; \cdot, {}^{-1}, 1]$ be a group. Then any equations of the form

$$ax = b, \quad ya = b \qquad (a, b \in G)$$

have unique solutions $x, y \in G$.

Proof. We deal first with the solution to equation $ax = b$. Uniqueness follows immediately from cancellation: $ax = ax' \,(= b) \Rightarrow x = x'$. As for existence, we show how to solve $ax = b$ for x:

$$ax = b \Rightarrow a^{-1}(ax) = a^{-1}b$$
$$\Rightarrow (a^{-1}a)x = a^{-1}b$$
$$\Rightarrow 1x = a^{-1}b$$
$$\Rightarrow x = a^{-1}b,$$

so we have found a solution (hence *the* solution) to $ax = b$, namely $x = a^{-1}b$.

The equation $ya = b$ is handled similarly. (Here $y = ba^{-1}$ is the unique solution.) $\square$

2.4 Products and Powers in Semigroups and Groups

In a (multiplicative) semigroup $[S; \cdot]$, the *product* of n elements $a_1, a_2, \ldots, a_n$ of S is denoted by

$$\prod_{i=1}^{n} a_i, \quad \text{or} \quad a_1 a_2 \cdots a_n,$$

and is defined recursively by

$$\prod_{i=1}^{n} a_i = a_1 \qquad \text{if } n = 1,$$
$$= \left(\prod_{i=1}^{n-1} a_i \right) a_n \quad \text{if } n > 1.$$

The above definition of (n-fold) product imposes a fixed order of associa-tion (left-to-right), hence is meaningful in a groupoid as well. In a semigroup, however, the order of association is immaterial, because of

Proposition 1 (*Generalized associative law*). If $\prod{}^{*}{}_{i=1}^{n} a_i$ denotes an arbi-trary (not necessarily left-to-right) association of the factors $a_1, a_2, \ldots, a_n$ in that order, then

$$\prod_{i=1}^{n}{}^{*} a_i = \prod_{i=1}^{n} a_i.$$

Proof. Exercise 8 of Section II.2. $\square$

If each a_i in the product $\prod_{i=1}^{n} a_i$ is the same element a, then we obtain the n-th *power* of a ("a to the n"):

$$a^n = \prod_{i=1}^{n} a = aa \cdots a \quad (n \text{ factors}).$$

Thus,

$$a^n = a \qquad \text{if } n = 1,$$
$$= (a^{n-1})a \quad \text{if } n > 1.$$

Proposition 2 (*Laws of exponents*). Let $[S; \cdot]$ be a semigroup, and let $a, b \in S$. Then for $m, n \in \mathbf{P}$,

(a) $a^{m+n} = a^m a^n$;
(b) $(a^m)^n = a^{mn}$;
(c) $a^n b^n = (ab)^n$ if S is commutative.

Proof. Example 4 and Exercise 6 of Section II.2. $\square$

For multiplicative monoids, we can extend the above definition of power to include exponent zero:

$$a^0 = 1.$$

For multiplicative groups, we can further extend the definition to include *negative* integral exponents: For $n \in \mathbf{P}$ we define a^{-n} by

$$a^{-n} = (a^n)^{-1}.$$

Proposition 2 can now be extended from semigroups to groups:

Proposition 3 (*Laws of exponents for multiplicative groups*). Let $[G; \cdot, {}^{-1}, 1]$ be a group, and let $a, b \in G$. Then for $m, n \in \mathbf{Z}$ (cf. Proposition 2),

(a) $a^{m+n} = a^m a^n$;
(b) $(a^m)^n = a^{mn}$;
(c) $a^n b^n = (ab)^n$ if G is abelian.

We leave the proof of these familiar laws as an exercise (Exercise 2).

Additive notation: sums and multiples. In an additive semigroup $[S; +]$, the n-fold product becomes the analogous n-fold *sum:*

$$\sum_{i=1}^{n} a_i, \quad \text{or} \quad a_1 + a_2 + \cdots + a_n,$$

and the n-fold power of a becomes the analogous n-fold *multiple* of a ("n times a"):

$$na = a + a + \cdots + a \quad (n \text{ terms}).$$

Propositions 2 and 3 hold for additive semigroups, with equal force of course. In particular, the familiar laws of exponents of Proposition 2 become the equally familiar (and abstractly equivalent) *laws of multiples:*

Proposition 2′. Let $[S; +]$ be a semigroup, and let $a, b \in S$. Then for $m, n \in \mathbf{P}$,

(a) $(m + n)a = ma + na$;
(b) $m(na) = (mn)a$;
(c) $na + nb = n(a + b)$ [Note: An additive semigroup (monoid or group) is always taken to be abelian.]

In an additive group $[G; +, -, 0]$ we can define zero and negative integral multiples of an element $a \in G$:

$$a0 = 0,$$
$$(-n)a = -na \quad (n \in \mathbf{P}).$$

Proposition 3, like Proposition 2, has its additive analog, which we omit.

3. SUBALGEBRAS (A UNIVERSAL ALGEBRA CONCEPT)

3.1 Definition and Examples

Definition. Let A be an Ω-algebra. A subset B of A is called a (Ω-) *subalgebra* of A if B is Ω-closed: for any operation $\omega \in \Omega$,

$$x_1, \ldots, x_n \in B \Rightarrow \omega(x_1, \ldots, x_n) \in B.$$

We write $B \leq A$ to denote that B is a subalgebra of A, and $B < A$ to denote that B is a *proper* subalgebra of A: $B \leq A$ and $B \neq A$.

Terminology. If Ω is the set of operations for a groupoid (semigroup, monoid, group, etc.), then we call an Ω-subalgebra a *subgroupoid* (*subsemigroup, submonoid, subgroup,* etc.).

Example 1. As additive groups, $\mathbf{Z} \leq \mathbf{Q} \leq \mathbf{R} \leq \mathbf{C}$. As multiplicative groups, $\mathbf{Q}^* \leq \mathbf{R}^* \leq \mathbf{C}^*$. $\square$

Remark. The subalgebra relation usually depends critically on the operations under consideration. For example, $\mathbf{Z}^* \leq \mathbf{Q}^*$ as multiplicative monoids, whereas $\mathbf{Z}^* \not\leq \mathbf{Q}^*$ as multiplicative groups. By rights, we should embellish our subalgebra notation $A \leq B$ to something like $A \leq_\Omega B$ in order to emphasize the dependence of $\leq$ on Ω; instead we choose to let context resolve any possible ambiguity.

Example 2. {odd integers} is a multiplicative submonoid of $\mathbf{Z}$. {even integers} is an additive subgroup of $\mathbf{Z}$ and a multiplicative subsemigroup (but not a multiplicative submonoid) of $\mathbf{Z}$. □

Example 3. {injections $A \to A$} is a submonoid ($= \{\circ, 1_A\}$-subalgebra) of A^A, as is {surjections $A \to A$} (Exercise 1). □

A general method for producing new subalgebras from old ones is provided by

Proposition 1. Let $\mathscr{B}$ be a collection of subalgebras of an Ω-algebra A. Then their intersection, $\cap\, \mathscr{B}$, is a subalgebra of A.

Proof. For any $\omega \in \Omega$ we have

$$x_1, \ldots, x_n \in \cap\, \mathscr{B}$$

$\Rightarrow x_1, \ldots, x_n \in B$ for all $B \in \mathscr{B}$ defn. of $\cap$

$\Rightarrow \omega(x_1, \ldots, x_n) \in B$ for all $B \in \mathscr{B}$ each $B \leq A$

$\Rightarrow \omega(x_1, \ldots, x_n) \in \cap\, \mathscr{B}$ defn. of $\cap$,

so that $\cap\, \mathscr{B}$ is Ω-closed. □

3.2 Subalgebras Generated by Subsets

This section can be skimmed lightly on a first reading.

Let A be an Ω-algebra and H a subset of A. Now there is at least one subalgebra of A that includes H, namely A itself. But far more interesting is the *least* subalgebra of A that includes H—this we denote by $[H]$. It may happen that $[H]$ is equal to all of A; in this case we say that A is *generated by* H, or that H is a *set of generators* for A. If $A = [H]$ for some finite subset H, then we say that A is *finitely generated* (by H).

Can we give a set-theoretic formula for $[H]$? Yes, according to

Proposition 1. Let A be an Ω-algebra and H a subset of A. Let $\mathscr{K}$ be defined as the set of all subalgebras of A that include H: $\mathscr{K} = \{B \leq A \mid H \subseteq B\}$. Then

$$[H] = \cap\, \mathscr{K}.$$

Proof. By Proposition 1 of Section 3.1, $\cap\, \mathscr{K}$ is a subalgebra of A. Also, $\cap\, \mathscr{K}$ includes H, since $H \subseteq B$ for every $B \in \mathscr{K}$.

Now let C be any subalgebra of A for which $H \subseteq C$. Then $C \in \mathcal{K}$ by the definition of $\mathcal{K}$, whence $\cap \mathcal{K} \subseteq C$. Thus $\cap \mathcal{K}$ is the *least* subalgebra of A that includes H; i.e., $\cap \mathcal{K} = [H]$. $\square$

Our next result shows how to construct $[H]$ recursively from H.

Proposition 2. Let A be an Ω-algebra and $H \subseteq A$. Let $S \subseteq A$ be defined recursively by

    1.   Any $h \in H$ is in S;
    2.   For any $\omega \in \Omega$,
$$x_1, \ldots, x_n \in S \Rightarrow \omega(x_1, \ldots, x_n) \in S;$$
    3.   (That's all.) There is nothing in S that is not given by clauses 1 and 2.

Then $S = [H]$.

Proof. By clause 1, $H \subseteq S$. By clause 2, $S \leq A$. Thus $[H] \subseteq S$, since $[H]$ is the *least* subalgebra of A that includes H.

In the other (more difficult) direction $(S \subseteq [H])$ we define recursively a collection S_m of subsets of S:

$$S_0 = H;$$
$$S_{m+1} = \{x \mid x = \omega(x_1, \ldots, x_n) \text{ for } \omega \in \Omega \quad \text{and} \quad x_1, \ldots, x_n \in S_m\}.$$

(Intuitively, S_m contains those elements of S that can be generated from H by m or fewer applications of operations from Ω.)

The claim now is that $S_m \subseteq [H]$ for all $m \in \mathbf{N}$. The proof is by induction on m.

Basis. $S_0 = H \subseteq [H]$.

Induction. Assume that $S_m \subseteq [H]$, and consider any $x \in S_{m+1}$. By definition of S_{m+1}, x can be expressed as $x = \omega(x_1, \ldots, x_n)$ for some $\omega \in \Omega$, $x_1, \ldots, x_n \in S_m$. By the induction hypothesis, each $x_i \in [H]$; hence $\omega(x_1, \ldots, x_n) \in [H]$, since $[H]$ is a subalgebra. Thus $x \in [H]$, which shows that $S_{m+1} \subseteq [H]$. This completes the proof by induction of the claim.

By clauses 1 and 2, $\cup_{m=0}^{\infty} S_m \subseteq S$; by clause 3, $S \subseteq \cup_{m=0}^{\infty} S_m$. Hence $S = \cup_{m=0}^{\infty} S_m$. Since $S_m \subseteq [H]$ for each $m \in \mathbf{N}$, it follows that $\cup_{m=0}^{\infty} S_m \subseteq [H]$; i.e., that $S \subseteq [H]$. $\square$

We conclude this section with some examples to show how simple the idea of subalgebra generation really is.

Notation. If $H = \{h_1, \ldots, h_n\}$ we write $[h_1, \ldots, h_n]$ instead of $[\{h_1, \ldots, h_n\}]$.

Example 1. We consider various subalgebras of $\mathbf{Z}$ generated by 2. With $\mathbf{Z}$ regarded

(a) as an additive group, $[2] = \{0, \pm 2, \pm 4, \pm 6, \ldots\}$;
(b) as an additive monoid, $[2] = \{0, 2, 4, 6, \ldots\}$;
(c) as an additive semigroup, $[2] = \{2, 4, 6, 8, \ldots\}$. $\square$

Example 2.

(a) As an additive group, $\mathbf{Z} = [1]$.
(b) As an additive semigroup, $\mathbf{Z} = [1, -1]$ (but not [1]—why not?). $\square$

Example 3.   As a multiplicative monoid, $\mathbf{P}$ is generated by {primes} (Exercise 3). $\square$

3.3 Subgroups

In this section we specialize the subalgebra concept to groups. The concepts and results that we obtain turn out to be among the most far-reaching in our study of algebraic systems, with important consequences for algebraic computing.

According to our general (universal algebra) definition of subalgebra, a subset S of a group $[G; \cdot, ^{-1}, 1]$ is a subgroup of G if

1. $x, y \in S \Rightarrow xy \in S$,
2. $x \in S \Rightarrow x^{-1} \in S$,
3. $1 \in S$.

Interestingly enough, these three closure rules can be encapsuled by a single closure rule, which involves the operation of division: $x/y = xy^{-1}$.

> **Proposition 1.**   Let S be a nonempty subset of a group $[G; \cdot, ^{-1}, 1]$. Then S is a subgroup of G if
>
> $(*)$ $x, y \in S \Rightarrow x/y \in S$.

Proof.   Let $S \neq \varnothing$ satisfy $(*)$. Since $S \neq \varnothing$, some $x \in G$ is in S; by $(*)$ we have $x/x = 1 \in S$. With $1 \in S$, we also have $1/x = x^{-1} \in S$. If $x, y \in S$, then $y^{-1} \in S$, whence by $(*)$ $x/y^{-1} = x(y^{-1})^{-1} = xy \in S$. Thus S is closed under 1, $^{-1}$, and $\cdot$.
$\square$

Using additive notation, Proposition 1 becomes: If S is a nonempty subset of an additive group $[G; +, -, 0]$, then S is a subgroup of G if

$$x, y \in S \Rightarrow x - y \in S.$$

Complexes, cosets, and Lagrange's Theorem.   We return briefly to the general setting of an arbitrary Ω-algebra A. By a *complex* of A we simply mean a nonempty subset of A. We can make the totality of complexes of A into an Ω-algebra $[P(A) - \{\varnothing\}; \Omega]$ by defining for any $\omega \in \Omega$ and complexes $X_1, X_2, \ldots, X_n \subseteq A$,

$$\omega(X_1, \ldots, X_n) = \{\omega(x_1, \ldots, x_n) \mid x_i \in X_i \text{ for } 1 \leq i \leq n\}.$$

Thus, for the case of a group $[G; \cdot, ^{-1}, 1]$ and complexes $S, T \subseteq G$, we have

$$ST = \{st \mid s \in S, t \in T\},$$
$$S^{-1} = \{s^{-1} \mid s \in S\},$$
$$1 = \{1\}.$$

An especially important case of a complex S of a group G arises when we take S to be a subgroup of G. For any $g \in G$, we then form the complex product $\{g\}S$, which we denote more simply by gS:

$$gS = \{gs \mid s \in S\};$$

gS is called the *left coset of S by g*. Similarly Sg is called the *right* coset of S by g:

$$Sg = \{sg \mid s \in S\}.$$

Of course gS and Sg are the same if G is abelian, but in general won't be if G is nonabelian—Exercise 7.

For the additive case, the (left) coset gS becomes

$$g + S = \{g + s \mid s \in S\}.$$

We denote by G/S ("$G \bmod S$") the set of all (left) cosets of S by elements of G:

$$G/S = \{gS \mid g \in G\} \qquad \text{(multiplicative)},$$
$$G/S = \{g + S \mid g \in G\} \qquad \text{(additive)}.$$

Note. We shall *never* interpret G/S as the complex $\{g/s \mid g \in G, s \in S\}$.

Our goal now is to establish some important properties of the system of cosets G/S.

Proposition 2. Let G be a group and S a subgroup. Then G/S is a partition of G.

Proof. Each coset $gS \in G/S$ is nonempty, since $g = g1 \in gS$. Also $\cup G/S = G$, again since $g \in gS$ for any $g \in G$. To complete the proof, we must show that the cosets of G/S are disjoint; i.e., that

$$gS \cap hS \neq \varnothing \Rightarrow gS = hS.$$

So assume that $gS \cap hS \neq \varnothing$. This means that $x \in gS$ and $x \in hS$ for some $x \in G$, which in turn means that

$$(*) \qquad\qquad x = gs_1 = hs_2 \qquad (s_1, s_2 \in S).$$

Now consider any $gs \in gS$ (we wish to show that $gs \in hS$). We can write

$$
\begin{aligned}
gs &= g(s_1 s_1^{-1})s & \text{inverse and identity laws} \\
&= (gs_1)s_1^{-1}s & \text{associativity} \\
&= (hs_2)s_1^{-1}s & \text{from } (*) \\
&= h(s_2 s_1^{-1}s) & \text{associativity.}
\end{aligned}
$$

But $s_2 s_1^{-1}s \in S$, since S is a subgroup of G. Thus $gs \in hS$, and we have shown that $gS \subseteq hS$. Symmetrically, $hS \subseteq gS$. $\square$

Let E_S denote the equivalence relation on G induced by G/S. Thus, $g\,E_S\,h$ means that g and h are in the same block (= coset) of G/S. We now inquire: When are two elements g and h E_S-equivalent? When are two cosets gS and hS of G/S the same? Both questions are sharply answered by

Proposition 3. Let S be a subgroup of a group G. Then

$$g\,E_S\,h \quad\Leftrightarrow\quad gS = hS \quad\Leftrightarrow\quad g^{-1}h \in S.$$

Proof. $(g\,E_S\,h \Rightarrow gS = hS)$ If $g\,E_S\,h$, then h is in the same coset as g. But the unique coset containing g is gS and the one containing h is hS, so that $gS = hS$.

$(gS = hS \Rightarrow g^{-1}h \in S)$ If $gS = hS$, then $gs = h1$ for some $s \in S$, whence $g^{-1}h = s \in S$.

$(g^{-1}h \in S \Rightarrow g\,E_S\,h)$ If $g^{-1}h \in S$, then $h = gs$ for some $s \in S$, whence $h \in gS$. But $g = g1 \in gS$, so that g and h lie in the same coset of G/S. Thus $g\,E_S\,h$. $\square$

For any subgroup S of G, we have $x \in S \Leftrightarrow x^{-1} \in S$. Hence, $g^{-1}h \in S \Leftrightarrow (g^{-1}h)^{-1} = h^{-1}g \in S$, from which we conclude the following

Corollary. If S is a subgroup of an *abelian* group G, then

$$g\,E_S\,h \quad\Leftrightarrow\quad gS = hS \quad\Leftrightarrow\quad gh^{-1} \in S.$$

In additive notation:

$$g\,E_s\,h \quad\Leftrightarrow\quad g + S = h + S \quad\Leftrightarrow\quad g - h \in S.$$

Remark. The criterion for coset equality provided by Proposition 3 and its corollary will henceforth be used routinely and without special explanation.

Our next proposition shows that the cosets of G/S all have the same number of elements.

Proposition 4. For every $g \in G$, $|gS| = |S|$.

Proof. We claim that the map $s \mapsto gs$ is a bijection $S \to gS$. First, it is clearly surjective. Second, if $gs = gs'$, then $s = s'$ by cancellation; hence $s \mapsto gs$ is injective. $\square$

Finally, we are in a position to establish Lagrange's Theorem—a theorem which, despite its simplicity, stands as one of the most far-reaching results of all group theory. (In Section 3.4 we shall encounter several of its consequences.)

Theorem 5 (*Lagrange's Theorem*). Let G be a finite group and S a subgroup of G. Then $|S|$ divides $|G|$.

Proof. By Proposition 2, $G/S = \{gS \mid g \in G\}$ is a partition of G. By Proposition 4, each block gS of G/S has the same number of elements as S. Hence we have the equation

$$|G| = |G/S| \times |S|.$$

Now $|G/S|$, the number of cosets in G/S, is of course an integer. Hence $|S|$ divides $|G|$. $\square$

Terminology. $|G/S|$ is called the *index* of S in G and is also denoted by $[G:S]$. (When G is infinite, $[G:S]$ may be finite or infinite depending on S, as we shall see below.)

Example 1. For $m \in \mathbf{P}$, the set

$$(m) = \{km \mid k \in \mathbf{Z}\}$$

is an additive subgroup of $\mathbf{Z}$. The induced equivalence relation $E_{(m)}$ then satisfies

$$a E_{(m)} b \quad \Leftrightarrow \quad a - b \in (m)$$

$$\Leftrightarrow \quad a \equiv_m b,$$

so that $E_{(m)} = \equiv_m$. Thus $\mathbf{Z}/\equiv_m$ and $\mathbf{Z}/(m)$ are one and the same, where the equivalence class $[a]$ of $\mathbf{Z}/\equiv_m$ is equal to the coset $a + (m)$ of $\mathbf{Z}/(m)$. By Theorem 1 of Section II.3.1, we have

$$\mathbf{Z}/(m) = \{a + (m) \mid a \in \mathbf{Z}_m\},$$

$$\text{where } a + (m) \neq b + (m) \text{ if } a \neq b \quad (a, b \in \mathbf{Z}_m).$$

Hence (m) has index m in $\mathbf{Z}$. $\square$

Example 2. The cosets of $\mathbf{Z}$ regarded as an additive subgroup of $\mathbf{R}$ are given by

$$\mathbf{R}/\mathbf{Z} = \{a + \mathbf{Z} \mid a \in \mathbf{R}\}.$$

Now

$$a + \mathbf{Z} = b + \mathbf{Z} \quad \Leftrightarrow \quad a - b \in \mathbf{Z},$$

so that each coset $a + \mathbf{Z}$ has a representative from the interval $0 \leq x < 1$, namely $a - n$ where n is the (unique) integer that satisfies $n \leq a < n + 1$. Thus,

$$\mathbf{R}/\mathbf{Z} = \{a + \mathbf{Z} \mid 0 \leq a < 1\}.$$

Also, $a - b$ is not an integer if $0 \leq a, b < 1$ and $a \neq b$, so that the cosets $a + \mathbf{Z}$ for $0 \leq a < 1$ are all distinct. Hence $\mathbf{Z}$ does not have finite index in $\mathbf{R}$. $\square$

3.4 Cyclic Groups; Order of Group Elements

Let $[G; \cdot, {}^{-1}, 1]$ be a group. We call a subalgebra of the form $[a]$ $(a \in G)$ a *cyclic subgroup*—the cyclic subgroup generated by a. If $G = [a]$ for some $a \in G$, we call G a *cyclic* group (generated by a).

What does the group $[a]$ look like?

Proposition 1. Let G be a group and $a \in G$. Then

$$[a] = \{a^i \mid i \in \mathbf{Z}\}.$$

Proof. Since $[a]$ contains a and is closed under $\cdot$, ${}^{-1}$, and 1, it follows that $[a]$

must also contain a^n and $(a^{-1})^n = a^{-n}$ for $n \in \mathbf{P}$, and $1 = a^0$. Hence $\{a^i \mid i \in \mathbf{Z}\}$ $\subseteq [a]$. But $\{a^i \mid i \in \mathbf{Z}\}$ is trivially checked to be a subalgebra of G, a subalgebra that contains $a = a^1$. Hence $[a] \subseteq \{a^i \mid i \in \mathbf{Z}\}$, since $[a]$, by definition, is the *least* subalgebra that contains a. $\square$

Definition. The *order*, $o(a)$, of an element $a \neq 1$ in a group $[G; \cdot, ^{-1}, 1]$ is defined by

$$o(a) = n \quad \text{if } a^n = 1 \text{ and } a^k \neq 1 \text{ for } 1 \leq k < n,$$
$$= \infty \quad \text{if } a^n \neq 1 \text{ for any } n \in \mathbf{P}.$$

For the case of an additive group $[G; +, -, 0]$,

$$o(a) = n \quad \text{if } na = 0 \text{ and } ka \neq 0 \text{ for } 1 \leq k < n,$$
$$= \infty \quad \text{if } na \neq 0 \text{ for any } n \in \mathbf{P}.$$

Proposition 2. Let $a \in G$ have order n. Then

(a) $a^k = a^{r_n(k)}$.
(b) $a^k = 1 \Leftrightarrow n \mid k$.

Proof. (a) Dividing k by n gives $k = nq + r_n(k)$, so that

$$a^k = a^{nq + r_n(k)}$$
$$= \left[(a^n)^q \right] \left[a^{r_n(k)} \right]$$
$$= a^{r_n(k)}, \quad \text{since } a^n = 1.$$

(b) ($\Leftarrow$) If $n \mid k$ then $k = mn$ for $m \in \mathbf{Z}$, whence $a^k = a^{mn} = (a^n)^m = 1$.
($\Rightarrow$) Let $a^k = 1$. Then $a^{r_n(k)} = 1$ by part (a). Since $0 \leq r_n(k) < n$, we must have $r_n(k) = 0$, otherwise we would contradict $o(a) = n$. Hence $n \mid k$, as required. $\square$

The concept of order allows us to give a more penetrating analysis of cyclic groups than we did in Proposition 1, from which a number of important corollaries flow.

Theorem 3. Let G be a group and $a \in G$.

(a) If $o(a) = n$, then

$$[a] = \{a^i \mid i \in \mathbf{Z}_n\},$$

where $a^i \neq a^j$ if $i \neq j$ $(i, j \in \mathbf{Z}_n)$.
(b) If $o(a) = \infty$, then

$$[a] = \{a^i \mid i \in \mathbf{Z}\},$$

where $a^i \neq a^j$ if $i \neq j$ $(i, j \in \mathbf{Z})$.

Proof. (a) By Proposition 1, $\{a^i \mid i \in \mathbf{Z}_n\} \subseteq [a] = \{a^i \mid i \in \mathbf{Z}\}$. For any $a^i \in [a]$, we have $a^i = a^{r_n(i)}$ (Proposition 2(a)), so that $[a] \subseteq \{a^i \mid i \in \mathbf{Z}_n\}$.

We complete the proof of part (a) by showing that $a^i \neq a^j$ if $i \neq j$ ($i,j \in \mathbf{Z}_n$). Assume, to the contrary, that $a^i = a^j$ with $0 \leq i < j < n$. Then

$$1 = a^i a^{-i} = a^j a^{-i} = a^{j-i}.$$

But $0 < j - i < n$, which contradicts $o(a) = n$ and forces the desired conclusion that the a^i's of $\{a^i \mid i \in \mathbf{Z}_n\}$ are all distinct.

(b) By Proposition 1, $[a] = \{a^i \mid i \in \mathbf{Z}\}$. The only point at issue here is whether $a^i \neq a^j$ if $i \neq j$ ($i,j \in \mathbf{Z}$). So assume to the contrary that $a^i = a^j$ with $i < j$ ($i,j \in \mathbf{Z}$). Then

$$1 = a^i a^{-i} = a^j a^{-i} = a^{j-i},$$

which contradicts $o(a) = \infty$. $\square$

Corollary 1. $o(a) = \|[a]\|$. (In words: The order of any group element a is the order of the subgroup generated by a.)

Remark. This result explains the use of the term "order" for $o(a)$. Indeed, some authors *define* the order of a by the order of $[a]$ and then prove our definition of order as a theorem.

Corollary 2. Let G be a finite group and $a \in G$. Then $o(a)$ divides $|G|$.

Proof. By Lagrange's Theorem, $\|[a]\|$ divides $|G|$. But by Corollary 1, $o(a) = \|[a]\|$.
$\square$

Corollary 3. Any group G of prime order p is cyclic.

Proof. Choose any $a \neq 1$ in G. By Corollary 2, $o(a) \mid p$. Hence $o(a) = p$, since p is prime, which means that $G = [a]$. $\square$

Corollary 4. Let G be a group of order n. Then $a^n = 1$ for any $a \in G$.

Proof. By Corollary 2, $n = o(a)k$ for some $k \in \mathbf{P}$, whence $a^n = (a^{o(a)})^k = 1$. $\square$

Corollary 5. Let G be a finite group. Then a nonempty subset S of G is a subgroup if S is closed under $\cdot$, the multiplication of G.

Proof. Exercise 9. $\square$

Now for some concrete examples of cyclic groups.

Example 1. $\mathbf{Z} = [1]$ is an infinite cyclic (additive) group. (1 has additive order ∞ in $\mathbf{Z}$.) $\square$

Example 2. $\mathbf{Z}_m = [1]$ is a cyclic (additive) group. (1 has additive order m in $\mathbf{Z}_m$.)
$\square$

Example 3. The complex number $\omega = e^{2\pi i/n}$ has multiplicative order n in C^*. (Check: First, $\omega^n = (e^{2\pi i/n})^n = e^{2\pi i} = 1$. Second, if $0 < j < n$, then $\omega^j = (e^{2\pi i/n})^j = e^{2\pi i(j/n)} \neq 1$, using the familiar property of complex numbers that $e^{2\pi i k} = 1$ if

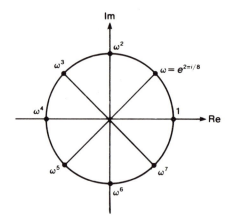

Fig. 2 Cyclic subgroup of $\mathbf{C}^*$ generated by $\omega = e^{2\pi i/8}$

and only if k is an integer.) Hence $[\omega] = \{1, \omega, \ldots, \omega^{n-1}\}$ is a cyclic subgroup of order n of the multiplicative group $\mathbf{C}^*$ (Fig. 2). $\square$

We observe that the generator of a cyclic group $[a]$ need not be unique. For example, the reader can check that the cyclic group $[\omega]$ generated by $\omega = e^{2\pi i/8}$ in Fig. 2 is also generated by ω^3 or ω^5. The following proposition is the key to determining which elements of a finite cyclic group are generators.

Theorem 4. Let $a \in G$ be an element of order n. Then

$$o(a^i) = n \quad \Leftrightarrow \quad (i, n) = 1.$$

Proof. ($\Rightarrow$) If $(i, n) = d > 1$, then $(a^i)^{n/d} = (a^n)^{i/d} = 1$, so that $o(a^i) \leq n/d < n$.

($\Leftarrow$) Let $o(a^i) = m$, and assume that $(i, n) = 1$. Now $(a^i)^n = (a^n)^i = 1$, so that $m = o(a^i)$ is $\leq n$.

For the other direction $(n \leq m)$, we use $(i, n) = 1$. By Theorem 2 of Section II.3.2, we can write

$$1 = (i, n) = si + tn \quad \text{for some } s, t \in \mathbf{Z}.$$

Then

$$a^m = (a^1)^m$$

$$= (a^{si+tn})^m$$

$$= ((a^i)^m)^s (a^n)^{tm}$$

$$= 1^s 1^{tm} = 1.$$

Hence $n = o(a)$ is $\leq m$, which gives $m = n$. $\square$

Definition. The *Euler phi-function* $\phi(n)$ is defined as the number of positive integers $\leq n$ and relatively prime to n. ($\phi(1) = 1$, $\phi(2) = 1$, $\phi(3) = 2$, $\phi(4) = 2$, $\phi(5) = 4$, $\phi(6) = 2$, etc.)

Corollary to Theorem 4. A cyclic group of order n has $\phi(n)$ distinct generators.

4. MORPHISMS (ANOTHER UNIVERSAL ALGEBRA CONCEPT)

4.1 The Morphism Concept

The idea of a *morphism*—a structure-preserving map from one algebraic system to another—is one of the most useful and beautiful concepts of all algebra.

We first need the notion of "similar" algebras. We call two algebras $[A; \Omega]$, $[A'; \Omega']$ *similar* if there is a bijection from Ω to Ω' with the property that corresponding operations $\omega \in \Omega$, $\omega' \in \Omega'$ have the same arity. In other words, two algebras A and A' are similar when they have corresponding nullary, unary, binary, etc., operations. In particular, any two groupoids, semigroups, monoids, groups, etc., are similar.

Remark. In practice we often denote corresponding operations ω, ω' of similar algebras A and A' by the same symbol ω, letting context decide whether ω is operating on A or on A'. We then speak of (similar) Ω-algebras A and A'.

With the idea of similar algebras in hand, we are ready for the main

Definition. Let $[A; \Omega]$ and $[A'; \Omega']$ be similar algebras. A function ϕ: $A \rightarrow A'$ is said to be a *morphism* (also called a *homomorphism*) from $[A; \Omega]$ to $[A'; \Omega']$ if for any $\omega \in \Omega$ and for any $a_1, \ldots, a_n \in A$,

(M) $$\phi(\omega(a_1, \ldots, a_n)) = \omega'(\phi(a_1), \ldots, \phi(a_n)).$$

Expressing the morphism relation (M) in words, the morphism ϕ *preserves*, or *respects*, corresponding operations of A and A': The same result is obtained whether (i) one performs an operation in A and maps the result via ϕ into A' (according to the left-hand side of (M)), or (ii) one first maps the arguments into A' (again via ϕ) and then performs the corresponding operation of A' (according to the right-hand side of (M)).

In terms of a mapping diagram, the morphism relation (M) is given by

$$
\begin{array}{ccc}
A^n & \xrightarrow{\ \omega\ } & A \\
\phi^n \downarrow & & \uparrow \phi \\
(A')^n & \xrightarrow{\ \omega'\ } & A'
\end{array}
$$

where ϕ^n is the *n*-fold Cartesian power of ϕ: $A \rightarrow A'$, the map $A^n \rightarrow (A')^n$ defined by

$$\phi^n: (a_1, \ldots, a_n) \mapsto (\phi(a_1), \ldots, \phi(a_n)).$$

If $\phi: A \to A'$ is a morphism of Ω-algebras, then by definition ϕ preserves all the operations of Ω. For the case of groups, then, a morphism ϕ from a group $[G; \cdot, {}^{-1}, 1]$ to a group $[G'; *, \ominus, 1']$ satisfies

1. $\phi(x \cdot y) = \phi(x) * \phi(y)$;
2. $\phi(x^{-1}) = \phi(x)\ominus$;
3. $\phi(1) = 1'$.

For groups, though, it turns out that 2 and 3 are actually consequences of 1 (hence only 1 need be checked). This is proved in

Proposition 1. Let $[G; \cdot, {}^{-1}, 1]$ and $[G'; \cdot, {}^{-1}, 1]$ be groups (where we now use the same symbols for the operations of both). If $\phi: G \to G'$ preserves multiplication,

$$\phi(xy) = \phi(x)\phi(y),$$

then ϕ also preserves inverses and the identity,

$$\phi(x^{-1}) = \phi(x)^{-1}, \qquad \phi(1) = 1.$$

(Thus if G and G' are groups, then a morphism $\phi: G \to G'$ of *semigroups* is a morphism of *groups*.)

Proof. First, we have

$$1\phi(1) = \phi(1) = \phi(1 \cdot 1) = \phi(1)\phi(1),$$

so that $\phi(1) = 1$ by cancellation. Second,

$$\phi(x)\phi(x^{-1}) = \phi(xx^{-1}) = \phi(1) = 1,$$

which shows that $\phi(x^{-1}) = \phi(x)^{-1}$. $\square$

Special kinds of morphisms. We classify morphisms according to their properties as functions: If $\phi: A \to A'$ is a morphism, then we call ϕ

1. an *isomorphism* if ϕ is bijective,
2. an *epimorphism* if ϕ is surjective,
3. a *monomorphism* if ϕ is injective.

CASE 1. *Isomorphism.* If A and A' are *isomorphic* (denoted $A \cong A'$), which is to say there exists an isomorphism $\phi: A \to A'$, then A and A' are abstractly identical; A' is essentially a copy of A which results from ϕ assigning different names to the elements of A. Consequently, any algebraic law satisfied by A must likewise be satisfied by A'. We illustrate this last statement.

Suppose that ϕ is an isomorphism from $[A; \cdot]$ to $[A'; *]$, and further suppose that $\cdot$ is associative. Then $*$ must be associative. Proof: For any

$a' = \phi(a),\, b' = \phi(b),\, c' = \phi(c)$ in A', we have

$$a' * (b' * c') = \phi(a) * (\phi(b) * \phi(c))$$

$$= \phi(a \cdot (b \cdot c)) \qquad\qquad \phi \text{ is a morphism}$$

$$= \phi((a \cdot b) \cdot c)) \qquad\qquad \cdot \text{ is associative}$$

$$= (\phi(a) * \phi(b)) * \phi(c) \qquad \phi \text{ is a morphism}$$

$$= (a' * b') * c'.$$

And so on for any other abstract property enjoyed by A. Indeed, we *define* the abstract properties of an algebraic system as being precisely those that are invariant under isomorphism.

Now, even though two isomorphic algebras have identical abstract properties, their concrete properties may be quite different. In particular, corresponding operations may differ considerably in their complexity of application. This is illustrated by the following classic

Example 1. Let ϕ be the map $x \mapsto \log x$ from $[\mathbf{R}^+\,;\,\cdot]$, the set of positive reals under multiplication, to $[\mathbf{R};\,+]$, the set of all reals under addition. We claim that ϕ is an isomorphism. Proof:

$$\phi(xy) = \log xy$$

$$= \log x + \log y$$

$$= \phi(x) + \phi(y),$$

so that ϕ is a morphism. Also, ϕ is bijective with inverse given by the exponential function, $x \mapsto \exp x$. Thus ϕ is an isomorphism, as claimed. The setup is described by the following mapping diagram:

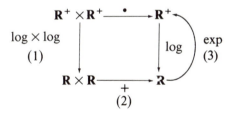

As the mapping diagram indicates, we can multiply two (positive) real numbers by the circuitous route of (1) taking logs of the two operands, (2) adding, and finally (3) exponentiating. Thus the mapping diagram renders a precise algebraic formulation of the conversion of multiplication to addition by the use of logarithms and antilogarithms—a common technique in pre-computer days. □

An application of the isomorphism concept to the characterization of cyclic groups is given by

Proposition 2.   Any two cyclic groups of the same order are isomorphic.

Proof. If $C_m = [a]$ and $\tilde{C}_m = [b]$ are two (multiplicative) cyclic groups of the same order m, then $o(a) = o(b) = m$. (The case $m = \infty$ is not excluded.) The map ϕ: $a^i \mapsto b^i$ then yields an isomorphism $C_m \to \tilde{C}_m$, trivially. $\square$

Corollary.   Let $C_m = [a]$ be a cyclic group of order m.

(a) If $m < \infty$, then C_m is isomorphic to the additive group $\mathbf{Z}_m$ of integers mod m.

(b) C_∞ is isomorphic to the additive group $\mathbf{Z}$ of integers.

Proof. (a) Since $\mathbf{Z}_m$ is an (additive) cyclic group of order m generated by 1, it follows from Proposition 2 that the map $\mathbf{Z}_m \to C_m$ defined by $k \mapsto a^k$ is an isomorphism.

(b) Since $\mathbf{Z}$ is an infinite order (additive) cyclic group, it follows again from Proposition 2 that the map $\mathbf{Z} \to C_\infty$ defined by $k \mapsto a^k$ is an isomorphism. $\square$

Now for our second major kind of morphism.

CASE 2.   *Epimorphism.* If ϕ: $A \to A'$ is an epimorphism, then we call A' a *homomorphic image* of A, and regard A' as an *abstraction*, or *model*, of A. To see the sense in which A' is a model of A we need interpret the morphism relation

(M) $\phi(\omega(a_1, \ldots, a_n)) = \omega'(\phi(a_1), \ldots, \phi(a_n)).$

This relation indicates that we can obtain at least partial information about the value $b = \omega(a_1, \ldots, a_n)$ in A by computing the corresponding value $b' = \omega'(\phi(a_1), \ldots, \phi(a_n))$ in A'. For by (M) we have that b, whatever it is, must be in $\phi^{-1}(b')$; i.e., we know b modulo the kernel relation of ϕ.

Of course knowing b' will not in general be sufficient to determine b uniquely, because of the many-to-one property of ϕ. A' can be regarded as a reflection of A giving imperfect information about A—akin to the view of a passing scene offered by a reflecting pool.

But the question arises, why bother performing operations in A' if we are interested in results in A? Here is where the telescoping property of an epimorphism comes into play. A' is simpler, less detailed, than A, hence is easier to deal with computationally. This observation will be elevated to a major theme of our algorithmic pursuits of Part Three.

Example 2.   The remainder mod m map, $a \mapsto r_m(a)$, is an epimorphism (of additive groups) from $[\mathbf{Z}; +, -, 0]$ to $[\mathbf{Z}_m; \oplus, \ominus, 0]$, as we now verify.

First, since $r_m(a) = a$ if $0 \le a < m$, the map r_m: $\mathbf{Z} \to \mathbf{Z}_m$ is surjective. As for the morphism properties,

$$r_m(a + b) = r_m(r_m(a) + r_m(b)) \qquad \text{Proposition 1 of Sec. 1.1}$$

$$= r_m(a) \oplus r_m(b) \qquad\qquad \text{defn. of } \oplus.$$

Since $[\mathbf{Z}; +, -, 0]$ and $[\mathbf{Z}_m; \oplus, \ominus, 0]$ are groups, we can appeal to Proposition 1 to conclude

$$r_m(-a) = \ominus r_m(a), \qquad r_m(0) = 0.$$

It is easily verified that r_m is also an epimorphism (of multiplicative monoids) from $[\mathbf{Z}; \cdot, 1]$ to $[\mathbf{Z}_m; \odot, 1]$.

Let us give an example of $\mathbf{Z}_m$'s role as an abstraction of $\mathbf{Z}$. In $\mathbf{Z}$, let $c = ab$ where $a = 17$ and $b = 14$, and consider the information about c given by the corresponding calculation mod 11. We have

$$r_{11}(c) = r_{11}(a) \odot r_{11}(b)$$
$$= 6 \odot 3 = 7.$$

So we know that $c \equiv_{11} 7$, i.e., that $c = 7 + 11k$ for some integer k; this is the partial information we have obtained about c by our modular (homomorphic image) computation. (Of course $k = 21$ by direct calculation.) $\square$

Example 3. The *projection maps*

$$p_1: (a,b) \mapsto a, \qquad p_2: (a,b) \mapsto b$$

are epimorphisms from the additive group $\mathbf{R} \times \mathbf{R}$ of points in the plane to the additive group $\mathbf{R}$ of points on the real line. $\square$

CASE 3. *Monomorphism.* If $\phi: A \to A'$ is a monomorphism, then ϕ regarded as a map $A \to \phi(A)$ is clearly an isomorphism. Thus A' includes what is essentially a copy of A. A monomorphism $A \to A'$ is sometimes called an *embedding* of A into A'.

Example 4. If A is an Ω-algebra and A' a subalgebra of A, then the inclusion map $x \mapsto x$ is a monomorphism $A' \to A$. $\square$

We single out yet another class of morphisms for special attention—morphisms that map algebras into themselves. We call a morphism $\phi: A \to A$ that maps an Ω-algebra A into itself an *endomorphism* of A. If ϕ is also bijective, hence an isomorphism $A \to A$, then we call it an *automorphism* of A.

Example 5. The map $n \mapsto 2n$ is an endomorphism of $\mathbf{Z}$ regarded as an additive group. The map $n \mapsto n^2$ is an endomorphism of $\mathbf{Z}$ regarded as a multiplicative monoid. The map $n \mapsto -n$ is an automorphism of $\mathbf{Z}$ regarded as an additive group. $\square$

For any Ω-algebra A, we denote by $\text{End}\, A$ the collection of all endomorphisms of A and by $\text{Aut}\, A$ the collection of all automorphisms of A. Both $\text{End}\, A$ and $\text{Aut}\, A$ have interesting algebraic structure:

Theorem 2. For any Ω-algebra A,

(a) $\text{End}\, A$ is a monoid;
(b) $\text{Aut}\, A$ is a group.

Proof. Exercise 5. $\square$

4.2 Structure-Preserving Properties of Morphisms

By definition, morphisms preserve operations. We now show that morphisms preserve much other significant algebraic structure as well.

First we show that epimorphisms preserve semigroup, monoid, and group properties: a homomorphic image of a semigroup is a semigroup, of a monoid is a monoid, of a group is a group.

Theorem 1. (a) Let ϕ be an epimorphism from $[S; \cdot]$ to $[S'; *]$. If $[S; \cdot]$ is a (commutative) semigroup, then $[S'; *]$ is a (commutative) semigroup.
(b) Let ϕ be an epimorphism from $[M; \cdot, 1]$ to $[M'; *, 1']$. If $[M; \cdot, 1]$ is a (commutative) monoid, then $[M'; *, 1']$ is a (commutative) monoid.

(c) Let ϕ be an epimorphism from $[G; \cdot, {}^{-1}, 1]$ to $[G'; *, \ominus, 1']$. If $[G; \cdot, {}^{-1}, 1]$ is an (abelian) group, then $[G'; *, \ominus, 1']$ is an (abelian) group.

Proof. (a) We must check that $*$ is associative. Since ϕ is surjective, the elements of S' can all be expressed in the form $\phi(a)$, $a \in S$. So for any $\phi(a), \phi(b), \phi(c) \in S'$ we have

$$\phi(a) * (\phi(b) * \phi(c))$$

$$= \phi(a \cdot (b \cdot c)) \qquad\qquad \phi \text{ is a morphism}$$

$$= \phi((a \cdot b) \cdot c) \qquad\qquad \cdot \text{ is associative}$$

$$= (\phi(a) * \phi(b)) * \phi(c) \qquad \phi \text{ is a morphism,}$$

so that $*$ is associative. Similar reasoning shows that if $\cdot$ is commutative, then $*$ is likewise commutative.

(b) Appealing to (a), we need only check that $1' = \phi(1)$ is an identity element for $*$. For any $a' = \phi(a)$ in M', we have

$$a' * 1' = \phi(a) * \phi(1)$$

$$= \phi(a1) \qquad\qquad \phi \text{ is a morphism}$$

$$= \phi(a) \qquad\qquad 1 \text{ is the identity element of } M$$

$$= a'.$$

Symmetrically, $1' * a' = a'$, and so $1' = \phi(1)$ is an identity element (hence *the* identity element) of M'.

(c) By part (b), $[G'; *, 1' = \phi(1)]$ is a monoid (commutative if G is). To complete the proof that G' is a group, we need only check that $\ominus$ satisfies the inverse laws. For any $\phi(a) \in G'$,

$$
\left.
\begin{aligned}
\phi(a) * \phi(a)^{\ominus} &= \phi(a) * \phi(a^{-1}) \\
&= \phi(aa^{-1})
\end{aligned}
\right\} \quad \phi \text{ is morphism}
$$

$$
= \phi(1) = 1' \qquad \text{inverse law of } G.
$$

Symmetrically, $\phi(a)^{\ominus} * \phi(a) = 1'$. $\quad\square$

(See Exercise 6 for a strengthened version of Theorem 1(c).)

Example 1. We can apply Theorem 1 to effortlessly deduce the additive group and multiplicative monoid structures of $\mathbf{Z}_m$ from the corresponding structures of $\mathbf{Z}$. By Example 2 of Section 4.1, $[\mathbf{Z}_m; \oplus, \ominus, 0]$ is a homomorphic image of $[\mathbf{Z}; +, -, 0]$ and $[\mathbf{Z}_m; \odot, 1]$ is a homomorphic image of $[\mathbf{Z}; \cdot, 1]$ (under the map $a \mapsto r_m(a)$). It immediately follows by Theorem 1(c) that $[\mathbf{Z}_m; \oplus, \ominus, 0]$ is an abelian group and by Theorem 1(b) that $[\mathbf{Z}_m; \odot, 1]$ is a commutative monoid. $\quad\square$

Our final collection of results about the structure-preserving properties of morphisms are universal in character.

Theorem 2 (*Morphisms preserve subalgebras*). Let A and A' be Ω-algebras, and $\phi: A \to A'$ a morphism. Then

(a) $B \leq A \Rightarrow \phi(B) \leq A'$;
(b) $C \leq A' \Rightarrow \phi^{-1}(C) \leq A$.

(In words: A morphism maps subalgebras into subalgebras, in both the forward and inverse direction.)

Proof. (a) We must show that $\phi(B)$ is Ω-closed. So let $\omega \in \Omega$ and $\phi(b_1), \ldots, \phi(b_n) \in \phi(B)$ (which means that each $b_i \in B$). Since ϕ is a morphism we have

$$
\omega(\phi(b_1), \ldots, \phi(b_n)) = \phi(\omega(b_1, \ldots, b_n)).
$$

Now, $\omega(b_1, \ldots, b_n)$ is in B because B is a subalgebra of A. Hence $\phi(\omega(b_1, \ldots, b_n))$ is in $\phi(B)$, which by the above equality means that $\omega(\phi(b_1), \ldots, \phi(b_n))$ is in $\phi(B)$. Thus $\phi(B)$ is Ω-closed.

(b) Here we must show that $\phi^{-1}(C)$ is Ω-closed. So let $\omega \in \Omega$ and $a_1, \ldots, a_n \in \phi^{-1}(C)$ (which means that each $\phi(a_i) \in C$). Since ϕ is a morphism we have

$$
\phi(\omega(a_1, \ldots, a_n)) = \omega(\phi(a_1), \ldots, \phi(a_n)).
$$

Now, $\omega(\phi(a_1), \ldots, \phi(a_n))$ is in C because C is a subalgebra of A', which by the above equality means that $\phi(\omega(a_1, \ldots, a_n))$ is in C. Hence $\omega(a_1, \ldots, a_n)$ is in $\phi^{-1}(C)$, and we have shown that $\phi^{-1}(C)$ is Ω-closed. $\quad\square$

Theorem 3 (*Morphisms preserve generators*). Let $\phi: A \to A'$ be a morphism, and let H be a subset of A. Then

$$\phi([H]) = [\phi(H)].$$

(In words: a morphism maps a set of generators for a domain subalgebra into a set of generators for the corresponding image subalgebra.)

Proof. We must show that $\phi([H])$ is the least subalgebra of B that includes $\phi(H)$. (By definition this least subalgebra is $[\phi(H)]$.)

First, $\phi([H])$ is a subalgebra of A' by Theorem 2(a). Also, since $[H] \supseteq H$, it follows that $\phi([H]) \supseteq \phi(H)$. To complete the proof, we must show that $\phi([H])$ is the *least* subalgebra of A' that includes $\phi(H)$ (this is the hard part).

So let C be any subalgebra of A' such that $C \supseteq \phi(H)$. (We must show that $C \supseteq \phi([H])$.) Since $C \supseteq \phi(H)$, it follows that

$$\phi^{-1}(C) \supseteq \phi^{-1}(\phi(H)) \supseteq H,$$

the last inclusion following by Exercise 6(a) of Section I.4.2. Thus $\phi^{-1}(C)$ is a subalgebra of A (Theorem 2(b)) that includes H. Hence

$$\phi^{-1}(C) \supseteq [H],$$

since $[H]$ is the *least* subalgebra of A that includes H. Finally, we have

$$C \supseteq \phi(\phi^{-1}(C)) \supseteq \phi([H]),$$

the first inclusion following by Exercise 6(b) of Section I.4.2. □

Corollary. Let $\phi: A \to A'$ be an epimorphism, and let H be a set of generators for A. Then $\phi(H)$ is a set of generators for A'.

Proof. $A' = \phi(A) = \phi([H]) = [\phi(H)]$. □

Theorem 4. Let ϕ and ψ be morphisms $A \to A'$, and let H be a set of generators for A. Then

$$\phi(x) = \psi(x) \text{ for all } x \in H \implies \phi = \psi.$$

(Thus, a morphism is uniquely determined by its effect on any set of generators.)

Proof. Exercise 8. □

EXERCISES

Section 1.

1. Find all ways of completing the following multiplication table to achieve a monoid.

·	a	b	c
a	c	b	a
b			b
c	a		

2. Is $[\mathbf{P}; \gcd]$ a semigroup? a monoid? What about $[\mathbf{P}; \mathrm{lcm}]$?

3. Draw up multiplication tables for S_2 and S_3, the symmetric groups on two and three letters.

4. Let S be the set of all strictly monotone and surjective functions $[0, 1] \rightarrow [0, 1] = \{x \in \mathbf{R} \mid 0 \le x \le 1\}$. (Strictly monotone means $x < y \Rightarrow f(x) < f(y)$.) Show that S under composition is a group.

Exercises 5–7 illustrate applications of group theory to geometry, specifically the geometry of $\mathbf{R}^2 = \mathbf{R} \times \mathbf{R}$, the Euclidean plane, or 2-space.

5. Let $T_{a,b}$ be the map ("transformation") $\mathbf{R}^2 \rightarrow \mathbf{R}^2$ defined by

$$T_{a,b}: \begin{pmatrix} x \\ y \end{pmatrix} \mapsto \begin{pmatrix} x' \\ y' \end{pmatrix} = \begin{pmatrix} x+a \\ y+b \end{pmatrix}$$

("translation by the vector $\begin{pmatrix} a \\ b \end{pmatrix}$").

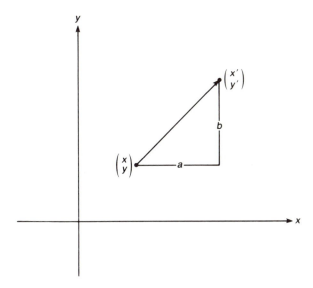

Show that the set $T(\mathbf{R}^2)$ of all such transformations,

$$T(\mathbf{R}^2) = \{ T_{a,b} \mid a, b \in \mathbf{R} \},$$

under composition is a group. ($T(\mathbf{R}^2)$ is called the group of all *translations* of $\mathbf{R}^2$.) Is this group abelian?

6. (*Matrix transformations.*) Let $M_2(\mathbf{R})$ be the set of all 2×2 *matrices*

$$A = \begin{pmatrix} a_{11} & a_{12} \\ a_{21} & a_{22} \end{pmatrix}$$

over $\mathbf{R}$ ($a_{11}, a_{12}, a_{21}, a_{22} \in \mathbf{R}$). For any such matrix A, we define a transformation T_A according to

$$T_A: \begin{pmatrix} x \\ y \end{pmatrix} \mapsto \begin{pmatrix} x' \\ y' \end{pmatrix} = \begin{pmatrix} a_{11}x + a_{12}y \\ a_{21}x + a_{22}y \end{pmatrix}.$$

(a) Show that $T_A \circ T_B = T_{AB}$, where AB is the *matrix product* of A and B defined by

$$\begin{pmatrix} a_{11}b_{11} + a_{12}b_{21} & a_{11}b_{12} + a_{12}b_{22} \\ a_{21}b_{11} + a_{22}b_{21} & a_{21}b_{12} + a_{22}b_{22} \end{pmatrix}.$$

(b) Show that T_I is the identity function on $\mathbf{R} \times \mathbf{R}$, where $I = \begin{pmatrix} 1 & 0 \\ 0 & 1 \end{pmatrix}$.

(c) Conclude that $\{T_A | A \in M_2(\mathbf{R})\}$ is a monoid under composition. Further conclude that $M_2(\mathbf{R})$ is a monoid under matrix multiplication.

7. (*Nonsingular matrix transformations.*) Define $NS_2(\mathbf{R}) \subseteq M_2(\mathbf{R})$ as the set of all *nonsingular* matrices $A \in M_2(\mathbf{R})$, i.e., 2×2 matrices A having nonzero *determinant*,

$$\det A = a_{11}a_{22} - a_{21}a_{12} \neq 0.$$

For $A \in NS_2(\mathbf{R})$ define $\tilde{A}$ by

$$\tilde{A} = \begin{pmatrix} a_{22}/D & -a_{12}/D \\ -a_{21}/D & a_{11}/D \end{pmatrix}$$

where $D = \det A$.

(a) Show that if $A \in NS_2(\mathbf{R})$, then T_A is invertible with inverse $T_{\tilde{A}}$.

(b) Conclude that $\{T_A | A \in NS_2(\mathbf{R})\}$ is a group under composition and that $NS_2(\mathbf{R})$ is a group under matrix multiplication.

Section 2.

1. Show the uniqueness of an identity element in any groupoid.

2. Prove Proposition 3 of Section 2.4 (laws of exponents for multiplicative groups). [*Hint:* Appeal to the corresponding laws for semigroups—Proposition 2.]

3. In a group G, prove that if $(ab)^n = a^n b^n$ for all $a, b \in G$ and positive integers n, then G is abelian, and conversely.

4. In a group G, prove that

$$(a_1 a_2 \cdots a_n)^{-1} = a_n^{-1} a_{n-1}^{-1} \cdots a_1^{-1}.$$

5. In a *finite* monoid $[M; \cdot, 1]$, prove that $uv = 1 \Rightarrow vu = 1$. (Thus, in a finite monoid either left or right invertibility implies two-sided invertibility.) [*Hint:* Since M is finite, successive powers of u cannot all be distinct.] Give a counterexample for the case of an infinite monoid.

6. (*Idempotents in finite semigroups.*) **Definition.** In a semigroup $[S; \cdot]$ an element s is an *idempotent* if $s^2 = s$.

   Let $[S; \cdot]$ be a finite semigroup. Show that for every $s \in S$, some power s^n $(n \in \mathbf{P})$ is an idempotent.

Exercises 7–9 deal with alternative characterizations of groups.

7. (*A weaker axiomatization of groups.*) Prove that a semigroup $[S; \cdot]$ is a group if
   1. There exists $1 \in S$ such that $1a = a$ for all $a \in S$ (thus 1 is postulated as a *left* identity only);
   2. Any $a \in S$ has a *left* inverse $b \in S$: $ba = 1$.

8. (*A solvability-of-equations axiomatization of groups*—cf. Theorem 1 of Section 2.3.) Prove that a semigroup $[S; \cdot]$ is a group if and only if for every $a, b \in S$, the equations $ax = b$, $ya = b$ have solutions in S.

9. (*An axiomatization of finite groups.*)   Prove that a finite monoid $[M; \cdot, 1]$ is a group if and only if 1 is the *only* idempotent of M. [*Hint:* Exercise 6.]

Section 3.

1. Prove that {injections $A \to A$}, {surjections $A \to A$} are submonoids of A^A.

2. Let S_X be the symmetric group (Example 3 of Section 1.3) on an infinite set X. Let $T \subseteq S_X$ be the set of all bijections that fix all but finitely many elements of X (thus, if $f \in T$, $f(x) = x$ for all but finitely many $x \in X$). Prove that T is a subgroup of S_X.

3. Show that as a multiplicative monoid, **P** is generated by {primes}.

4. (a) Is the additive group $\mathbf{Z} \times \mathbf{Z}$ finitely generated?
   (b) Is the additive group **Q** finitely generated?

5. Determine the smallest group (with respect to composition) of functions defined on $\mathbf{R} - \{0, 1\}$ that contains $f(x) = 1/x$ and $g(x) = 1/(1 - x)$.

6. Let S be a subgroup of G with $S \neq G$. Prove that $G - S$ is a set of generators for G (i.e., $G = [G - S]$).

7. Consider the two functions f_1, f_2 defined by

$$f_1: 1 \mapsto 1,\ 2 \mapsto 2,\ 3 \mapsto 3,$$

$$f_2: 1 \mapsto 2,\ 2 \mapsto 1,\ 3 \mapsto 3.$$

   Verify that $S = \{f_1, f_2\}$ is a subgroup of the symmetric group S_3, and display the left and right cosets, gS and Sg, for $g \in S_3$.

8. Show that each coset of $\mathbf{R}^+$ regarded as a multiplicative subgroup of $\mathbf{C}^*$ has a unique representative having absolute value of unity. Does $\mathbf{R}^+$ have finite index in $\mathbf{C}^*$?

9. Prove Corollary 5 of Theorem 3, Section 3.4.

10. Let G be group of even order. Show that G contains an element $x \neq 1$ that is its own inverse: $x^{-1} = x$.

11. Show that there are no nonabelian groups of order < 6.

Exercises 12–16 deal with cyclic groups.

12. Write down all the generators of the additive cyclic groups (a) $\mathbf{Z}_{10}$, (b) $\mathbf{Z}_{11}$, (c) $\mathbf{Z}_{12}$.

13. How many generators does an *infinite* cyclic group $G = [a]$ have?

14. Prove that every subgroup of a cyclic group $G = [a]$ is cyclic. [*Hint:* If S is a subgroup of G, let m be the smallest positive integer such that $a^m \in S$.]

15. Let $G = [a]$ be a cyclic group of order n. Show that for every divisor d of n there is a unique subgroup of order d. What are the subgroups of $\mathbf{Z}_{12}$?

16. Let C_m denote a cyclic group of order m. Prove that $C_m \times C_n$ is cyclic if and only if $(m, n) = 1$. [*Hint:* If $(m, n) = 1$, exhibit a generator of order mn for $C_m \times C_n$.] Illustrate this result for $\mathbf{Z}_3 \times \mathbf{Z}_4$.

17. (a) Establish the following multiplicative property of the Euler phi function:

$$\phi(mn) = \phi(m)\phi(n) \text{ if } (m, n) = 1.$$

   [*Hint:* Exercise 16.]
   (b) Show that $\phi(p^n) = p^n - p^{n-1}$ (p prime).

(c)   Now derive the formula

$$\phi(n) = n \prod_{p \mid n} (1 - 1/p)$$

where the product is over all distinct prime divisors p of n.

18. (*A universal algebra result.*)   Let $\mathbb{S}_A$ be the set of all subalgebras of an Ω-algebra A. Show that $\mathbb{S}_A$ is a lattice (Section I.3.3) under the subalgebra relation $\leq$.

Section 4.

1. Let G be an abelian group. Show that the map ϕ_n: $a \mapsto a^n$ is an endomorphism of G.

2. (*Complex conjugation.*)   Show that the map ϕ: $a + ib \mapsto a - ib$ is an automorphism
   (i)   of $\mathbf{C}$ regarded as an additive group;
   (ii) of $\mathbf{C}^*$ regarded as a multiplicative group.

3. (*Inner automorphisms.*)
   (a)   Let G be a group, and let $a \in G$. Show that the map

   $$\phi_a: x \mapsto axa^{-1}$$

   is an automorphism of G. (Such an automorphism is called an *inner* automorphism of G.)
   (b)   Show that the set $\{\phi_a \mid a \in G\}$ of all inner automorphisms of G is a subgroup of the group Aut G of all automorphisms of G. (For the group property of Aut G, see Exercise 5(b).)

4. (*Cayley's Theorem—the representation of any monoid by a monoid of functions under composition.*)   Let $[M; \cdot, 1]$ be a monoid. Show that the map ϕ: $M \to M^M$ defined by

   $$\phi(a) = f_a \quad \text{where } f_a(x) = ax$$

   is a monomorphism of monoids from $[M; \cdot, 1]$ to $[M^M; \circ, 1_M]$. What if M is a group?

5. Let A, B, C be Ω-algebras.
   (a)   Prove that if ϕ: $A \to B$, ψ: $B \to C$ are morphisms, then $\psi\phi$ is a morphism. Conclude that End A is a monoid.
   (b)   Prove that if ϕ: $A \to B$ is an isomorphism, then so is ϕ^{-1}. [*Hint:* $x = \phi\phi^{-1}(x)$.] Conclude that Aut A is a group.

6. Let ϕ be an epimorphism from $[S; \cdot]$ to $[S'; *]$. Prove: If $[S; \cdot, ^{-1}, 1]$ is a group, then so is $[S'; *, \ominus, 1']$ (with $\ominus$ and $1'$ appropriately defined).

7. Prove that the homomorphic image of a cyclic group is cyclic.

8. Prove Theorem 4 of Section 4.2.

9. Let C_p be a cyclic group of prime order p. Show that Aut $C_p \cong C_{p-1}$. [*Hint:* Theorem 4 of Section 4.2.]

IV

RINGS, INTEGRAL DOMAINS, AND FIELDS

In the previous chapter we restricted our attention to algebraic systems based on a single binary operation. In this chapter we deal with algebraic systems based on two binary operations—addition and multiplication.

1. THE RING CONCEPT

1.1 Basic Definitions and Examples

Definition. A *ring* $[R; +, -, 0, \cdot, 1]$ is an algebraic system that satisfies:

1. $[R; +, -, 0]$ is an abelian group   (the "additive group" of R);
2. $[R; \cdot, 1]$ is a monoid   (the "multiplicative monoid" of R);
3. $a(b + c) = ab + ac$, $(b + c)a = ba + ca$ (left and right distributive laws).

A ring is called *commutative* if its multiplication is commutative: $ab = ba$. (Interestingly enough, ring addition *must* be commutative—Exercise 1.)

The multiplicative identity element 1 of R is variously called the *unity, unit element*, or *one* of R.

Example 1. The familiar number systems **Z**, **Q**, **R**, and **C** are all rings under the usual operations $+, -, 0, \cdot, 1$. $\square$

John D. Lipson, Elements of Algebra and Algebraic Computing. ISBN 0-201-04115-4.

A subalgebra S of a ring R—i.e., a subset of R closed under the ring operations $+, -, 0, \cdot, 1$ of R— is called a *subring*. For the above numerical rings we have the hierarchy of (proper) subrings $\mathbf{Z} < \mathbf{Q} < \mathbf{R} < \mathbf{C}$.

Example 2.   The set $\mathbf{Z}_m = \{0, 1, \ldots, m - 1\}$ of integers mod m is a ring under mod m operations $\oplus, \ominus, 0, \odot, 1$. Since $[\mathbf{Z}_m; \oplus, \ominus, 0]$ is an abelian group and $[\mathbf{Z}_m; \odot, 1]$ is a monoid, we need only check the left or right distributive law—Exercise 2. $\square$

Notation.   We shall often denote the mod m operations $\oplus, \ominus, \odot$ simply by $+, -, \cdot$ if there is no danger of confusion as to whether operations are to be interpreted in $\mathbf{Z}_m$ or $\mathbf{Z}$. For example, if it is understood that we are in $\mathbf{Z}_9$, then we could write $3 + 5 \cdot (-7) = 3 + 5 \cdot 2 = 4$.

Our next proposition shows that the familiar laws of ordinary arithmetic relating addition and multiplication apply to rings generally.

Proposition 1.   In a ring R,

(a) $a0 = 0a = 0$ (i.e., the *additive* zero is also a *multiplicative* zero, or *annihilator*);
(b) $a(-b) = (-a)b = -ab$;
(c) $(-a)(-b) = ab$.

Remark.   We know from the outset that the distributive laws must play a vital role in the proof of the above proposition, since they are the only postulated relationships between the additive and multiplicative parts of a ring.

Proof of Proposition 1.

(a) $a0 = a(0 + 0)$ additive identity law
      $ = a0 + a0$ distributive law.

Also,

$a0 = a0 + 0$ additive identity law,

whence $a0 = 0$ by cancellation in the additive group of R. Similarly, $0a = 0$.

(b) $ab + a(-b)$
      $ = a(b + (-b))$ distributive law
      $ = a0$ additive identity law
      $ = 0$ part (a),

and so $a(-b)$ is the additive inverse of ab; i.e., $a(-b) = -ab$. Symmetrically, $(-a)b = -ab$.

(c)   Exercise 3. $\square$

We call a ring R *trivial* if it contains but a single element: $R = \{0\}$. In this ring the result of all operations is 0; in particular, $1 = 0$.

Proposition 2. If R is a nontrivial ring (i.e., if R has more than one element), then $1 \neq 0$.

Proof. Let a be a nonzero element of R. Then $1a = a \neq 0$, whereas $0a = 0$ by Proposition 1(a). Hence $1 \neq 0$. $\square$

We continue with our examples of rings by looking at some important *derived* rings—rings that are defined in terms of an already existing "ground" ring.

Example 3 (*Formal power series*). Let R be a commutative ring (e.g., **Z**, **Q**, **R**, **C**, or $\mathbf{Z}_m$). We define $R[[x]]$ as the set of all expressions

$$a(x) = a_0 + a_1 x + a_2 x^2 + \cdots$$

where the *coefficients* a_i ($i \in \mathbf{N}$) of $a(x)$ lie in R. We also write $a(x)$ as $\sum_{i=0}^{\infty} a_i x^i$. Since we are regarding $a(x)$ as a formal expression, i.e., as a string of symbols, the summation symbol does not represent a sum of values in R; hence convergence is not an issue. Each such $a(x)$ is called a *formal power series over R in an indeterminate x*. Equality of formal power series is of course defined formally: $a(x) = b(x)$ means $a_i = b_i$ for all $i \in \mathbf{N}$.

Each $a_i x^i$ is called a *term* (of *degree i*) of $a(x)$. If $a_i \neq 0$ we say that the term $a_i x^i$ occurs in $a(x)$. We take the usual liberties of notation in denoting $(-a_i)x^i$ by $-a_i x^i$, $1x^i$ by x^i, ax^1 by ax, ax^0 by a, and in eliminating terms with zero coefficients. Thus we would write $1 - x + 2x^3 + \cdots$ rather than $1x^0 + (-1)x^1 + 0x^2 + 2x^3 + \cdots$. In keeping with our notational liberties, we identify $a \in R$ with the "constant" power series $a = a + 0x + 0x^2 + \cdots \in R[[x]]$ and regard the indeterminate x as an element of $R[[x]]$: $x = 0 + x + 0x^2 + \cdots$.

Finally, we endow $R[[x]]$ with a ring structure by defining operations $+, -, \cdot, 0 \, [= 0(x)], 1 \, [= 1(x)]$ according to:

$$c(x) = a(x) + b(x) \Leftrightarrow c_i = a_i + b_i \qquad (i = 0, 1, 2, \dots),$$

$$= a(x)b(x) \quad \Leftrightarrow c_i = a_0 b_i + a_1 b_{i-1} + \cdots + a_i b_0$$

$$= \Sigma_{j=0}^i a_j b_{i-j},$$

$$= -a(x) \quad \Leftrightarrow c_i = -a_i,$$

$$= 0 \qquad \Leftrightarrow c_i = 0,$$

$$= 1 \qquad \Leftrightarrow c_i = 1 \quad \text{if } i = 0$$

$$= 0 \quad \text{otherwise,}$$

where the right-hand side operations are those of R. The reader should note, in particular, that the so-called *convolution rule* $c_i = \Sigma_{j=0}^i a_j b_{i-j}$ for multiplication of formal power series is nothing more than a compact formula for the result of expanding and collecting like terms in

$$(a_0 + a_1 x + a_2 x^2 + \cdots)(b_0 + b_1 x + b_2 x^2 + \cdots).$$

It is a routine matter to verify that $R[[x]]$ under the above operations is a ring. As an example, let us check out the left distributive law

$$\underbrace{a(x)[b(x) + c(x)]}_{d(x)} = \underbrace{a(x)b(x) + a(x)c(x)}_{e(x)}.$$

We have

$$d_i = \sum_{j=0}^{i} a_j(b_{i-j} + c_{i-j}) \qquad \text{defn. of } +, \cdot \text{ in } R[[x]]$$

$$= \sum_{j=0}^{i} a_j b_{i-j} + \sum_{j=0}^{i} a_j c_{i-j} \qquad \begin{array}{l}\text{generalized distributivity} \\ \text{in } R \text{ (Exercise 5)}\end{array}$$

$$= e_i \qquad \text{defn. of } +, \cdot \text{ in } R[[x]],$$

which establishes left distributivity. $\square$

Example 4   (*Polynomials*).   Let $R[[x]]$ be a formal power series ring, as defined in the previous example. We define the *degree* of $a(x) = \sum_{i=0}^{\infty} a_i x^i$ as follows:

$$\deg a(x) = -\infty \qquad \text{if } a(x) = 0,$$

$$= n \qquad \begin{array}{l}\text{if for some } n \in \mathbf{N}, \\ a_n \neq 0 \text{ and } a_i = 0 \text{ for } i > n,\end{array}$$

$$= +\infty \qquad \text{otherwise.}$$

We then define $R[x]$, the set of all *polynomials* over R in an indeterminate x, by

$$R[x] = \{a(x) \in R[[x]] \mid \deg a(x) < \infty\}.$$

If $a(x) \in R[x]$ has degree n, then we usually write $a(x) = \sum_{i=0}^{n} a_i x^i$ (rather than $\sum_{i=0}^{\infty} a_i x^i$).

The claim now is that $R[x]$ under the formal power series operations $+, -, 0, \cdot, 1$ is a subring of $R[[x]]$. Verification: If $a(x), b(x) \in R[x]$ have degrees m and n respectively ($m, n < \infty$), then the definitions of formal power series operations yield

$$\deg[a(x) + b(x)] \leq \max(m, n) < \infty,$$

$$\deg a(x)b(x) \quad \leq m + n < \infty,$$

$$\deg -a(x) \quad = m < \infty,$$

$$\deg 0, \deg 1 \quad < \infty,$$

so that $R[x]$ is a subring of $R[[x]]$ as claimed. (Note that the above degree relationships hold if either or both of $a(x), b(x)$ is 0—degree $= -\infty$.) $R[x]$ is called the *polynomial ring* over R in an indeterminate x. □

Example 5 (*Bivariate polynomial rings*). Let R be a commutative ring. Since $R[x]$ is a ring, we can let $R' = R[x]$ in the polynomial ring $R'[y]$ in order to obtain the *bivariate* polynomial ring $R[x][y]$, the ring of all polynomials $a(x,y)$ in y whose coefficients are polynomials in x. Thus, if $a(x,y) \in R[x][y]$, then

$$a(x,y) = \sum_{j=0}^{n} \left(\sum_{i=0}^{m} a_{ij} x^i \right) y^j \quad \text{(form 1)}.$$

By finitely many applications of the commutative ring laws, we can write

$$a(x,y) = \sum_{i=0}^{m} \left(\sum_{j=0}^{n} a_{ij} y^j \right) x^i \quad \text{(form 2)}.$$

Or, we can treat the indeterminates x and y in a simultaneous fashion, and write

$$a(x,y) = \sum_{k=0}^{m+n} \left(\sum_{i+j=k} a_{ij} x^i y^j \right) \quad \text{(form 3)}.$$

Expressions of form 1 constitute $R[x][y]$; expressions of form 2 constitute $R[y][x]$. We shall denote the set of all expressions of form 3 by $R[x,y]$.

Now as polynomial rings, $R[x][y]$, $R[y][x]$, and $R[x,y]$ are essentially the same, for an expression of one form can be transformed to either one of the other forms by finitely many applications of the commutative ring laws. Nevertheless, we choose to distinguish $R[x][y]$, $R[y][x]$, and $R[x,y]$ according to their respective "canonical forms"—forms 1, 2 and 3, respectively.

As a particular application of this distinction, we define the degree in y of $a(x,y)$, $\deg_y a(x,y)$, as the (univariate) degree of $a(x,y)$ regarded as a polynomial in $R[x][y]$. Similarly, $\deg_x a(x,y)$ is the univariate degree of $a(x,y)$ regarded as a polynomial in $R[y][x]$. Finally, the *total degree* of $a(x,y)$, $\deg a(x,y)$, is directly related to $R[x,y]$; it is the maximum integer $k = i + j$ such that $a_{ij} \neq 0$ (the term $a_{ij} x^i y^j$ is said to have [total] degree $i + j$).

Let us illustrate all this with an example:

$$\begin{aligned} a(x,y) &= (5x^2 - 3x + 4)y + (2x^2 + 1) &&\in \mathbf{Z}[x][y], \\ &= (5y + 2)x^2 + (-3y)x + (4y + 1) &&\in \mathbf{Z}[y][x], \\ &= 5x^2 y + (2x^2 - 3xy) + (4y) + 1 &&\in \mathbf{Z}[x,y]; \end{aligned}$$

here we have $\deg_y a(x,y) = 1$, $\deg_x a(x,y) = 2$, and $\deg a(x,y) = 3$. □

Example 6 (*Function rings*). Let R be a ring and consider R^R, the set of all functions $R \to R$. We endow R^R with a ring structure in a natural way: if

$f, g \in R^R$, then

$$(f + g)(x) = f(x) + g(x),$$
$$(fg)(x) = f(x)g(x),$$
$$(-f)(x) = -f(x),$$
$$1(x) = 1,$$
$$0(x) = 0,$$

where the right-hand side operations are those of R. (The verification that R^R under the above operations is a ring is trivial.)

Thus, if we take R to be $\mathbf{R}$, the reals, then we obtain the ring of all real-valued functions of a real variable. For example, if $f(x) = x + 2$ and $g(x) = x^2 - 1$, then $(f + g)(x) = x^2 + x + 1$ and $(fg)(x) = x^3 + 2x^2 - x - 2$. (Note that multiplication of functions is quite different from composition of functions.) □

Example 7 (*Polynomial function rings*). Let $a(x) = \sum_{i=0}^{n} a_i x^i$ be a polynomial in $R[x]$, R commutative. We define $a: R \rightarrow R$, the *polynomial function* induced by $a(x)$, by the familiar process of *polynomial evaluation:* for $\alpha \in R$,

$$a(\alpha) = \sum_{i=0}^{n} a_i \alpha^i = a_n \alpha^n + a_{n-1} \alpha^{n-1} + \cdots + a_0.$$

A function $f \in R^R$ is called a *polynomial function* if $f = a$ for some $a(x) \in R[x]$. The collection P_R of all polynomial functions on R,

$$P_R = \{a: R \rightarrow R \mid a \text{ is induced by } a(x) \in R[x]\},$$

is a subring of R^R, which is essentially the content of

Proposition 3. Let $a(x), b(x) \in R[x]$, and let $s(x) = a(x) + b(x)$ and $t(x) = a(x)b(x)$. Then as polynomial functions $s, t \in R^R$,

$$s = a + b, \qquad t = ab.$$

(Thus the sum and product of two polynomial functions is again a polynomial function.)

Proof. If $s(x) = a(x) + b(x)$, then for $\alpha \in R$,

$$s(\alpha) = \sum_i s_i \alpha^i \qquad \text{defn. of polynomial evaluation}$$

$$= \sum_i (a_i + b_i)\alpha^i \qquad \text{defn. of } + \text{ in } R[x]$$

$$= \sum_i a_i \alpha^i + \sum_i b_i \alpha^i \qquad \text{distributivity in } R$$

$$= a(\alpha) + b(\alpha) \qquad \text{defn. of polynomial evaluation}$$

$$= (a + b)(\alpha) \qquad \text{defn. of } + \text{ in } R^R,$$

so that $s = a + b$ in R^R. The proof that $t(x) = a(x)b(x)$ in $R[x]$ implies $t = ab$ in R^R proceeds similarly (Exercise 4). $\square$

Example 8 (*Matrix rings*).   Let R be a commutative ring, and let $M_n(R)$ be the set of all $n \times n$ matrices over R:

$$M_n(R) = \{A = (a_{ij}) \mid a_{ij} \in R\},$$

where $A = B$ means $a_{ij} = b_{ij}$ $(1 \leq i, j \leq n)$. Then $M_n(R)$ is a ring under the usual matrix operations:

$$C = A + B \Leftrightarrow c_{ij} = a_{ij} + b_{ij},$$

$$= AB \quad \Leftrightarrow c_{ij} = \sum_{k=1}^{n} a_{ik} b_{kj},$$

$$= -A \quad \Leftrightarrow c_{ij} = -a_{ij},$$

$$= 0 \quad \Leftrightarrow c_{ij} = 0,$$

$$= 1 \quad \Leftrightarrow c_{ii} = 1, c_{ij} = 0 \ (i \neq j).$$

The verification that $M_n(R)$ is a ring, as claimed, is straightforward but tedious. The reader might verify the left or right distributive law to see how the ring properties of $M_n(R)$ follow from those of R.

A noteworthy property of $M_n(R)$ is that it is always non-commutative (if $n > 1$), even though R is commutative. For example, in $M_2(\mathbf{Z})$ we have

$$\begin{pmatrix} 1 & 0 \\ 0 & 0 \end{pmatrix} \begin{pmatrix} 0 & 1 \\ 0 & 0 \end{pmatrix} = \begin{pmatrix} 0 & 1 \\ 0 & 0 \end{pmatrix}, \quad \text{whereas} \quad \begin{pmatrix} 0 & 1 \\ 0 & 0 \end{pmatrix} \begin{pmatrix} 1 & 0 \\ 0 & 0 \end{pmatrix} = \begin{pmatrix} 0 & 0 \\ 0 & 0 \end{pmatrix}. \ \square$$

Example 9 (*Direct products, sums, of rings*).   Let R and S be rings. Then their Cartesian product, $R \times S = \{(a, b) \mid a \in R, b \in S\}$, is a ring under component-wise operations as is trivially verified. This ring is called the *direct product* of R and S. (It is occasionally called the *direct sum* of R and S, in which case it is denoted by $R \oplus S$.) $\square$

Integral multiples; the characteristic of a ring.   The *integral multiples* $n \cdot a$ of $a \in R$ are defined from group theory, regarding R as an additive group: for $n \in \mathbf{P}$,

$$n \cdot a = a + a + \cdots + a \quad (n \text{ terms}),$$

$$(-n) \cdot a = -n \cdot a,$$

$$0 \cdot a = 0 \in R.$$

We distinguish between the ring multiple ra, where $r, a \in R$, and the integral multiple $n \cdot a$, where $n \in \mathbf{Z}$, $a \in R$.

Proposition 4 (*Laws of integral multiples*). Let R be a ring. Let $a, b \in R$ and $m, n \in \mathbf{Z}$. Then

(a) $(m + n) \cdot a = m \cdot a + n \cdot a$;
(b) $m \cdot (n \cdot a) = (mn) \cdot a$;
(c) $n \cdot (a + b) = n \cdot a + n \cdot b$;
(d) $n \cdot (ab) = (n \cdot a)b$;
(e) $n \cdot a = (n \cdot 1)a$.

Proof. (a), (b), and (c) are nothing more than the laws of multiples in the additive group of R (see Proposition 3 of Section III.2.4); (d) and (e) are easily proved consequences of the distributive laws (Exercise 5(c)). $\square$

Definition. The *characteristic* of a ring R, char R, is the order of 1 in the additive group of R.

Thus, if R has finite characteristic m, then $m \cdot 1 = 0$ and $n \cdot 1 \neq 0$ for $1 \leq n < m$. If the characteristic of R is not finite, then we say that R has *characteristic zero* (rather than characteristic infinity). Examples: $\mathbf{Z}, \mathbf{Q}, \mathbf{R}, \mathbf{C}$ all have characteristic zero; $\mathbf{Z}_m$ has characteristic m.

Of course a finite ring must have finite characteristic. But the converse is not true: $\mathbf{Z}_m[x]$ is an infinite ring with finite characteristic.

Proposition 5. Let R be a ring of characteristic m. Then
$$m \cdot a = 0 \quad \text{for any } a \in R.$$

Proof. $m \cdot a = (m \cdot 1)a$ Proposition 4(e)
$\qquad\quad = 0a$ char $R = m$
$\qquad\quad = 0. \square$

1.2 Subrings

We have already encountered the concept of subring as a special case of the universal concept of subalgebra: A subset R' of a ring R is a subring of R ($R' \leq R$) if R' is closed under all the ring operations $+, -, 0, \cdot, 1$ of R. If R' is a subring of R we also say that R is an *extension ring* of R' ($R \geq R'$).

Example 1. In Section 1.1 we encountered the following subring hierarchies:

(a) $\mathbf{Z} < \mathbf{Q} < \mathbf{R} < \mathbf{C}$;
(b) $R < R[x] < R[[x]]$;
(c) $P_R = \{\text{polynomial functions } R \to R\} \leq R^R$. (We shall see later that for certain rings $P_R = R^R$.)

We can simplify somewhat the subring criterion (closure under all the ring operations $+, -, 0, \cdot, 1$) by appealing to Proposition 1 of Section III.3.3:

Proposition 1. A subset S of a ring R is a subring of R if for any $a, b \in S$
$$a - b \in S, \quad ab \in S, \quad 1 \in S.$$

Example 2. Let $\alpha = \sqrt[3]{2}$. Then $S = \{a + b\alpha + c\alpha^2 \,|\, a, b, c \in \mathbf{Q}\}$ is a subring of

R. For $1 = 1 + 0\alpha + 0\alpha^2$ is in S, and if $x = a + b\alpha + c\alpha^2$ and $y = d + e\alpha + f\alpha^2$ are in S, then so are

$$x - y = (a - d) + (b - e)\alpha + (c - f)\alpha^2,$$
$$xy = (ad + 2bf + 2ce) + (ae + bd + 2cf)\alpha + (af + be + cd)\alpha^2. \quad \square$$

Let S be a subset of a ring R. From universal algebra we can speak of the *subring generated by* S: $[S] =$ the smallest subring of R that includes S.

Example 3 *(Unital subring)*. Let R be any ring. Since 1 is contained in any subring, it follows that $[1]$ is the smallest subring of R. $[1]$ is called the *unital subring* of R. We claim now that

$$(*) \qquad\qquad [1] = \{n \cdot 1 \mid n \in \mathbf{Z}\}.$$

For by closure under $+$, $-$, and 0, $[1]$ must contain along with 1 the elements $n \cdot 1$ and $-n \cdot 1 = (-n) \cdot 1$ for $n \in \mathbf{P}$, and $0 = 0 \cdot 1$. Hence $\{n \cdot 1 \mid n \in \mathbf{Z}\} \subseteq [1]$. But $\{n \cdot 1 \mid n \in \mathbf{Z}\}$ contains $1 = 1 \cdot 1$ and is easily shown to be a subring. Hence $[1] \subseteq \{n \cdot 1 \mid n \in \mathbf{Z}\}$, since $[1]$ is the *smallest* subring that contains 1.

We observe from $(*)$ that the unital subring of $\mathbf{Z}$ is $\mathbf{Z}$ itself; i.e., $\mathbf{Z} = [1]$. $\square$

1.3 Morphisms of Rings

Let $[R; +, -, 0, \cdot, 1]$ and $[R'; \oplus, \ominus, 0', \odot, 1']$ be rings. According to our general (universal) definition, a function $\phi: R \to R'$ is a morphism of rings if ϕ preserves the ring operations $+, -, 0, \cdot, 1$ of R. Proposition 1 of Section III.4.1 provides the following modest simplification:

Proposition 1. Let R and R' be rings. Then a function $\phi: R \to R'$ is a morphism of rings if

1. $\phi(a + b) = \phi(a) \oplus \phi(b)$;
2. $\phi(ab) = \phi(a) \odot \phi(b)$;
3. $\phi(1) = 1'$.

For rings we have the following ring-theoretic analogue of Theorem 1 of Section III.4.2.

Theorem 2. Let ϕ be an epimorphism from $[R; +, -, 0, \cdot, 1]$ to $[R'; \oplus, \ominus, 0', \odot, 1']$. Then $[R'; \oplus, \ominus, 0', \odot, 1']$ is a ring (commutative if R is).

Proof. Theorem 1 (parts (b) and (c)) of Section III.4.2 gives us everything except the distributive law in R', which we now verify. For any $\phi(a), \phi(b), \phi(c) \in R'$,

$$\phi(a) \odot (\phi(b) \oplus \phi(c))$$

$= \phi(a(b + c))$	ϕ is a morphism
$= \phi(ab + ac)$	distributive law in R
$= (\phi(a) \odot \phi(b)) \oplus (\phi(a) \odot \phi(c))$	ϕ is a morphism. $\square$

Thus: *A homomorphic image of a (commutative) ring is a (commutative) ring.*

Example 1.   According to Example 2 of Section III.4.1, the remainder mod m map, $a \mapsto r_m(a)$, is an epimorphism from the ring $\mathbf{Z}$ to the ring $\mathbf{Z}_m$. Or we can take the converse viewpoint of Theorem 2 to infer the ring structure of $\mathbf{Z}_m$: Since r_m is an epimorphism from $[\mathbf{Z}; +, -, 0, \cdot, 1]$ to $[\mathbf{Z}_m; \oplus, \ominus, 0, \odot, 1]$ and since $[\mathbf{Z}; +, -, 0, \cdot, 1]$ is a commutative ring, so is $[\mathbf{Z}_m; \oplus, \ominus, 0, \odot, 1]$. $\square$

Example 2 *(Evaluation morphism).*   Let E and R be commutative rings with $R \le E$. For $\alpha \in E$, we define the $(\alpha\text{-})$evaluation map $\phi_\alpha : R[x] \to E$ by

$$\phi_\alpha : a(x) \mapsto a(\alpha).$$

Then ϕ_α is a morphism of rings ("evaluation morphism").

Proof.   If $s(x) = a(x) + b(x)$ and $t(x) = a(x)b(x)$, then Proposition 3 of Section 1.1 gives

$$s(\alpha) = a(\alpha) + b(\alpha), \quad t(\alpha) = a(\alpha)b(\alpha) \qquad \text{for any } \alpha \in E.$$

Hence

$$\phi_\alpha(a(x) + b(x)) = \phi_\alpha(a(x)) + \phi_\alpha(b(x)),$$

$$\phi_\alpha(a(x)b(x)) = \phi_\alpha(a(x))\phi_\alpha(b(x)).$$

Also $\phi_\alpha(1) = 1$, trivially, completing the proof.

We note that if $E = R$ then ϕ_α is an *epi*morphism $R[x] \to R$, since any $a \in R$ is the image under ϕ_α of the "constant polynomial" $a = a + 0x + 0x^2 + \cdots \in R[x]$. $\square$

Example 3.   We have observed that the derived rings $R[x]$ and $R[[x]]$ include the ground ring R as a subring. We now show this to be essentially the case with the other derived rings that we have encountered, the function ring R^R and the matrix ring $M_n(R)$ (Examples 6 and 8 of Section 1.1). More precisely, we show that each of these rings contains a subring that is *isomorphic* with R.

For the case of $M_n(R)$, we define *scalar multiplication* as follows: if $r \in R$ and $A = (a_{ij}) \in M_n(R)$, then $rA = (ra_{ij})$. Next we define the set

$$S = \{rI \mid r \in R\}$$

of all *scalar matrices*. It is now easy to show that S is a subring of $M_n(R)$ that is isomorphic to R under the map $r \mapsto rI$.

For the case of R^R, let f_r be the *constant function* defined by $f_r(a) = r$ for all $a \in R$. Then the set C of all constant functions in R^R,

$$C = \{f_r \mid r \in R\},$$

is readily checked to be a subalgebra of R^R that is isomorphic with R under the map $r \mapsto f_r$. $\square$

Ring adjunction.   Let R be a ring and E an extension ring of R. We are often interested, for a given $\alpha \in E$, in the smallest extension ring of R that includes α;

this is the subring $[R \cup \{\alpha\}]$ of E, which we also denote by $R[\alpha]$ ("R adjoined by α").

For the commutative case, an explicit description of $R[\alpha]$ (one that justifies the notation $R[\alpha]$) is provided by the following proposition, the proof of which makes use of one of our universal algebra results about morphisms.

Proposition 3. Let E and R be commutative rings with $E \geq R$, and let $\alpha \in E$. Then

$$R[\alpha] = \{a(\alpha) | a(x) \in R[x]\}.$$

Proof. Let $S = \{a(\alpha) | a(x) \in R[x]\}$. Now any subring of E that includes $R \cup \{\alpha\}$ must contain any $a(\alpha) \in S$ by closure. Thus $S \subseteq R[\alpha]$.

In the other (more interesting) direction, we consider the evaluation morphism (Example 2) $\phi_\alpha : R[x] \to E$,

$$\phi_\alpha : a(x) \mapsto a(\alpha).$$

By our universal algebra result of Theorem 2(a) of Section III.4.2, $\phi_\alpha(R[x]) = S$ is a subring of E. Moreover, S is a subring that contains each $a = \phi_\alpha(a)$ in R and $\alpha = \phi_\alpha(x)$. Thus $R[\alpha] \subseteq S$, since $R[\alpha]$ by definition is the *smallest* subalgebra that contains each $a \in R$ and α. $\square$

When $\alpha \in E$ satisfies a polynomial over R, then some simplification in $R[\alpha]$ is possible. We shall investigate this situation in Chapter VI (Section 1) for the important case R a field; here we shall be content with the following

Example 4. Let $\alpha = \sqrt[3]{2}$. We claim that

$$Q[\alpha] = \{a + b\alpha + c\alpha^2 | a, b, c \in Q\}.$$

Proof of claim. Denoting $\{a + b\alpha + c\alpha^2 | a, b, c \in Z\}$ by S, we note that $S \subseteq Q[\alpha]$ by closure of the subring $Q[\alpha]$. But S is a subring of R (Example 2 of Section 1.2) that contains each $a = a + 0\alpha + 0\alpha^2 \in Q$ and $\alpha = 0 + 1\alpha + 0\alpha^2$. Hence $Q[\alpha] \subseteq S$, since $Q[\alpha]$ is the smallest subalgebra of R that contains each $a \in Q$ and α. $\square$

2. INTEGRAL DOMAINS AND FIELDS

From this point on we focus our attention on *commutative* rings and the special kinds of commutative rings that are of paramount importance for applications.

2.1 Zerodivisors and Units; Integral Domains and Fields

In a commutative ring R, there are two kinds of elements that we should like to define: *zerodivisors* and *units*.

An element $a \neq 0$ in R is said to be a *zerodivisor* if $ab = 0$ for some $b \neq 0$ in R. (Symmetrically, b is also a zerodivisor.)

An element $u \neq 0$ in R is said to be a *unit* if u is invertible: $uv = 1$ for some $v \in R$ $(v = u^{-1})$.

Example 1.   In $\mathbf{Z}_8$, 2 is a zerodivisor, since $2 \times 4 = 0$ while $2 \neq 0$, $4 \neq 0$. In $\mathbf{Z}_8$, 3 is a unit, since $3 \times 3 = 1$. $\square$

We denote the set of all units of a ring R by $U(R)$.

Proposition 1. $U(R)$ is a multiplicative group ("the group of units of R").

Proof.   Easy. $\square$

We now show that a ring element cannot simultaneously be a zerodivisor and a unit.

Proposition 2.   Let R be a commutative ring and $u \in R$ be a unit. Then u is not a zerodivisor.

Proof.   Assume that $uv = 0$ (we must show that v is necessarily equal to zero to conclude that u is not a zerodivisor). We then have $v = u^{-1}uv = u^{-1}0 = 0$. $\square$

The concepts of zerodivisor and unit lead to two important classes of commutative rings, one class distinguished by the absence of zerodivisors, the other by the presence of units.

Definition.   An *integral domain* is a nontrivial commutative ring with no zerodivisors: $ab = 0 \Rightarrow a = 0$ or $b = 0$.

Definition.   A *field* is a nontrivial commutative ring in which every nonzero element is a unit.

Corollary (to Proposition 2).   A field is an integral domain.

Fields are important for obvious operational reasons. In a field we can perform all four rational operations: addition, subtraction, multiplication, and division by nonzero elements $(a/b = ab^{-1})$. It is this capability that underlies many numerical (real and complex) algorithms, such as the solution of linear systems of equations by elimination.

Integral domains are important for less obvious reasons. Suffice it to say that many familiar facts from the algebra of real and complex numbers depend in a crucial way on the integral domain property of absence of zerodivisors. We shall see this particularly vividly in Section 3.3.

Example 2.   The prototypical example of an integral domain (from which the name is derived) is the ring $\mathbf{Z}$ of integers. We note that $\mathbf{Z}$ is not a field: 1 and -1 are the only units. $\square$

Example 3.   The most familiar examples of fields are provided by the number rings $\mathbf{Q}$, $\mathbf{R}$, and $\mathbf{C}$. Early in one's mathematical education, one learns the simple formula for inverses in $\mathbf{Q}$: for a/b with $a \neq 0$ (hence $a/b \neq 0$),

$$(a/b)^{-1} = b/a.$$

Later, one learns the more complex (forgive the pun) formula for inverses in $\mathbf{C}$: for $a + ib$ with $a, b \in \mathbf{R}$ not both zero (hence $a + ib \neq 0$),

$$(a + ib)^{-1} = \frac{a}{a^2 + b^2} + i\frac{-b}{a^2 + b^2}$$

[this formula can be derived by expressing $(a + ib)^{-1}$ as $\dfrac{a - ib}{(a + ib)(a - ib)}$]. $\square$

Example 4 (*Prime fields*). For p a prime, we have already determined that every element of $\mathbf{Z}_p^*$ is a unit (Theorem 1 of Section III.1.3). On the other hand, if $m = ab$ is composite ($1 < a, b < m$), then $ab = 0$ in $\mathbf{Z}_m$. Thus we have proved

Theorem 3.

(a) If p is prime, then $\mathbf{Z}_p$ is a field.
(b) If m is composite, then $\mathbf{Z}_m$ is a ring with zerodivisors.

Example 5. The addition and multiplication tables below provide an example of a four element field.

+	0	1	α	β
0	0	1	α	β
1	1	0	β	α
α	α	β	0	1
β	β	α	1	0

$\cdot$	1	α	β
1	1	α	β
α	α	β	1
β	β	1	α

Thus we have exhibited a finite *nonprime* field. (These examples suggest the problem of characterizing all finite fields, the nonprime as well as the prime ones. This we shall do in Chapter VI.) $\square$

A useful alternative characterization of integral domains is provided by

Theorem 4. D is an integral domain if and only if the following *cancellation law* holds:

$(*)$ $\qquad\qquad\qquad [ab = ac \quad \text{and} \quad a \neq 0] \Rightarrow b = c.$

Proof. (Only if) Let D be an integral domain: $ab = 0 \Rightarrow a = 0$ or $b = 0$. Then

$$ab = ac \text{ and } a \neq 0$$

$$\Rightarrow a(b - c) = 0 \text{ and } a \neq 0$$

$$\Rightarrow b - c = 0 \qquad D \text{ has no zerodivisors,}$$

giving $b = c$.

(If) Let D be a commutative ring for which the cancellation law ($*$) holds. For any $a \neq 0$ we have

$$ab = 0 \Rightarrow ab = a0 \qquad a0 = 0$$

$$\Rightarrow b = 0 \qquad \text{cancellation law,}$$

so that D is free of zerodivisors. $\square$

In rings with zerodivisors the cancellation law will not in general hold. For example, in $\mathbf{Z}_{12}$ we have $3 \times 2 = 3 \times 6$, yet $2 \neq 6$.

The ring of integers, $\mathbf{Z}$, gives us an example of an infinite integral domain that is not a field. However, in the finite case the concepts of integral domain and field coincide:

Theorem 5.   A finite integral domain is a field.

Proof.   Let D be a finite integral domain with distinct nonzero elements

$$d_1, d_2, \ldots, d_n.$$

For any $x \neq 0$ in D, consider

$$xd_1, xd_2, \ldots, xd_n.$$

Now each $xd_i \neq 0$, since D is an integral domain. Moreover, $xd_i \neq xd_j$ if $i \neq j$, otherwise we would have $d_i = d_j$ by cancellation of $x \neq 0$ (Theorem 4). Hence $xd_1, xd_2, \ldots, xd_n$ are distinct and nonzero, hence must exhaust the nonzero elements $d_1, d_2, \ldots, d_n$ of D. In particular, we must have $1 = xd_j$ for some d_j, whence $d_j = x^{-1}$. $\square$

Integral domains having finite characteristic are restricted according to

Theorem 6.   If an integral domain has finite characteristic m, then m is prime.

Proof.   Suppose to the contrary that an integral domain D has composite characteristic $m = ab$, with $1 < a$, $b < m$. Then $a \cdot 1, b \cdot 1 \neq 0$, but $(a \cdot 1)(b \cdot 1) = (ab) \cdot 1 = m \cdot 1 = 0$, contradicting the fact that D has no zerodivisors. $\square$

Morphism considerations.   The following proposition shows that a ring morphism preserves the (partial) operation of inversion.

Proposition 7.   Let $\phi: R \to R'$ be a morphism of rings, and let $a \in R$ be a unit. Then $\phi(a^{-1}) = \phi(a)^{-1}$.

Proof. $\phi(a)\phi(a^{-1}) = \phi(aa^{-1}) = \phi(1) = 1$; symmetrically, $\phi(a^{-1})\phi(a) = 1$. Thus $\phi(a^{-1}) = \phi(a)^{-1}$ as required. $\square$

Corollary.   If $a \in R$ is a unit, then so is $\phi(a) \in R'$ (i.e., ring morphisms map units into units). ▪

We observe that the analog of Theorem 2 of Section 1.3 does *not* hold for integral domains: a homomorphic image of an integral domain need not be an integral domain, as evidenced by the epimorphism $a \mapsto r_m(a)$ from $\mathbf{Z}$ to $\mathbf{Z}_m$; if m is composite, then $\mathbf{Z}_m$ is not an integral domain. For integral domains and fields, however, we do have the following modest

Proposition 8.   Let $\phi: R \to R'$ be an isomorphism of rings.

(a) If R is an integral domain, then so is R'.
(b) If R is a field, then so is R'.

Proof.   Exercise 5. $\square$

2.2   Field of Quotients

We address ourselves to the following problem: given an integral domain D, construct a field F that includes D. (Thus in F every nonzero element of D will be invertible.) Of course if D is finite, then according to Theorem 5 of the previous section we already have a field. So our problem is interesting only if D is infinite and not already a field.

We shall show in the following theorem that the familiar construction whereby the rationals are obtained as quotients of integers applies to integral domains in general.

Theorem 1   (*Construction of the field of quotients of an integral domain*). Let D be an integral domain. Define a relation $\sim$ on $D \times D^*$ ($D^* = D - \{0\}$) according to

(1) $(a, b) \sim (c, d) \quad \Leftrightarrow \quad ad = bc.$

Then
   1. $\sim$ is an equivalence relation on $D \times D^*$.

Let $Q(D) = (D \times D^*)/\sim$, and denote the equivalence class $[(a, b)]$ by the "fraction" a/b. Thus, according to (1), we have

(2) $a/b = c/d \quad \Leftrightarrow \quad ad = bc$

(and we are reminded of the familiar rule for equality of rational numbers; e.g., $2/3 = 8/12$, since $2 \times 12 = 3 \times 8$). Define $+$ and $\cdot$ on $Q(D)$ (recalling the familiar formulas for addition and multiplication of rational numbers) by

$$a/b + c/d = (ad + bc)/bd,$$
$$(a/b)(c/d) = (ac)/(bd).$$

Then
   2. $+$ and $\cdot$ are well-defined.

3. $Q(D)$ is a field, in which
$$0 = 0/1,$$
$$1 = 1/1,$$
$$-(a/b) = (-a)/b,$$
$$(a/b)^{-1} = b/a \qquad (a \neq 0).$$

4. $\tilde{D} = \{a/1 \mid a \in D\}$ is a subdomain of $Q(D)$ that is isomorphic with D under the map $a \mapsto a/1$.

Identifying $a \in D$ with $a/1 \in \tilde{D}$, we finally have

5.   If F is a field that includes D, then F includes $Q(D)$ (thus, $Q(D)$ is the *smallest* field that includes D).

Proof of Theorem 1   (highlights).

PART 1.   Trivial.

PART 2.   Since $+$ and $\cdot$ are defined on equivalence classes, there *is* something to prove: we must show that the results of $+$ and $\cdot$ are independent of the choice of representatives of equivalence classes. Thus, to conclude that $+$ is well-defined, we must show that if $a/b = a'/b'$ and $c/d = c'/d'$, then $a/b + c/d = a'/b' + c'/d'$. So assume $a/b = a'/b'$ and $c/d = c'/d'$. By (2) we then have

(3) $$ab' = a'b, \qquad cd' = c'd.$$

By the definition of $+$ on $Q(D)$,

$$a/b + c/d = (ad + bc)/(bd),$$
$$a'/b' + c'/d' = (a'd' + b'c')/(b'd').$$

We then compute

$$(ad + bc)(b'd')$$
$$= (ab')(dd') + (cd')(bb')$$
$$= (a'b)(dd') + (c'd)(bb') \qquad \text{by (3)}$$
$$= (a'd' + b'c')(bd);$$

by (2) it follows that $(ad + bc)/(bd) = (a'd' + b'c')/(b'd')$.

The proof that $\cdot$ is well-defined follows similarly.

PART 3.   It is straightforward albeit tedious to verify the associative, commutative, and distributive laws for $+$ and $\cdot$ on $Q(D)$. We complete the proof that $Q(D)$ is a field.

$0/1$ is the additive identify element.
Check: $a/b + 0/1 = (a + 0)/(b1) = a/b$.

1/1 is the multiplicative identity element.
Check: $(a/b)(1/1) = (a1)/(b1) = a/b$.

$-(a/b) = (-a)/b$.
Check: $a/b + (-a)/b = (ab + (-a)b)/b^2 = 0/b^2 = 0/1$.

Hence $Q(D)$ is a commutative ring.

As for multiplicative inverses for nonzero elements, if $a/b \neq 0/1$, then $a \neq 0$, whence $b/a \in Q(D)$. Then

$$(a/b)(b/a) = (ab)/(ba) = 1/1,$$

so that $(a/b)^{-1} = b/a$. Thus, every nonzero element of $Q(D)$ is a unit, which makes the commutative ring $Q(D)$ a field.

PART 4.   Trivial.

PART 5.   Let F be a field that includes D. For every $a, b \in D$ ($b \neq 0$), F must contain ab^{-1}, since $F \supseteq D$ is a field. But we have agreed to identify $a \in D$ with $a/1 \in Q(D)$, so that

$$ab^{-1} = (a/1)(b/1)^{-1}$$
$$= (a/1)(1/b)$$
$$= a/b.$$

Thus F contains a/b for every $a, b \in D$ ($b \neq 0$); i.e., $F \supseteq Q(D)$, as required. $\square$

The most familiar example of a field of quotients (indeed, the one that motivated the abstract concept) is the field $\mathbf{Q} = Q(\mathbf{Z})$ of rational numbers. We shall encounter other important examples in Section 3.

If two integral domains are isomorphic, we would expect their respective fields of quotients to be likewise isomorphic. This is indeed the case.

Proposition 2.   Let D and D' be integral domains with $D \cong D'$. Then $Q(D) \cong Q(D')$.

Proof.   If $\phi: D \rightarrow D'$ is an isomorphism, then the map

$$a/b \mapsto \phi(a)/\phi(b)$$

is readily shown to be an isomorphism $Q(D) \rightarrow Q(D')$. (Exercise 6.) $\square$

3.   POLYNOMIALS AND FORMAL POWER SERIES

In this section we investigate the algebraic properties of polynomials and formal power series, paying special attention to those properties related to zerodivisors and units.

3.1 Algebra of Polynomials and Formal Power Series

Let R be a commutative ring. In our examples of rings, we have already encountered the ring $R[[x]]$ of formal power series (Example 3 of Section 1.1)

$$a(x) = \sum_{i=0}^{\infty} a_i x^i \qquad (a_i \in R)$$

and the ring $R[x]$ (subring of $R[[x]]$) of polynomials (Example 4 of Section 1.1)

$$a(x) = \sum_{i=0}^{n} a_i x^i \qquad (a_i \in R, n \in \mathbf{N}).$$

For polynomials we have defined the degree:

$$\deg a(x) = n \text{ if } a(x) = a_n x^n + \text{l.d.t.} \qquad (a_n \neq 0)$$

$$\deg 0 = -\infty$$

(l.d.t. = lower degree terms). a_n is called the *leading coefficient* of $a(x)$; a_0 is called the *constant term* of $a(x)$.

For power series we introduce the notion of *order:*

$$\operatorname{ord} a(x) = n \quad \text{if} \quad a(x) = a_n x^n + \text{h.d.t.} \quad (a_n \neq 0)$$

$$\operatorname{ord} 0 = +\infty$$

(h.d.t. = higher degree terms). a_n is called the *low order* coefficient of $a(x)$; as in the polynomial case, a_0 is called the *constant term* of $a(x)$.

The following proposition shows a dual relationship between degree and order.

Proposition 1.

(a) In $R[x]$,

$$\deg a(x)b(x) \leq \deg a(x) + \deg b(x),$$

$$\deg[a(x) + b(x)] \leq \max[\deg a(x), \deg b(x)].$$

(b) In $R[[x]]$,

$$\operatorname{ord} a(x)b(x) \geq \operatorname{ord} a(x) + \operatorname{ord} b(x),$$

$$\operatorname{ord}[a(x) + b(x)] \geq \min[\operatorname{ord} a(x), \operatorname{ord} b(x)].$$

Proof. Simply appeal to the definitions of $+$ and $\cdot$ in $R[x]$ and $R[[x]]$. $\square$

Remark. By assigning $\deg 0 = -\infty$ and $\operatorname{ord} 0 = +\infty$, the relationships of Proposition 1 apply if either of $a(x)$, $b(x)$ is zero.

Theorem 2. If D is an integral domain, then so are (a) $D[x]$, (b) $D[[x]]$.

Proof. We know that $D[x]$ and $D[[x]]$ are commutative rings; at issue here is the absence of zerodivisors.

(a) Let $a(x) = a_m x^m + \text{l.d.t.}$ $(a_m \neq 0)$, $b(x) = b_n x^n + \text{l.d.t.}$ $(b_n \neq 0)$. (Hence $a(x), b(x) \neq 0$.) Then

$$(*) \qquad\qquad a(x)b(x) = a_m b_n x^{m+n} + \text{l.d.t.}$$

Now $a_m b_n \neq 0$, since D has no zerodivisors. Thus $a(x)b(x) \neq 0$, and we have shown that $D[x]$ does not contain zerodivisors.

(b) (For $D[[x]]$ we focus attention on the lowest degree terms.) Let $a(x) = a_m x^m + \text{h.d.t.}$ $(a_m \neq 0)$, $b(x) = b_n x^n + \text{h.d.t.}$ $(b_n \neq 0)$. (Hence $a(x), b(x) \neq 0$.) Then

$$(**) \qquad\qquad a(x)b(x) = a_m b_n x^{m+n} + \text{h.d.t.}$$

and the conclusion that $a(x)b(x) \neq 0$ follows just as in (a). $\square$

From $(*)$ and $(**)$ we conclude the following

Corollary. Let D be an integral domain. Then the degree function in $D[x]$ and order function in $D[[x]]$ obey (cf. Proposition 1)

(a) $\deg a(x)b(x) = \deg a(x) + \deg b(x)$;
(b) $\operatorname{ord} a(x)b(x) = \operatorname{ord} a(x) + \operatorname{ord} b(x)$.

Example 1. In $\mathbf{Z}_6[x]$, $(3x + 2)(2x + 4) = 4x + 2$. But $\mathbf{Z}_6$ is not an integral domain, so we have not refuted the above corollary. $\square$

We now investigate the units of $D[x]$ and $D[[x]]$.

NOTE.   Henceforth, unless otherwise qualified,

D denotes a integral domain,

F denotes a field.

Theorem 3. $a(x)$ is a unit of $D[x] \Leftrightarrow a(x)$ is a unit of D.

Proof. One direction $(\Leftarrow)$ is trivial. As for the other direction $(\Rightarrow)$, let $a(x) \in D[x]$ be a unit. Then $a(x)b(x) = 1$ for some $b(x) \in D[x]$, and we have

$$0 = \deg 1 = \deg a(x)b(x) = \deg a(x) + \deg b(x).$$

The only solution to this equation is $\deg a(x) = \deg b(x) = 0$, whence $a(x) = a$ is a unit in D. $\square$

Corollary. The units of $F[x]$ are the nonzero constant polynomials of F.

When we turn to $D[[x]]$, we obtain many more units compared with $D[x]$.

Theorem 4. $a(x) = a_0 + a_1 x + \cdots$ is a unit in $D[[x]] \Leftrightarrow a_0$ is a unit in D.

Proof. $(\Rightarrow)$ If $a(x)b(x) = 1$ for $b(x) = b_0 + b_1 x + \cdots \in D[[x]]$, then equating coefficients gives $a_0 b_0 = 1$; i.e., a_0 is a unit in D.

($\Leftarrow$) (Constructive.) Look at the condition for the existence of $b(x) = b_0 + b_1 x + \cdots \in D[[x]]$ satisfying $a(x)b(x) = 1$. Equating coefficients gives

$$a_0 b_0 = 1,$$
$$a_0 b_1 + a_1 b_0 = 0,$$
$$\vdots$$
$$a_0 b_n + a_1 b_{n-1} + \cdots + a_n b_0 = 0,$$
$$\vdots$$

The existence of $b(x)$ is then tantamount to the existence of a solution $b_0, b_1, b_2, \ldots$ to the above system of equations.

Now by assumption a_0 is a unit. We can therefore solve the above equations in turn for the coefficients b_i of $b(x)$ ($i = 0, 1, \ldots, n, \ldots$):

$$b_0 = a_0^{-1},$$
$$b_1 = -a_0^{-1}(a_1 b_0),$$
$$\vdots$$
$$b_n = -a_0^{-1}(a_1 b_{n-1} + a_2 b_{n-2} + \cdots + a_n b_0),$$
$$\vdots$$

By construction, $b(x)$ satisfies $a(x)b(x) = 1$; i.e., $b(x) = a(x)^{-1}$. $\square$

Corollary. $a(x) = a_0 + a_1 x + \ldots$ is a unit in $F[[x]] \Leftrightarrow a_0 \neq 0$.

Example 2. Let $a(x) = 1 + x + x^2 + \cdots \in Q[[x]]$. Then equating coefficients in $a(x)b(x) = 1$ (or substituting into the above formula) yields $b(x) = a(x)^{-1} = 1 - x$. $\square$

We note that $F[[x]]$, though containing many more units than $F[x]$, is still not a field. For example, x is not a unit in $F[[x]]$. With the idea in mind of extending $F[[x]]$ to a field, we introduce the set $F\langle x \rangle$ of *extended* formal power series

$$a(x) = \sum_{i=-\infty}^{\infty} a_i x^i$$

where $a_i \neq 0$ for only finitely many $i < 0$. This is to say that for every $a(x) \in F\langle x \rangle$ there is an integer n (which depends on $a(x)$) such that $a_i = 0$ for all $i < n$. Thus we may write $a(x) = \sum_{i=n}^{\infty} a_i x^i$ ($n \in Z$).

Addition and multiplication are defined on $F\langle x \rangle$ by the usual rule of collecting like powers of x:

(*)
$$s(x) = a(x) + b(x) \Leftrightarrow s_i = a_i + b_i;$$

$$p(x) = a(x)b(x) \quad \Leftrightarrow p_i = \sum_{j+k=i} a_j b_k.$$

It is not obvious that the sum and product of two extended formal power series are themselves extended formal power series. We therefore establish

Proposition 5. If $a(x), b(x) \in F\langle x \rangle$, then

$$a(x) + b(x) \in F\langle x \rangle, \qquad a(x)b(x) \in F\langle x \rangle.$$

Proof. If $a(x), b(x) \in F\langle x \rangle$, then $a(x) = \sum_{i=m}^{\infty} a_i x^i$, $b(x) = \sum_{i=n}^{\infty} b_i x^i$ for some $m, n \in \mathbf{Z}$. If $s(x) = a(x) + b(x)$, then from (*), $s_i = 0$ for $i < \min(m, n)$. Hence $s(x) \in F\langle x \rangle$.

If $p(x) = a(x)b(x)$, we must examine $p_i = \sum_{j+k=i} a_j b_k$. Since $a_j = 0$ for $j < m$ and $b_k = 0$ for $k < n$, it follows that $p_i = 0$ for $i < m + n$. Also, for any i,

$$p_i = a_m b_{i-m} + a_{m+1} b_{i-m-1} + \cdots + a_{i-n} b_n$$

is a *finite* summation, hence well-defined. $\square$

Proposition 6. $F[[x]]$ is a subring of $F\langle x \rangle$.

Proof (sketch). The proof that $F\langle x \rangle$ is a commutative ring is similar to the proof that $F[[x]]$ is a commutative ring. We omit the easy but tedious details.

Clearly $F[[x]] \subseteq F\langle x \rangle$. Also, the ring operations of $F\langle x \rangle$ are seen to reduce to those of $F[[x]]$ for the case of operands in $F[[x]]$. Hence $F[[x]]$ is a subring of $F\langle x \rangle$, as required. $\square$

$F\langle x \rangle$ is distinguished by

Proposition 7. $F\langle x \rangle$ is a field.

Proof. Let $a(x) \in F\langle x \rangle$, $a(x) \neq 0$. (We must establish that $a(x)$ is a unit in $F\langle x \rangle$.) Then there exists $n \in \mathbf{Z}$ such that $a_i = 0$ for $i < n$. Let n be the greatest such integer; i.e., $a_i = 0$ for $i < n$ and $a_n \neq 0$. (Since $a(x) \neq 0$, such an integer exists.) We then have

$$a(x) = a_n x^n + a_{n+1} x^{n+1} + a_{n+2} x^{n+2} + \cdots$$

$$= x^n \underbrace{\left(a_n + a_{n+1} x + a_{n+2} x^2 + \cdots \right)}_{b(x)}$$

where $b(x) = b_0 + b_1 x + b_2 x^2 + \ldots$ is an element of $F[[x]]$ and has $b_0 = a_n \neq 0$. By the corollary to Theorem 4, $b(x)$ is a unit in $F[[x]]$ (hence *a fortiori* a unit in $F\langle x \rangle$), and we have

$$a(x)^{-1} = x^{-n} b(x)^{-1}. \quad \square$$

Example 3.   Invert

(a) $a(x) = x^3 + x^4 + x^5 + \cdots \in \mathbf{Q}\langle x \rangle$;
(b) $a(x) = x^{-3} + x^{-2} + x^{-1} + \cdots \in \mathbf{Q}\langle x \rangle$.

Solution:

(a) $a(x) = x^3 + x^4 + x^5 + \cdots$

$\qquad = x^3(1 + x + x^2 + \cdots)$;

$\quad a(x)^{-1} = x^{-3}(1 - x)$ (cf. Example 2)

$\qquad = x^{-3} - x^{-2}$.

(b) $a(x) = x^{-3} + x^{-2} + x^{-1} + \cdots$

$\qquad = x^{-3}(1 + x + x^2 + \cdots)$;

$\quad a(x)^{-1} = x^3(1 - x)$

$\qquad = x^3 - x^4$.

* * *

We now take stock of the various extensions that we have of a polynomial domain $F[x]$. Thus far in this section we have focussed our attention on power series extensions of $F[x]$:

$$F[x] \subset F[[x]] \subset F\langle x \rangle.$$

Algebraically and operationally, the main point of distinction among these integral domains as we proceed from $F[x]$ through to $F\langle x \rangle$ is the increasing existence of units. The only units of $F[x]$ are the nonzero constant polynomials; in $F[[x]]$ all power series with nonzero constant terms are units; and in $F\langle x \rangle$, every nonzero (extended) power series is a unit—$F\langle x \rangle$ is a field.

But $F[x]$ and $F[[x]]$ have other field extensions besides $F\langle x \rangle$. Since $F[x]$ is an integral domain, it has a field of quotients (Theorem 1 of Section 2.2), which we denote by $F(x)$:

$$F(x) = \{a(x)/b(x) \mid a(x), b(x) \in F[x], \ b(x) \neq 0\}.$$

$F(x)$ is called the *field of rational functions over F*.

$F[[x]]$ is also an integral domain, so it too has a field of quotients, which we denote by $F((x))$:

$$F((x)) = \{a(x)/b(x) \mid a(x), b(x) \in F[[x]], \ b(x) \neq 0\}.$$

We call elements of $F((x))$ *power series* rational functions in order to distinguish them from the (polynomial) rational functions of $F(x)$.

Our various extensions of $F[x]$ now look thus:

$$F[x] \; \subset \; F[[x]] \; \subset \; F\langle x \rangle$$
$$\cap \quad ? \quad \cap \quad ?$$
$$F(x) \; \subset \; F((x))$$

We have posed two questions in the above diagram, the answers to which we shall leave as interesting exercises.

Question 1: What is the relationship between $F\langle x \rangle$ and $F((x))$? (We note that $F\langle x \rangle$ and $F((x))$ are both fields that include $F[[x]]$.) See Exercise 4.

Question 2: When is a power series $a(x) \in F[[x]]$ equal [in $F((x))$] to some $h(x)/d(x) \in F(x)$? Thus we ask when a power series expression is expressible as a rational function. [Note: According to the criterion for equality of fractions in a field of quotients, $a(x) = a(x)/1 = h(x)/d(x) \Leftrightarrow h(x) = a(x)d(x)$.] For example, $1 + x + x^2 + \cdots \in \mathbf{Q}[[x]]$ is equal in $\mathbf{Q}((x))$ to $1/(1-x)$, where $1/(1-x) \in \mathbf{Q}(x)$. On the other hand, it can be shown that $1 + x + x^2/2! + x^3/3! + \cdots$ is not expressible as a rational function in $\mathbf{Q}(x)$ (a fact that should not surprise any serious student of calculus). The answer to Question 2 is the subject of Exercise 5.

Remark. The importance of Question 2 is simply this: A power series $a(x) \in F[[x]]$ is in general an *infinite* expression, whereas its expression as a rational function $h(x)/d(x) \in F(x)$ (if it exists) is a *finite* expression—a "closed form" version of $a(x)$. In many areas of applied mathematics, combinatorics and probability theory in particular, the rational function version of a power series is regarded as a (rational) *generating function* for the coefficients of its power series "expansion."

3.2 The Division Property of Polynomials

The following theorem underlies the working of the familiar process of polynomial "long division." The general setting of an arbitrary commutative ring R for the coefficients causes no real difficulty.

Theorem 1 (*Division Property of $R[x]$*). Let $a(x) = a_m x^m + \cdots + a_0$ and $b(x) = b_n x^n + \cdots + b_0$ be polynomials of degree m and n in $R[x]$, and let b_n be a unit in R. Then there exist *unique* $q(x), r(x) \in R[x]$ (the *quotient* and *remainder* polynomials, respectively) that satisfy the following *Division Property*:

$$a(x) = b(x)q(x) + r(x), \qquad \deg r(x) < \deg b(x).$$

Proof (*Existence*). Our existence proof is very much motivated by how polynomial long division works. Symbolically, we have

$$b_n x^n + \cdots + b_0 \overline{\smash{\big)}\ \begin{aligned} & b_n^{-1} a_m x^{m-n} \\ & a_m x^m + a_{m-1} x^{m-1} + \cdots + a_0 \\ & \underline{a_m x^m + \text{l.d.t.}} \\ & \qquad\qquad \bar{a}_{m-1} x^{m-1} + \text{l.d.t.} \end{aligned}}$$ (iterate)

Now for a formal proof of existence of $q(x)$ and $r(x)$ (cf. the proof of Example 5, Section II.2). Let $p(m)$ be the assertion of Theorem 1 (apart from uniqueness). We now establish $p(m)$ for all $m \in \mathbf{N}$ by course-of-values induction.

Basis ($m = 0$). If $n = 0$, then $a(x) = a$ and $b(x) = b$ are constant polynomials. Then $q = b^{-1}a$, $r = 0$ trivially satisfy the Division Property. If $n > 0$, then $q(x) = 0$, $r = a(x)$ satisfy the Division Property.

Induction. Assume $p(k)$ for $0 \le k < m$, and let $a(x)$ have degree m.

CASE 1. $\deg b(x) > m$. Then $q(x) = 0$, $r = a(x)$ satisfy the Division Property.

CASE 2. $\deg b(x) \le m$ (the interesting case). As the result of performing one step of polynomial long division, we have

$$a(x) = b_n^{-1} a_m x^{m-n} b(x) + \bar{a}(x)$$

where (key observation) the partial remainder $\bar{a}(x)$ has degree $< m$. The induction hypothesis then lets us divide $\bar{a}(x)$ by $b(x)$ to obtain

$$\bar{a}(x) = q(x)b(x) + r(x), \qquad \deg r(x) < \deg b(x).$$

This gives $a(x)$ as

$$a(x) = \left[b_n^{-1} a_m x^{m-n} + q(x) \right] b(x) + r(x),$$

which has yielded a quotient and remainder satisfying the Division Property. Thus we have established $p(m)$, which completes the proof of the existence part of Theorem 1.

(*Uniqueness*.) Suppose

$$\begin{aligned} a(x) &= b(x)q(x) + r(x) \\ &= b(x)\bar{q}(x) + \bar{r}(x) \end{aligned}$$

where $\deg r(x), \deg \bar{r}(x) < n = \deg b(x)$. Subtracting gives

$$(*) \qquad\qquad b(x)\left[q(x) - \bar{q}(x) \right] = \bar{r}(x) - r(x).$$

Assume to the contrary that $q(x) \ne \bar{q}(x)$. Then $q(x) - \bar{q}(x)$ can be written as $d_l x^l + \text{l.d.t.}$ ($l \ge 0$) where $d_l \ne 0$, so that $b(x)[q(x) - \bar{q}(x)] = b_n d_l x^{n+l} + \text{l.d.t.}$ Since b_n is a unit, $b_n d_l \ne 0$ (Proposition 2 of Section 2.1). Hence $b(x)[q(x) - \bar{q}(x)]$ has degree $n + l \ge n$. But $\bar{r}(x) - r(x)$ has degree $< n$. This contradiction to $(*)$ forces $q(x) = \bar{q}(x)$ and consequently $r(x) = \bar{r}(x)$, as required. $\square$

We note that over $F[x]$ (F a field) the Division Property holds provided only that the divisor polynomial $b(x)$ is nonzero.

Example 1. In $\mathbf{Z}_8[x]$ compute $q(x)$ and $r(x)$ when $a(x) = 5x^4 + 2x^3 + 4x^2 + 7x + 2$ is divided by $b(x) = 3x^2 + 5$.

Solution. Noting that 3 is a unit in $\mathbf{Z}_8$, we are assured of the existence of $q(x)$ and $r(x)$ satisfying the Division Property. We now carry out the polynomial division algorithm:

$$
\begin{array}{r}
7x^2 + 6x\ + 3 \\
3x^2 + 5\,\overline{)5x^4 + 2x^3 + 4x^2\ + 7x + 2} \\
\underline{5x^4\qquad\quad + 3x^2} \\
2x^3 +\ x^2 + 7x + 2 \\
\underline{2x^3\qquad\ + 6x} \\
x^2 +\ x + 2 \\
\underline{x^2\qquad + 7} \\
x + 3
\end{array}
$$

Thus $q(x) = 7x^2 + 6x + 3$, $r(x) = x + 3$. □

Remainders and equivalence mod m(x). Let $m(x) = x^n + \text{l.d.t.}$ be a *monic* (leading coefficient = unity) polynomial in $R[x]$. By Theorem 1 the Division Property holds for any $a(x) \in R[x]$:

$$a(x) = m(x)q(x) + r(x), \qquad \deg r(x) < \deg m(x).$$

We embellish the notation for $r(x)$ in the above to $r_{m(x)}(a(x))$. We refer to $r_{m(x)}$ as the remainder mod $m(x)$ function and to $r_{m(x)}(a(x))$ as $a(x) \bmod m(x)$. If $r_{m(x)}(a(x)) = 0$, we write $m(x) \mid a(x)$ ("$m(x)$ divides $a(x)$").

We then define *equivalence mod* $m(x)$, $\equiv_{m(x)}$, on $R[x]$ by

$$a(x) \equiv_{m(x)} b(x) \quad\Leftrightarrow\quad m(x) \mid (a(x) - b(x)).$$

It is trivially verified that $\equiv_{m(x)}$ is an equivalence relation on $R[x]$.

Of course $r_{m(x)}$ and $\equiv_{m(x)}$ are analogous to r_m and $\equiv_m$ on $\mathbf{Z}$. Completely analogous to Lemma 1 and Theorem 1 of Section II.3.1, we have the following results:

Lemma 1. Let $m(x)$ be a monic polynomial in $R[x]$ (R a commutative ring). Then

(a) $a(x) \equiv_{m(x)} r_{m(x)}(a(x))$;

(b) $a(x) = r_{m(x)}(a(x)) \quad\Leftrightarrow\quad \deg a(x) < \deg m(x)$;

(c) $a(x) \equiv_{m(x)} b(x) \quad\Leftrightarrow\quad r_{m(x)}(a(x)) = r_{m(x)}(b(x))$.

We define
$$R[x]_{m(x)} = \{a(x) \in R[x] \mid \deg a(x) < \deg m(x)\}.$$

Theorem 2.

(a) For any $[a(x)] \in R[x]/\equiv_{m(x)}$,
$$[a(x)] = [r_{m(x)}(a(x))].$$
(b) For any $a(x), b(x) \in R[x]_{m(x)}$,
$$a(x) \neq b(x) \Rightarrow [a(x)] \neq [b(x)].$$

Proofs of Lemma 1 *and Theorem* 2. Exercise 7. □

From Theorem 2 we conclude: each equivalence class $[a(x)]$ has a unique representative in $R[x]_{m(x)}$, namely $r_{m(x)}(a(x))$.

At this point we could define a ring structure on $R[x]_{m(x)}$ much as we did on Z_m in Example 2 of Section III.1.1. We choose, instead, to do so as an application of some results that we shall establish in the next chapter (Section V.2.1).

3.3 Polynomials as Functions

Much of high school algebra and early calculus is devoted to the study of polynomials as functions—rules for computing (typically real) values. In this section we shall see that many of the familiar facts about polynomial functions (especially those facts about roots of polynomials) depend in a critical way on the absence of zerodivisors, i.e., on the integral domain property of the coefficients.

Roots of polynomials. Let $a(x) \in R[x]$ and $\alpha \in R$, R a commutative ring. If $a(\alpha) = 0$, then we call α a *root*, or *zero*, of $a(x)$.

The following theorem establishes a useful relationship between roots and linear factors.

Theorem 1   (*Factor Theorem*). Let $a(x) \in R[x]$, $\alpha \in R$. Then
$$\alpha \text{ is a root of } a(x) \iff (x - \alpha) \mid a(x).$$

Proof. ($\Rightarrow$) Let $a(\alpha) = 0$. Since the leading coefficient of $x - \alpha$ is a unit, we can divide $a(x)$ by $x - \alpha$ to obtain
$$(*) \qquad\qquad a(x) = (x - \alpha)q(x) + r(x)$$
where $\deg r(x) < \deg(x - \alpha) = 1$. Hence $r(x) = r \in R$. From $(*)$ we have
$$r = a(\alpha) - (\alpha - \alpha)q(\alpha) = 0 - 0 = 0,$$
so that $(x - \alpha) \mid a(x)$.

($\Leftarrow$)   If $a(x) = (x - \alpha)q(x)$, then $a(\alpha) = 0q(\alpha) = 0$. □

Corollary (*Remainder Theorem*). Let $a(x) \in R[x]$, $\alpha \in R$. Then the remainder on dividing $a(x)$ by $x - \alpha$ is $a(\alpha)$.

Proof. By the Factor Theorem, $x - \alpha$ divides $a(x) - a(\alpha)$. Thus we have $a(x) = (x - \alpha)q(x) + a(\alpha)$, making $a(\alpha)$ the unique remainder when $a(x)$ is divided by $x - \alpha$. $\square$

If α is a root of $a(x)$, then by the Factor Theorem $a(x)$ can be expressed as

$$a(x) = (x - \alpha)^m q(x), \qquad m \geq 1, q(\alpha) \neq 0.$$

The positive integer m is called the *multiplicity* of the root α of $a(x)$. If $m = 1$, then α is called a *simple* root; if $m > 1$, then α is called a *multiple* root.

We now recall a familiar result from the algebra of real and complex numbers: a polynomial of degree n can have at most n roots, counting multiplicities (a root of multiplicity m counts as m roots). But consider the quadratic equation $x^2 + 12$ over $\mathbf{Z}_{16}$. It has four roots: 2, 6, 10, 14. Now $\mathbf{Z}_{16}$, unlike the real or complex number system, is a ring with zerodivisors. It therefore behooves us to take careful note of where the absence of zerodivisors comes into the proof of

Theorem 2. Let D be an integral domain. A nonzero polynomial $a(x) \in D[x]$ of degree n has at most n roots in D, counting multiplicities.

Proof. The proof is by course-of-values induction on $n = \deg a(x)$.

Basis $(n = 0)$. A nonzero polynomial of degree 0 is a (nonzero) constant of D, hence trivially has no roots.

Induction. Assume the statement of the theorem for all polynomials of degree $< n$, and let $a(x) \in D[x]$ have degree n. If $a(x)$ has a root α of multiplicity m $(m \geq 1)$, then

$$a(x) = (x - \alpha)^m q(x)$$

where $\deg q(x) = n - m$. If $a(x)$ has another root $\beta \neq \alpha$, then

$$0 = a(\beta) = (\beta - \alpha)^m q(\beta).$$

Since $\beta \neq \alpha$, we must have $q(\beta) = 0$ *because D has no zerodivisors.* Hence the only roots of $a(x)$ are α (multiplicity m) and those of $q(x)$. By the induction hypothesis $q(x)$ has at most $n - m$ roots. Thus $a(x)$ has at most $(n - m) + m = n$ roots. $\square$

In particular, we note that a nonzero polynomial of degree n in $D[x]$ has at most n distinct roots in D.

Formal derivatives and multiple roots. Let R be a commutative ring, and let $a(x) = a_n x^n + a_{n-1} x^{n-1} + \cdots + a_0$ be a polynomial over R. We define the (*formal*) derivative of $a(x)$,

$$a'(x) = na_n x^{n-1} + (n-1)a_{n-1} x^{n-2} + \cdots + a_1.$$

In terms of Σ-notation, if $a(x) = \Sigma_i a_i x^i$, then

$$a'(x) = \sum_i (i + 1)a_{i+1} x^i.$$

This definition of formal derivative is seen to agree with the formula for the derivative of a polynomial that we know from calculus. Moreover, we can now establish for formal derivatives the well-known properties of analytic derivatives (with no recourse to continuity arguments which are just not applicable to arbitrary commutative rings).

Proposition 3.   Let $a(x), b(x) \in R[x]$. Then

(a) $c(x) = a(x) + b(x) \Rightarrow c'(x) = a'(x) + b'(x)$;
(b) $c(x) = a(x)b(x) \quad \Rightarrow c'(x) = a(x)b'(x) + a'(x)b(x)$;
(c) $c(x) = a(b(x)) \quad\quad \Rightarrow c'(x) = a'(b(x))b'(x)$.

Proof.   (a) Trivial.
   (b) We denote the derivative of $f(x) = \Sigma_i f_i x^i$ by $f'(x) = \Sigma_i f'_i x^i$. (Thus $f'_i = (i+1)f_{i+1}$.) With $c(x) = a(x)b(x)$ and $d(x) = a(x)b'(x) + a'(x)b(x)$, our task is to show that $c'(x) = d(x)$, i.e., that $c'_i = d_i$.
   Now

$$d_i = \sum_{j+k=i} a_j b'_k + \sum_{j+k=i} a'_j b_k$$

$$= \sum_{j+k=i} a_j(k+1)b_{k+1} + \sum_{j+k=i} (j+1)a_{j+1}b_k$$

$$= \sum_{j+k=i+1} a_j k b_k + \sum_{j+k=i+1} j a_j b_k$$

$$= \sum_{j+k=i+1} (j+k)a_j b_k$$

$$= (i+1) \sum_{j+k=i+1} a_j b_k$$

$$= (i+1)c_{i+1}$$

$$= c'_i.$$

(c) Exercise 13. $\square$

Let α be a root of $a(x) \in R[x]$. As already noted, the Factor Theorem yields the factorization

$$(*) \qquad\qquad a(x) = (x-\alpha)^m q(x), \qquad m \geq 1, q(\alpha) \neq 0.$$

The following theorem gives a useful criterion for α to be a multiple root ($m > 1$ in $(*)$).

Theorem 4.   Let α be a root of $a(x) \in R[x]$. Then α is a multiple root of $a(x)$ if and only if $a'(\alpha) = 0$.

Proof.   Factoring $a(x)$ as in $(*)$, we can differentiate $a(x)$ via Proposition 3 to obtain

$$a'(x) = (x-\alpha)^m q'(x) + m(x-\alpha)^{m-1} q(x).$$

If $m > 1$, then $a'(\alpha) = 0$, trivially. If $m = 1$ then $a'(\alpha) = q(\alpha)$, and $q(\alpha)$ is $\neq 0$ according to ($*$). $\square$

Interpolation. Given a collection of n "sample points" $(\alpha_k, \beta_k) \in F \times F$ ($k = 0, \ldots, n - 1$; α_k's distinct), the *interpolation problem* over F is to find a polynomial $u(x) \in F[x]$ such that $u(\alpha_k) = \beta_k$ ($k = 0, \ldots, n - 1$). The "interpolating polynomial" $u(x)$ is said to interpolate to the points (α_k, β_k).

The most familiar setting for this problem is the case $F = \mathbf{R}$, where we regard the polynomial as a curve that passes through the points (α_k, β_k) in the plane. This viewpoint is illustrated in Fig. 1.

A solution to the interpolation problem, a problem first solved by the great Sir Isaac Newton (1642–1729), is the content of

Theorem 5 (*Interpolation*). Given any n points $(\alpha_k, \beta_k) \in F \times F$ ($k = 0, \ldots, n - 1$; α_k's distinct), there exists a unique polynomial $u(x) \in F[x]$ having degree $< n$ such that $u(\alpha_k) = \beta_k$ ($k = 0, \ldots, n - 1$).

Proof (*Existence*). For the existence of $u(x)$, we present an ingenious formula attributed to Lagrange, called *Lagrange's Interpolation Formula*:

$$u(x) = \sum_{k=0}^{n-1} \beta_k L_k(x)$$

where

$$L_k(x) = \frac{(x - \alpha_0) \cdots (x - \alpha_{k-1})(x - \alpha_{k+1}) \cdots (x - \alpha_n)}{(\alpha_k - \alpha_0) \cdots (\alpha_k - \alpha_{k-1})(\alpha_k - \alpha_{k+1}) \cdots (\alpha_k - \alpha_n)}.$$

The claim is that $u(\alpha_k) = \beta_k$ ($k = 0, \ldots, n - 1$). Proof: By inspection, $L_k(\alpha_j) = 0$ if $k \neq j$ (due to the factor $\alpha_j - \alpha_j$ of $L_k(x)$), while $L_k(\alpha_k) = 1$. The result $u(\alpha_k) = \beta_k$ follows immediately. By construction, $\deg u(x) < n$.

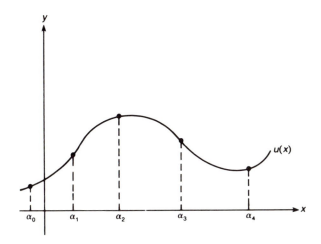

Fig. 1 Interpolating polynomial

(*Uniqueness*—an application of Theorem 2.) Suppose that $u(x), \bar{u}(x)$ both have degree $< n$ and $u(\alpha_k) = \bar{u}(\alpha_k) = \beta_k$ ($k = 0, \ldots, n-1$). Then $d(x) = u(x) - \bar{u}(x)$ is a polynomial that has degree $< n$ but has n distinct roots, namely the n α_k's. By Theorem 2 this is impossible unless $d(x) = 0$. Hence $u(x) = \bar{u}(x)$, proving uniqueness. □

Example 1. Compute the polynomial $u(x) \in \mathbf{Z}_{11}[x]$ that interpolates to

$$
\begin{array}{c|ccc}
\alpha_k & 0 & 2 & 5 \\
\hline
\beta_k & 3 & 9 & 4
\end{array}.
$$

Solution. Substitute into Lagrange's Interpolation Formula:

$$
u(x) = 3\frac{(x-2)(x-5)}{(0-2)(0-5)} + 9\frac{(x-0)(x-5)}{(2-0)(2-5)} + 4\frac{(x-0)(x-2)}{(5-0)(5-2)}
$$

$$
= 2x^2 + 10x + 3 \quad \text{(performing coefficient calculations in } \mathbf{Z}_{11}\text{)}.
$$

Check: $u(0) = 3$, $u(2) = 9$, $u(5) = 4$. □

4. DIVISIBILITY; EUCLIDEAN DOMAINS

This section presents a case study in how useful and revealing the process of abstraction can be. In Section 4.1, we generalize divisibility and related notions from the concrete settings of integers and polynomials to the abstract setting of an arbitrary integral domain. In Section 4.2, we then introduce a special class of integral domains, called *Euclidean domains*, that captures the important divisibility properties both of integers and of polynomials over a field.

As usual,

$$D \text{ denotes an integral domain,}$$

$$F \text{ denotes a field.}$$

4.1 Divisibility Concepts in Integral Domains

Divisibility. The generalization of the divisibility relation from integers and polynomials to an arbitrary integral domain is immediate: For $a, b \in D$,

$$a \mid b \quad \text{means} \quad b = ac \text{ for some } c \in D.$$

Example 1.
(a) In $\mathbf{Z}$: $3 \mid 6$, $3 \nmid 7$.
(b) In $\mathbf{Z}[x]$: if $a(x) = 2x + 4$ and $b(x) = 2x^2 + 3x - 2$, then $b(x)/a(x) = x - \frac{1}{2}$. Hence

  (i) $a(x) \mid b(x)$ in $\mathbf{Q}[x]$;

  (ii) $a(x) \nmid b(x)$ in $\mathbf{Z}[x]$.

(c) In any field: $a \mid b$ provided $a \neq 0$, for we have $b = a(a^{-1}b)$. (The divisibility relation is not very interesting in a field.) □

Associates. $a, b \in D^*$ are called *associates*, denoted $a \sim b$ ("*a tilda b*"), if $a = ub$ where u is a unit in D.

The association relation is closely related to divisibility:

Proposition 1. $a \sim b \iff a|b$ and $b|a$.

Proof. ($\Rightarrow$) If $a \sim b$, then $a = ub$, so that $b|a$. Since u is a unit, $b = u^{-1}a$, so that $a|b$.

($\Leftarrow$) If $a|b$ and $b|a$, then $b = xa$ and $a = yb$ for $x, y \in D$. Therefore $b = xyb$, so that $xy = 1$ by cancellation of $b \neq 0$. Thus y (and x) is a unit, which means $a \sim b$. $\square$

A divisor a of b is called a *proper* divisor of b if a is neither a unit nor an associate of b.

Proposition 2.   Let $a \sim b$ and $c \sim d$. Then,
$$a|c \Rightarrow b|d.$$

Proof. If $a \sim b$ and $c \sim d$, then $a = ub$ and $c = vd$ for units u and v. Now,

$$a|c \Rightarrow c = aq$$
$$\Rightarrow vd = ubq$$
$$\Rightarrow d = b(v^{-1}uq),$$

and so $b|d$. $\square$

In words, Proposition 2 states that if a divides c then any associate of a divides any associate of c. Consequently, any element can be replaced by any of its associates without affecting divisibility relationships. Our goal now is to bring the useful notion of *distinguished* associate into the picture.

Proposition 3.   The relation $\sim$ is an equivalence relation on D^*.

Proof. Easy—Exercise 2. $\square$

Thus, $\sim$ partitions D^* into equivalence classes of associated elements: for any $a \in D^*$, the $\sim$-equivalence class containing a is given by

$$[a] = \{ua \mid u \in U(D)\},$$

where $U(D)$ is the group of units of D.

By a *distinguished associate* of $a \in D^*$, we simply mean a distinguished representative of the $\sim$-equivalence class $[a]$. We shall agree on the following:

(a) In $\mathbf{Z}$, $U(\mathbf{Z}) = \{1, -1\}$. Hence $[m] = \{m, -m\}$. We choose the distinguished associate of $[m]$ to be positive.

(b) In $F[x]$, $U(F[x]) = F^*$. Hence $[a(x)] = \{ca(x) \mid c \in F^*\}$. We choose the distinguished associate of $[a(x)]$ to be monic: if $a(x) = a_n x^n + \text{l.d.t.}$ $(a_n \neq 0)$, its unique monic associate is $a_n^{-1}a(x)$.

(c) In any integral domain D, $[u] = U(D)$ if u is a unit. We choose unity as the distinguished representative of $[u]$.

Greatest common divisors. A *greatest common divisor* (g.c.d.) of $a, b \in D$ (a, b not both zero) is an element $g \in D$ that satisfies:

1. $g \mid a$ and $g \mid b$ (g is a *common divisor* of a and b);
2. $c \mid a$ and $c \mid b \Rightarrow c \mid g$ (g is the *greatest* common divisor of a and b).

Example 2. In **Z**, the common divisors of 12 and 18 are $\pm 1, \pm 2, \pm 3, \pm 6$; the greatest common divisors of 12 and 18 are accordingly ± 6. $\square$

The next proposition shows that a g.c.d. (whenever it exists) is unique up to associates.

> **Proposition 4.** If g is a g.c.d. of $a, b \in D$, then so is any associate of g. Conversely, if g, h are g.c.d.'s of $a, b \in D$, then $g \sim h$.

Proof. By definition, any g.c.d. g of $a, b \in D$ satisfies 1 and 2 above. But by Proposition 2, so does any associate of g.

As for the converse, suppose g, h are both g.c.d.'s of $a, b \in D$. By 2, $g \mid h$ and $h \mid g$, whence $g \sim h$ (Proposition 1). $\square$

Since a g.c.d. is determined only up to associates—up to multiplication by an arbitrary unit—it is convenient to achieve uniqueness by choosing a distinguished associate for any g.c.d. We can then speak of *the* g.c.d. of a and b, which we denote by $\gcd(a, b)$ or simply by (a, b):

(a) In **Z** we take (a, b) to be > 0.
(b) In $F[x]$ we take $(a(x), b(x))$ to be monic.
(c) In any integral domain, if a g.c.d. of a and b is a unit, then we take (a, b) to be 1. We then call a and b *relatively prime*. (Examples: In **Z**, $(4, 9) = 1$; in **R**$[x]$, $(x^2 + 1, x^2 - 1) = 1$.)

Primes. We have already classified the nonzero elements of an integral domain D into units and nonunits. We further subdivide the nonunits into *prime* and *composite* elements.

A nonunit $p \in D^*$ is called *prime* if whenever $p = ab$ ($a, b \in D^*$), either a or b is a unit. Thus, a prime has no proper divisors. A nonunit that is not prime is called *composite*. Thus, a composite element $c \in D^*$ has a factorization of the form $c = ab$ ($a, b \in D^*$) where neither a nor b is a unit.

Example 3. 7 is prime in **Z**, since the only factorizations of 7 are 1×7 and $(-1) \times (-7)$. On the other hand, $6 = 2 \times 3$ is composite, since neither 2 nor 3 is a unit in **Z**. $\square$

> **Proposition 5.** If p is a prime, then so is any associate of p.

Proof. Easy—Exercise 3. $\square$

Thus with primes, as with g.c.d.'s, we may restrict ourselves to distinguished associates, e.g., positive primes in the case of the integers.

Turning to polynomial domains, we introduce the following

Terminology. If $a(x)$ is prime (composite) in $D[x]$, we say that $a(x)$ is *irreducible* (*reducible*) *over D*.

In $F[x]$ (F a field), the units are all given by the constant (degree zero) polynomials F^*. Hence $a(x) \in F[x]$ is reducible over F if $a(x) = b(x)c(x)$ for polynomials $b(x), c(x) \in F[x]$ having degree ≥ 1. In $D[x]$ (D an integral domain), the latter condition is sufficient but not necessary for reducibility over D (see Example 4(a) below).

Example 4.

(a) $2x + 2$ is irreducible over **Q**;
    $2x + 2 = 2(x + 1)$ is reducible over **Z**, since neither 2 nor $x + 1$ is a unit in $\mathbf{Z}[x]$.
(b) $x^2 - 2 = (x + \sqrt{2})(x - \sqrt{2})$ is reducible over **R**;
    $x^2 - 2$ is irreducible over **Q** (since $\sqrt{2} \notin \mathbf{Q}$).
(c) $x^2 + 1 = (x + i)(x - i)$ is reducible over **C**;
    $x^2 + 1$ is irreducible over **R** (or **Q**, or **Z**). $\square$

Let $a(x) \in F[x]$ be a quadratic or cubic. If $a(x)$ is reducible, then $a(x) = b(x)c(x)$ where $b(x), c(x)$ have degree ≥ 1. *Hence one of* $b(x)$ *and* $c(x)$ *must be linear* (in either the quadratic or cubic case). Hence by the Factor Theorem (Theorem 1 of Section 3.3), $a(x)$ must have a root in F. We have just proved the somewhat specialized but often useful

Proposition 6. A quadratic or cubic polynomial $a(x) \in F[x]$ is reducible over F if and only if $a(x)$ has a root in F.

Example 5. Let $a(x) = x^3 + 2x + 2$, regarded as a polynomial (a) over $\mathbf{Z}_5$, (b) over $\mathbf{Z}_3$.

(a) $a(x)$ is reducible over $\mathbf{Z}_5$, since $a(1) = 0$ (evaluation over $\mathbf{Z}_5$).
(b) $a(x)$ is irreducible over $\mathbf{Z}_3$, since $a(0), a(1), a(2) \neq 0$ (evaluation over $\mathbf{Z}_3$). $\square$

4.2 Euclidean domains

Because of the striking similarity of integers and polynomials (with coefficients in a field) with respect to divisibility, it behooves us to introduce a level of abstraction that includes them both—to kill two birds with one stone. This is accomplished by the following

Definition. A *Euclidean domain* is an integral domain D together with a "degree" function $d: D^* \to \mathbf{N}$ ($d(a) =$ the (Euclidean) degree of a) such that

1. $d(ab) \geq d(a)$ $(a, b \neq 0)$.
2. (*Division Property.*) For every a and b in D ($b \neq 0$), there exist a

"quotient" q and "remainder" r in D that satisfy the *Division Property:*

$$a = bq + r, \qquad d(r) < d(b) \text{ or } r = 0.$$

We denote the (unique or preferred) remainder r satisfying the above Division Property by $r_b(a)$ ("the remainder when a is divided by b").

Example 1. $\mathbf{Z}$ with $d(a) = |a|$. Postulate 2, in particular, is satisfied by the Division Property of $\mathbf{Z}$: for every $a, b \in \mathbf{Z}$, $b \neq 0$, there exist (unique) q, r such that

$$a = bq + r, \qquad 0 \leq r < |b|.$$

We note that if $r \neq 0$, then $r - b$ also satisfies postulate 2. Thus the axioms for a Euclidean domain do not guarantee a unique remainder (or quotient). For $\mathbf{Z}$ we obtain uniqueness by taking the preferred remainder to be nonnegative. $\square$

Example 2. $F[x]$ with $d(a) = \deg a(x)$. Postulate 1 is a consequence of the corollary to Theorem 2 of Section 3.1. Postulate 2 is satisfied by the Division Property of $F[x]$: for every $a(x), b(x) \in F[x]$, $b(x) \neq 0$, there exist unique $q(x), r(x)$ such that

$$a(x) = b(x)q(x) + r(x), \qquad \deg r(x) < \deg b(x).$$

In contrast to $\mathbf{Z}$, for $F[x]$ we do have uniqueness for the quotient and remainder satisfying postulate 2 (Theorem 1 of Section 3.2). $\square$

A trivial (uninteresting) example of a Euclidean domain is

Example 3. Any field F with $d(a) = 1$ if $a \neq 0$. Since F is a field, it is always the case that the remainder on division by a nonzero field element is zero. $\square$

A couple of non-Euclidean examples:

Example 4. $\mathbf{Q}$ with $d(a) = |a|$ is not a Euclidean domain even if we extend the range of d to include the positive rationals. For postulate 1 does not hold; e.g., $d(2 \times \frac{1}{2}) = d(1) < d(2)$. $\square$

Example 5. $\mathbf{Z}[x]$ with $d(a) = \deg a(x)$ is not a Euclidean domain because postulate 2 does not hold. For example, the (unique) quotient when $a(x) = 3x^5$ + l.d.t. is divided (in $\mathbf{Q}[x]$) by $b(x) = 2x^3 + $ l.d.t. is $\frac{3}{2}x^2 + $ l.d.t., which is not in $\mathbf{Z}[x]$. $\square$

Greatest Common Divisors in Euclidean Domains. Our first major divisibility property for Euclidean domains is that any Euclidean domain is a *g.c.d. domain* —an integral domain in which every pair of elements has a g.c.d., a g.c.d. which, moreover, is expressible in a special useful form. We have already encountered this result for $\mathbf{Z}$ (Theorem 2 of Section II.3.2); the generalization to Euclidean domains will grant us the same result for $F[x]$.

> **Theorem 1** (*A Euclidean domain is a g.c.d. domain*). Let D be a Euclidean domain, and let $a, b \in D$ (not both zero). Then a and b have a g.c.d. g

expressible as a linear combination of a and b:

$$g = sa + tb \qquad (s, t \in D).$$

We first establish the preliminary

Lemma 1.   With $a, b \in D$ (not both zero), define

$$S(a, b) = \{sa + tb \mid s, t \in D\}.$$

Choose $g \in S(a, b)$ $(g \neq 0)$ to have minimum degree: $d(g) \leq d(x)$ for all $x \in S(a, b)$. Then

$$S(a, b) = (g),$$

where $(g) = \{rg \mid r \in D\}$ is the set of all multiples of g.

Proof.   First we note that $S(a, b)$ contains nonzero elements, since at least one of $a = 1a + 0b$ and $b = 0a + 1b$ in $S(a, b)$ is nonzero. We can therefore choose $g \in S(a, b)$ as in the statement of the lemma.

As for the inclusion $S(a, b) \subseteq (g)$, divide $x \in S(a, b)$ by g to obtain

$$x = gq + r \quad \text{where } d(r) < d(g) \text{ or } r = 0.$$

(Our goal is to show $r = 0$.) Since $x, g \in S(a, b)$, we have

$$x = s_x a + t_x b, \qquad g = s_g a + t_g b,$$

whence

$$r = x - gq = (s_x - s_g q)a + (t_x - t_g q)b$$

is in $S(a, b)$. Hence $r = 0$, otherwise we would have $d(r) < d(g)$ in contradiction to g having minimum degree in $S(a, b)$. Thus $S(a, b) \subseteq (g)$. The reverse inclusion holds trivially. $\square$

Proof of Theorem 1.   With $S(a, b)$ and g as in Lemma 1, the claim is that g is a g.c.d. of a and b.

Proof of claim: Since $a, b \in S(a, b) = (g)$, a and b are multiples of g, which means that g is a common divisor of a and b. And if $c \mid a$ and $c \mid b$, then $c \mid s_g a + t_g b$ trivially. Thus $g = s_g a + t_g b$ is a *greatest* common divisor of a and b.

$\square$

According to Proposition 4 and the ensuing discussion of the previous section, we can obtain a *unique* g.c.d. for any a, b in a Euclidean domain by choosing a distinguished associate: In $\mathbf{Z}$, we take $\gcd(a, b) > 0$; in $F[x]$ we take $\gcd(a(x), b(x))$ to be monic; and in any Euclidean domain, we take $\gcd(a, b) = 1$ if a and b are relatively prime, i.e., have a unit for their g.c.d.

In Chapter VII we shall deal with computational aspects of g.c.d.'s in Euclidean domains.

Prime factorization in Euclidean domains.   Our second major divisibility property of Euclidean domains is that any Euclidean domain is a *unique factorization domain*—an integral domain in which every nonunit can be (essentially) uniquely expressed as a product of primes. (This generalizes what we proved for the integers in Section II.3.3.)

We shall need the following lemmas, which provide the key to proving facts by induction in abstract Euclidean domains. Let D be a Euclidean domain with degree function d.

Lemma 3.

(a) $d(1) \leq d(a)$ for any $a \in D^*$.
(b) $d(1) = d(a) \Leftrightarrow a$ is a unit.

Proof.   (a) $d(1) \leq d(1a = a)$ by postulate 1 for a Euclidean domain.

(b) ($\Leftarrow$)   If a is a unit, then $d(a) \leq d(aa^{-1} = 1)$, so that $d(a) = d(1)$ by part (a).

($\Rightarrow$)   Let $d(a) = d(1)$. Dividing 1 by a gives $1 = aq + r$ where $d(r) < d(a)$ or $r = 0$. But $d(r) < d(1) (= d(a))$ is impossible by part (a). Hence $r = 0$, whence a is a unit as required. $\square$

Lemma 4.   In D^*, if $a = bc$ where c is a nonunit, then $d(b) < d(a)$.

Proof.   By postulate 1 for a Euclidean domain, $d(b) \leq d(a)$. The point at issue here is *strict* inequality. Dividing b by a gives

$$b = aq + r \quad \text{where } d(r) < d(a) \text{ or } r = 0.$$

But $r = 0$ means that $b = aq = bcq$, which gives $cq = 1$ by cancellation of b—a contradiction to c being a nonunit. Hence the possibility that $r = 0$ is excluded, and we have

$$r = b - aq$$
$$= b - bcq$$
$$= b(1 - cq),$$

so that $d(b) \leq d(r)$. Thus we have $d(b) \leq d(r) < d(a)$; i.e., $d(b) < d(a)$ as required. $\square$

Corollary.   If b is a proper divisor of a, then $d(b) < d(a)$.

Proof.   If $b|a$ and $a\nmid b$, then $a = bc$ where c is a nonunit. $\square$

Theorem 3A   (*Existence of a factorization into primes*).   Any $a \in D^*$ is either a unit or can be expressed as a finite product of primes: $a = p_1 p_2 \cdots p_n$.

Proof.   By induction on $d(a)$ (note the utility of Lemmas 3 and 4 in the following).

Basis. $d(a) = d(1)$. Then by Lemma 3, a is a unit.

Induction.   Assume the statement of the theorem for any $x \in D^*$ such that $d(x) < d(a)$. If a is prime or a unit, then we are done. If a is composite, then we can write $a = xy$ where neither x nor y is a unit. By Lemma 4, we have

$d(x) < d(a)$ and $d(y) < d(a)$, whence by the induction hypothesis we can express x and y as products of primes:

$$x = p_1 p_2 \cdots p_m \text{ and } y = q_1 q_2 \cdots q_n.$$

This yields a as a product of primes:

$$a = xy = p_1 p_2 \cdots p_m q_1 q_2 \cdots q_n. \quad \square$$

For our second factorization theorem, we shall need the corollaries that flow from the following useful

Lemma 5.   Let a and b be relatively prime in a Euclidean domain D. Then

(a) $a \mid bc \Rightarrow a \mid c$;
(b) $a \mid c$ and $b \mid c \Rightarrow ab \mid c$.

Remark.   The stipulation that a and b be relatively prime is necessary. In **Z**, for example, $4 \mid (2 \times 6)$, whereas $4 \nmid 6$; $4 \mid 12$ and $6 \mid 12$, whereas $(4 \times 6) \nmid 12$.

Proof of Lemma 5.   Since a and b are relatively prime, by Theorem 1 we have

$$sa + tb = 1 \quad \text{for some } s, t \in D,$$

whence

$$(*) \qquad\qquad\qquad sac + tbc = c.$$

(a) If $a \mid bc$, then $bc = ka$ for some $k \in D$. From $(*)$ we then obtain $a(sc + tk) = c$, so that $a \mid c$ as required.

(b) If $a \mid c$ and $b \mid c$, then $c = ma = nb$ for some $m, n \in D$. From $(*)$ we then have $sanb + tbma = c$, giving $ab(sn + tm) = c$. Thus $ab \mid c$ as required. $\square$

Corollary 1 (to part (a)).   If p is a prime in a Euclidean domain D, then

$$p \mid ab \Rightarrow p \mid a \text{ or } p \mid b.$$

Proof.   Suppose $p \mid ab$ and $p \nmid a$. Since p is a prime, its only divisors are associates and units. Since $p \nmid a$, the only common divisors of a and p are units, which means that a and p are relatively prime. By Lemma 5(a), $p \mid b$. $\square$

Corollary 2.   If p is a prime in a Euclidean domain D, then

$$p \mid a_1 a_2 \cdots a_n \Rightarrow p \mid a_i \text{ for some } 1 \le i \le n.$$

Proof.   Corollary 1 + induction. $\square$

Theorem 3B (*Uniqueness of prime factorization*).   In a Euclidean domain D, every nonunit can be expressed as a product of primes in essentially one way: if

$$(*) \qquad\qquad\qquad a = p_1 p_2 \cdots p_s = q_1 q_2 \cdots q_t$$

where the p_i's, q_j's are primes, then $s = t$ and there exists a reordering of the q_j's such that

$$p_1 \sim q_1, p_2 \sim q_2, \ldots, p_s \sim q_s.$$

Proof. By Theorem 3A, any nonunit $a \in D^*$ can be expressed as a product of primes; it is uniqueness that is at issue. The proof proceeds by induction on $d(a)$.

Basis. $d(a) = d(1)$. Then a is a unit and there is nothing further to be proved.

Induction. Assume the statement of the theorem for all $x \in D^*$ such that $d(x) < d(a)$. Let a be expressed as products of primes in ostensibly two ways:

$$a = p_1 p_2 \cdots p_s = q_1 q_2 \cdots q_t$$

with $s \leq t$.

We dispense first with the case $s = 1$. If $s = 1$, then $a = p_1 = q_1 q_2 \cdots q_t$, forcing $t = 1$ since p_1 is a prime. Thus the theorem holds for this (rather trivial) case.

For the case $s > 1$, we observe that $p_1 | a$, which is to say that

$$p_1 | q_1 q_2 \cdots q_t.$$

By Corollary 2 to Lemma 5, we have

$$p_1 | q_i \quad \text{for some } 1 \leq i \leq t,$$

and by reordering of the q_j's, we may assume that $p_1 | q_1$. But q_1 is prime, so that $p_1 \sim q_1$ (otherwise the prime q_1 would have p_1 as a proper divisor), say $p_1 = u q_1$. Dividing a by p_1 gives

$$a/p_1 = p_2 \cdots p_s = q_2' q_3 \cdots q_t$$

where $q_2' = u^{-1} q_2$ is a prime (Proposition 5 of Section 4.1).

Now $d(a/p_1) < d(a)$ by Lemma 4, which means that the induction hypothesis can be applied to a/p_1. This gives $s - 1 = t - 1$ and

$$p_2 \sim q_2' \sim q_2, p_3 \sim q_3, \ldots, p_s \sim q_s$$

for some reordering of the q_j's $(2, 3, \ldots, t)$. This, combined with $p_1 \sim q_1$, completes the induction step of the proof. $\square$

As an immediate consequence of Theorems 3A and 3B (cf. the corollary to Theorems 1A and 1B of Section II.3.3), we have the

Corollary. In a Euclidean domain, any nonunit a has a prime decomposition of the form

$$a = u p_1^{e_1} p_2^{e_2} \cdots p_r^{e_r}$$

where u is a unit, the p_i's are the distinct nonassociated prime factors of a, and each exponent $e_i > 0$ is uniquely determined by the $\sim$-equivalence class of p_i. (Example: $360 = 1 \times 2^3 \times 3^2 \times 5^1 = (-1) \times (-2)^3 \times 3^2 \times 5^1$.)

* * *

To summarize our findings about Euclidean domains: they are integral domains that have a divisibility theory essentially like $\mathbf{Z}$. First, greatest common divisors always exist in a special useful form; second, every nonunit can be factored into a product of primes in an essentially unique way.

But it is certainly fair to ask at this point, does *every* integral domain enjoy a divisibility theory like $\mathbf{Z}$? Interestingly enough, the answer is no. In Exercise 12, an example is developed to show that an integral domain can indeed have quite "nonstandard" divisibility properties. The particular example of Exercise 12 goes back to Richard Dedekind (1831–1916), an early investigator of nonstandard divisibility results.

EXERCISES

Section 1

1. Prove that the commutativity of ring addition follows from the remaining ring postulates. [*Hint:* Apply the distributive laws in two different ways to the element $(a + b)(1 + 1)$.]

2. Verify the distributive law in $\mathbf{Z}_m$:
$$a \odot (b \oplus c) = (a \odot b) \oplus (a \odot c).$$

3. Prove the familiar rule of signs $(-a)(-b) = ab$ in an arbitrary ring.

4. Complete the proof of Proposition 3 of Section 1.1: Show that if $t(x) = a(x)b(x)$ in $R[x]$ then $t = ab$ in R^R.

5. Prove the following generalized distributive laws in an arbitrary ring:
   (a) $(\sum_{i=1}^{n} a_i)b = \sum_{i=1}^{n} a_i b; \; a(\sum_{i=1}^{n} b_i) = \sum_{i=1}^{n} ab_i.$
   (b) $(\sum_{i=1}^{m} a_i)(\sum_{i=1}^{n} b_i) = \sum_{i=1}^{m} \sum_{j=1}^{n} a_i b_j.$
   (c) $n \cdot (ab) = (n \cdot a)b,$
       $n \cdot a = (n \cdot 1)a \quad (n \in \mathbf{Z}).$

6. (*The Binomial Theorem.*) Let R be a commutative ring. Prove the *binomial theorem* in R:
$$(a + b)^n = \sum_{k=0}^{n} \binom{n}{k} \cdot a^k b^{n-k}.$$
[Recall: The binomial coefficient $\binom{n}{k} = \dfrac{n!}{k!(n-k)!}$.]

7. In $R[x]$ what is $[R \cup \{x^2\}]$?

8. Prove that the homomorphic image of a (commutative) ring is a (commutative) ring (cf. Exercise 6 of Section III.4).

9. Let $\phi: R \to S$ be a morphism of rings. Prove that the image under ϕ of the unital subring of R is the unital subring of S.

Section 2

1. (a) Show that every nonzero element of $\mathbf{Z}_m$ is either a unit or a zerodivisor.
   (b) What is $U(\mathbf{Z}_m)$?

2. In an integral domain D, prove: $a^2 = 1 \Rightarrow a = \pm 1$. Does this result hold in a commutative ring with zerodivisors?

3. True or false: If D is an integral domain, then so is the direct product $D \times D$.

4. If a commutative ring R has prime characteristic p, is R necessarily an integral domain? (Cf. Theorem 6 of Section 2.1.) What if R has characteristic zero?

5. Prove Proposition 8 of Section 2.1.

6. Prove Proposition 2 of Section 2.2.

7. (*Quadratic extensions.*)   Let D be an integral domain and E an extension domain of D. Let $d \in D$ have a square root, $\sqrt{d}$, in E but not in D. (For example, $D = \mathbf{Z}$, $E = \mathbf{R}$, $d = 2$; or $D = \mathbf{R}$, $E = \mathbf{C}$, $d = -1$.) We now examine

$$D[\sqrt{d}] = \{a + b\sqrt{d} \mid a, b \in D\},$$

the *quadratic extension* of D obtained by adjoining a square root of d to D.

(a) Verify that $D[\sqrt{d}]$ is, as the notation indicates, the smallest subdomain of E that contains D and $\sqrt{d}$. Also show that every element $\alpha \in D[\sqrt{d}]$ has a *unique* expression of the form $\alpha = a + b\sqrt{d}$.

Define the *norm* of $\alpha = a + b\sqrt{d}$ by $N(\alpha) = a^2 - b^2 d$.

(b) (The reason for introducing the norm.) Establish the following multiplicative property for the norm: $N(\alpha\beta) = N(\alpha)N(\beta)$. Conclude that α is a unit in $D[\sqrt{d}]$ if and only if $N(\alpha)$ is a unit in D.

(c) Further conclude that if D is a field, then so is $D[\sqrt{d}]$.

(d) Determine the units of
    (i) $\mathbf{Z}[\sqrt{2}]$;   (ii) $\mathbf{Z}[\sqrt{-1}]$;   (iii) $\mathbf{Z}[\sqrt{-5}]$.

(e) Show that $Q(D[\sqrt{d}]) = Q(D)[\sqrt{d}]$.

Exercises 8–10 deal with an important class of domains called *ordered domains*, in which the familiar concepts of "positive" and "negative" elements and "order" are defined.

8. (*Ordered domains.*)

Definition.   An *ordered domain* is an integral domain in which there is a special subset $D^+ \subset D$, called the set of *positive* elements of D, that satisfies
    1. $a, b \in D^+ \Rightarrow a + b, ab \in D^+$;
    2. For each $a \in D$, exactly one of the following holds:

$$a \in D^+, a = 0, \text{ or } -a \in D^+ \qquad (\textit{trichotomy}).$$

(a) In an ordered domain prove that $a^2 \in D^+$ if $a \neq 0$. Conclude that $1 \in D^+$.

(b) Show that the only way to make the integers $\mathbf{Z}$ and reals $\mathbf{R}$ into ordered domains is the familiar one. (A reassuring result.)

(c) Show that there is no way of making the complex field $\mathbf{C}$ into an ordered domain. Also show that there is no way of making a finite field into an ordered field.

Definition.   In an ordered domain D, the binary relation $<$ ("less than") is defined by

$$a < b \Leftrightarrow b - a \in D^+.$$

(Thus $a > 0 \Leftrightarrow a \in D^+$.)

(d) Verify the following familiar laws of order:

(i) For any $a, b \in D$, exactly one of the following holds:
$$a < b, \ a = b, \ \text{or } a > b \qquad (\textit{trichotomy});$$
(ii) $a < b$ and $b < c \Rightarrow a < c$ (*transitivity*);
(iii) $b < c \Rightarrow a + b < a + c$ (*isotonicity for sums*);
(iv) $a > 0$ and $b < c \Rightarrow ab < ac$ (*isotonicity for positive factors*).

Definition. In an ordered domain D, the *absolute value* $|a|$ of $a \in D$ is defined by a if $a \geq 0$, $-a$ otherwise.

(e) Verify the familiar laws of absolute value:
(i) $|ab| = |a| |b|$;
(ii) $|a + b| \leq |a| + |b|$.

9. Let D be an ordered domain. Show that its field of quotients, $Q = Q(D)$, becomes an ordered domain if we define Q^+ by

$$(*) \qquad\qquad\qquad a/b \in Q^+ \ \Leftrightarrow \ ab \in D^+.$$

Verify that $D^+ \subseteq Q^+$ (i.e., verify that the order of D is *extended* to Q). Further argue that this extension of the order of D to Q is unique, in that $(*)$ provides the only ordering of Q^+ that satisfies $D^+ \subseteq Q^+$. [*Hint:* $a/b = ab(1/b)^2$.]

10. (*A well-ordering characterization of the integers.*) In Section II.1 we essentially described $\mathbf{Z}$ as an ordered domain in which the set $\mathbf{P}$ of positive integers is *well-ordered:* every nonempty subset of $\mathbf{P}$ has a least element. In this exercise we show that this description provides a (unique) characterization of $\mathbf{Z}$ in the sense that if D is any other ordered domain in which the set D^+ of positive elements is well-ordered (with respect to the order $<$ of D), then $D \cong \mathbf{Z}$.

     So let D be an ordered domain in which the set D^+ positive elements is well-ordered. For convenience we denote the unity of D by e (to distinguish it from the unity 1 of $\mathbf{Z}$).

(a) Show that e is the least element of D^+. [*Hint:* Suppose to the contrary that $0 < m < e$ for some $m \in D^+$. Let m be the least such element. Now consider m^2.]

(b) (The crux of the proof.) Show that $D^+ = \{n \cdot e \mid n \in \mathbf{P}\}$. [*Hint:* Suppose to the contrary this is false. Let S be the (nonempty) set of all elements of D^+ not of this form, and let s be the least element of S. Now consider the element $s - e$.]

(c) Conclude that $D = \{n \cdot e \mid n \in \mathbf{Z}\}$.

(d) (The finale.) Show that the obvious map $\phi \colon n \mapsto n \cdot e$ is an isomorphism $\mathbf{Z} \to D$.

Section 3

1. Compute the inverse in $\mathbf{Q}[[x]]$ of $a(x) = 1 + x + 2x^2 + 3x^3 + 5x^4 + \cdots$ where $a_k = a_{k-1} + a_{k-2}$ ($k \geq 2$).

2. Express as extended formal power series in $F\langle x \rangle$:
(a) $1/(x^3 - x^4)$;
(b) $(1 + x + x^3)/(x^2 - x^4)$.

3. (a) Express $\dfrac{\frac{1}{2} + \frac{2}{9}x^3}{\frac{4}{7}x + \frac{5}{3}x^2}$ as an element of $\mathbf{Z}(x)$.

(b) (*Generalization.*) Let D be an integral domain. Show that $D(x) = Q(D)(x)$.

4. Prove that $Q(F[[x]]) = F\langle x \rangle$.

5. Let $a(x) = \sum_{k=0}^{\infty} a_k x^k$ be a power series in $F[[x]]$. Prove: $a(x)$ is equal in $F((x))$ to a rational function $h(x)/d(x) \in F(x)$ if and only if the a_k's ultimately satisfy a *linear recurrence:* for some $l, n \in \mathbb{N}$ and $d_1, d_2, \ldots, d_n \in F$,
$$a_k = d_1 a_{k-1} + d_2 a_{k-2} + \cdots + d_n a_{k-n} \quad \text{for all } k > l.$$
[*Hint:* Show that if $a(x)$ is in $F(x)$, then $a(x)$ must be expressible in the form
$$a(x) = \frac{h(x)}{d(x)} = \frac{h_0 + h_1 x + \cdots + h_m x^m}{1 - d_1 x - \cdots - d_n x^n}$$
for $h(x), d(x) \in F[x]$.]

6. (*Application of Exercise 5.*)   Derive a rational generating function for the coefficients of the following power series:
    (a) $a(x) = 1 + x + 2x^2 + 3x^3 + 5x^4 + \cdots$ where $a_k = a_{k-1} + a_{k-2}$ $(k \geq 2)$.
    (b) $a(x) = 3x^2 + 3x^3 + 12x^4 + 42x^5 + 117x^6 + \cdots$ where $a_k = a_{k-1} - 3a_{k-2} - 7a_{k-3} - 6a_{k-4}$ $(k \geq 5)$.

7. Prove Lemma 1 and Theorem 2 of Section 3.2 (cf. the proofs of the analogous integer results of Section II.3.1).

8. Show that $x + p$ is a factor of $x^3 + (2p - 2q + 1)x^2 + (p^2 - 3pq + q^2 + p - q)x - pq(p - q + 1)$.

9. Let $a(x) \in R[x]$, where R is a commutative ring. Show that if $a(x^2)$ has a factor $x - 1$, then it also has a factor $x^2 - 1$.

10. Find all $p \in R$ such that the remainder on dividing
$$a(x) = px^3 - (p - 1)x^2 - (1 - 5p)x + p^2 + 5p + 45$$
by $x + 2$ is -21.

11. Let F be a field. Show that the natural correspondence between $F[x]$ and P_F (the set of all polynomial functions $F \rightarrow F$),
$$a(x) \mapsto [a: F \rightarrow F: \alpha \mapsto a(\alpha)],$$
is a bijection if and only if F is an infinite field. [*Hint:* If F is finite, say $|F| = n$, consider $x^n - x$.]

12. Show that if F is finite, then $P_F = F^F$ (i.e., every function $F \rightarrow F$ is a *polynomial function*).

13. (*Differentiation of composites.*)   Let $a(x), b(x) \in R[x]$, R a commutative ring. Prove:
    (a) $c(x) = a(x)^n \Rightarrow c'(x) = na(x)^{n-1}$;
    (b) $c(x) = a(b(x)) \Rightarrow c'(x) = a'(b(x))b'(x)$ (Proposition 3(c) of Section 3.3).

14. Let $a(x) \in R[x]$, $\alpha \in R$, where R is a commutative ring. Let $q(x)$ be the quotient when $a(x)$ is divided by $x - \alpha$. Show that the remainder when $q(x)$ is divided by $x - \alpha$ is $a'(\alpha)$. (Thus $a(\alpha), a'(\alpha)$ can be obtained by successive divisions by $x - \alpha$.)

Section 4

1. In an integral domain can a unit have a proper divisor?

2. Prove Proposition 3 of Section 4.1.

3. Prove Proposition 5 of Section 4.1.

4. Factor into irreducibles the following polynomials in $\mathbf{Z}_3[x]$.
   (a) $x^2 + 1$;   (b) $x^3 + x + 1$;   (c) $x^3 + 2x + 2$;
   (d) $x^4 + x^3 + x^2 + 1$;   (e) $x^4 + x^3 + x + 1$.

5. In a Euclidean domain, prove: $a \sim b \Rightarrow d(a) = d(b)$.

6. Let D be a Euclidean domain, and let $a, b \in D$. Show that if $sa + tb = 1$ for some $s, t \in D$, then $(a, b) = 1$. (Cf. Exercise 2 of Section II.3.)

7. Let D be a Euclidean domain, and let $a, b, c \in D$ where a and b are relatively prime.
   (a) Show that the equation
   $$(*) \quad ax + by = c$$
   has a solution $x, y \in D$.
   (b) Show that if x_0, y_0 is a particular solution to $(*)$, then all solutions $u, v \in D$ are given by
   $$u = x_0 - tb, \quad v = y_0 + ta \quad \text{for arbitrary } t \in D.$$

8. (*Generalization of Exercise 7.*) We now remove the condition of Exercise 7 that a and b be relatively prime. Let g be a g.c.d. of a and b.
   (a) If $g \nmid c$, show that $ax + by = c$ has no solutions.
   (b) If $g | c$, show the equivalence of the equations $ax + by = c$ and $a'x + b'y = c'$ where $a' = a/g$, $b' = b/g$, and $c' = c/g$. Conclude that $ax + by = c$ has solutions over D.

9. Let D be a Euclidean domain, p a prime in D. Prove that $\sqrt{p} \notin Q(D)$. Conclude that $\sqrt{2}$, $\sqrt{3}$, $\sqrt{5}$, etc., are *irrational*. [*Hint:* Assume, to the contrary, that $\sqrt{p} = a/b$. Then consider $b^2 p = a^2$ in the light of unique factorization.]

10. (*Least common multiples.*) The *least common multiple* (l.c.m.) l of a, b in a Euclidean domain D is characterized by the properties (cf. Exercise 3 of Section II.3):
    1. $a | l$ and $b | l$ (l is a *common* multiple of a and b);
    2. $a | c$ and $b | c \Rightarrow l | c$ (l is the *least* common multiple of a and b).
    (a) Show that the l.c.m. of a and b is unique up to associates.

    Notation. We denote the (unique) distinguished associate of any l.c.m. of a and b by $\text{lcm}(a, b)$ or by $[a, b]$.

    (b) Prove that if $(a, b) = 1$, then $[a, b] \sim ab$.
    (c) Prove that $[ka, kb] \sim k[a, b]$ ($k \in D$).
    (d) Conclude that $[a, b] \sim ab/(a, b)$.

11. (*G.c.d.'s and l.c.m.'s by prime decomposition.*) Let a and b in a Euclidean domain have prime decompositions
    $$a = u p_1^{e_1} p_2^{e_2} \cdots p_n^{e_n} \quad (e_i \geq 0),$$
    $$b = v p_1^{f_1} p_2^{f_2} \cdots p_n^{f_n} \quad (f_i \geq 0),$$
    where $p_1, \ldots, p_n$ are nonassociated prime factors of either a or b, and u, v are units. Define
    $$g = p_1^{g_1} p_2^{g_2} \cdots p_n^{g_n} \quad \text{where } g_i = \min(e_i, f_i),$$
    $$l = p_1^{l_1} p_2^{l_2} \cdots p_n^{l_n} \quad \text{where } l_i = \max(e_i, f_i).$$
    Show (i) that g is a g.c.d. of a and b and (ii) that l is an l.c.m. of a and b.

12. (*A non-Euclidean integral domain.*)   In this exercise we investigate divisibility properties of $\mathbf{Z}[\sqrt{-5}\,] = \{a + b\sqrt{-5} \mid a, b \in \mathbf{Z}\}$. $\mathbf{Z}[\sqrt{-5}\,]$ is evidently a subring of $\mathbf{C}$. (You will find it useful to review the concept of *norm* defined in Exercise 7 of Section 2.)

(a) Show that the units of $\mathbf{Z}[\sqrt{-5}\,]$ are ± 1 (just like $\mathbf{Z}$).

(b) (*Failure of unique factorization in* $\mathbf{Z}[\sqrt{-5}\,]$.)   Consider the two factorizations of 9 in $\mathbf{Z}[\sqrt{-5}\,]$:

$$9 = 3 \times 3 = (2 + \sqrt{-5}\,)(2 - \sqrt{-5}\,).$$

Show that $3, 2 + \sqrt{-5}, 2 - \sqrt{-5}$ are primes, no two of which are associates. Conclude that unique factorization does not hold in $\mathbf{Z}[\sqrt{-5}\,]$.

(c) (*Failure of the g.c.d. property in* $\mathbf{Z}[\sqrt{-5}\,]$.)   Show that 9 and $3(2 + \sqrt{-5}\,)$ do not have a g.c.d. in $\mathbf{Z}[\sqrt{-5}\,]$.

13. (a) In any integral domain D, show that if p satisfies

$$p \mid ab \Rightarrow p \mid a \text{ or } p \mid b,$$

then p is prime. (Cf. Corollary 1 to Lemma 5 of Section 4.2.)

(b) Show that the converse to (a) need not hold in a non-Euclidean integral domain. [*Hint:* Exercise 12(b).]

QUOTIENT ALGEBRAS

A fundamental problem of algebra is to determine the homomorphic images (i.e., the abstractions, or simplified versions) of a given Ω-algebra A. The problem as stated might seem hopeless. Where do we find such homomorphic images? What do they look like? Yet interestingly enough, the problem is not hopeless. Indeed, Universal Algebra offers a remarkably comprehensive solution, which revolves around the notion of a *quotient algebra*—a quotient set A/E of an Ω-algebra A by a special kind of equivalence relation called a *congruence relation*, with operations defined on A/E in a natural way.

In Section 1 we investigate the theory of congruence relations and quotient algebras from a universal algebra viewpoint. In Sections 2 and 3 we specialize the setting to rings.

1. UNIVERSAL QUOTIENT ALGEBRAS

1.1 Congruence Relations

A *congruence relation* on an Ω-algebra, like a subalgebra of an Ω-algebra or a morphism of Ω-algebras, is a set-theoretic construction (in this case an equivalence relation) that respects the operations of Ω:

John D. Lipson, Elements of Algebra and Algebraic Computing. ISBN 0-201-04115-4.

Definition. A *congruence relation* E on an Ω-algebra A is an equivalence relation on A that satisfies the following *substitution property:* for any $\omega \in \Omega$,

$$a_i \, E \, b_i \text{ for } i = 1, \ldots, n \Rightarrow \omega(a_1, \ldots, a_n) \, E \, \omega(b_1, \ldots, b_n).$$

In words, the above substitution property (s.p.) states that if corresponding operands of an operation ω are equivalent then so are the values. The substitution property can be regarded as a generalization of Euclid's axiom on magnitudes that "equals added to or multiplied by equals gives equals."

A partition π induced by a congruence relation is called a *partition with the substitution property*, or *s.p. partition* for short. In terms of partitions, the above substitution property becomes

$$[a_i] = [b_i] \text{ for } i = 1, \ldots, n \Rightarrow [\omega(a_1, \ldots, a_n)] = [\omega(b_1, \ldots, b_n)].$$

As always, one passes from equivalence to equality when one passes from the set A to the quotient set (partition) A/E.

For the important special case of a binary operation, say multiplication, the substitution property becomes

$$a \, E \, c \text{ and } b \, E \, d \Rightarrow ab \, E \, cd;$$

or in terms of a partition we have the following setup:

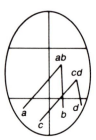

Example 1. The classical example of a congruence relation is our old friend equivalence mod m on $\mathbf{Z}$ (Example 3 of Section I.3.2), which we now call *congruence* mod m and generalize to the setting of an arbitrary commutative ring R.

Let $m \in R$. We define *congruence* mod m on R by

$$a \equiv_m b \quad \Leftrightarrow \quad a - b = km \qquad (k \in R).$$

(Thus, $a \equiv_m b \Leftrightarrow m \,|\, a - b$.) We now claim that congruence mod m is indeed a congruence relation on the ring R.

Proof of claim: $\equiv_m$ is easily checked to be an equivalence relation. We now check that $\equiv_m$ has the substitution property with respect to $+$, $-$, and $\cdot$. (The substitution property is vacuously satisfied for nullary operations, in this case 0

and 1.) So let $a \equiv_m c$ and $b \equiv_m d$. By the definition of $\equiv_m$,

$$a = c + km, \quad b = d + lm \quad (k, l \in R).$$

We then have

$$a + b = c + d + (k + l)m,$$
$$-a = -c + (-k)m,$$
$$ab = cd + (cl + dk + klm)m,$$

so that

$$a + b \equiv_m c + d,$$
$$-a \equiv_m -c,$$
$$ab \equiv_m cd,$$

completing the proof of the claim.

In Example 1 of Section III.3.3, we showed that the partition induced by $\equiv_m$ on $\mathbf{Z}$ is the collection $\mathbf{Z}/(m)$ of cosets of (m) regarded as an additive subgroup of $\mathbf{Z}$. Thus we have shown that $\mathbf{Z}/(m)$ is an *s.p.* partition of $\mathbf{Z}$, regarding $\mathbf{Z}$ as either an additive group or a ring. $\square$

Example 2. We return to Example 2 of Section III.3.3 and investigate the equivalence relation $E_{\mathbf{Z}}$ on $\mathbf{R}$ induced by the subgroup $\mathbf{Z}$ of the additive group of $\mathbf{R}$,

$$a E_{\mathbf{Z}} b \quad \Leftrightarrow \quad a - b \in \mathbf{Z}.$$

Now $E_{\mathbf{Z}}$ has the substitution property with respect to $+$ and $-$ (verify). But it does not have the substitution property with respect to $\cdot$; e.g., $1/3\, E_{\mathbf{Z}}\, 4/3$, but $(1/3)^2\, E_{\mathbf{Z}}\, (4/3)^2$. We conclude, therefore, that $E_{\mathbf{Z}}$ is a congruence relation on $\mathbf{R}$ when $\mathbf{R}$ is regarded as an additive group, but not when $\mathbf{R}$ is regarded as a ring.

$\square$

In the next section we shall see that the quotient set of an algebra by a congruence relation can be naturally endowed with algebraic structure.

1.2 The Quotient Algebra/Morphism Theorems of Universal Algebra

The two theorems of this section establish that congruence relations and homomorphic images go hand in hand, that they are but different facets of the same fundamental idea, namely abstraction. The first theorem shows that the quotient set A/E of an Ω-algebra A by a congruence relation E can be turned into an Ω-algebra ("quotient algebra"), one which is a homomorphic image of A. The second theorem shows that, conversely, every homomorphic image of an Ω-algebra A can be captured up to isomorphism by some quotient algebra of A.

Theorem 1 (*Quotient Algebras*). Let A be an Ω-algebra and E a congruence relation on A. Then

(a) The quotient set A/E is an Ω-algebra if we define $\omega \in \Omega$ on A/E by

$$\omega([a_1], \ldots, [a_n]) = [\omega(a_1, \ldots, a_n)];$$

(b) A/E is a homomorphic image of A under the natural map $\nu: a \mapsto [a]$.

Proof.   (a) The only point at issue in claiming that A/E is an Ω-algebra is that the operations $\omega \in \Omega$ are well-defined. Since these operations are defined on equivalence classes, we must show that the values obtained are independent of the choice of representatives for those equivalence classes. For any $\omega \in \Omega$, we have

$$[a_i] = [b_i] \quad (i = 1, \ldots, n)$$

$\Rightarrow [\omega(a_1, \ldots, a_n)] = [\omega(b_1, \ldots, b_n)]$ s.p. of A/E

$\Rightarrow \omega([a_1], \ldots, [a_n]) = \omega([b_1], \ldots, [b_n])$ defn. of ω on A/E,

and so ω is well-defined.

(b) For any $\omega \in \Omega$ and $a_1, \ldots, a_n \in A$,

$$\nu(\omega(a_1, \ldots, a_n))$$

$= [\omega(a_1, \ldots, a_n)]$ defn. of ν

$= \omega([a_1], \ldots, [a_n])$ defn. of ω on A/E

$= \omega(\nu(a_1), \ldots, \nu(a_n))$ defn. of ν,

so that $\nu: A \to A/E$ is a morphism, hence trivially an epimorphism. $\square$

Taking stock, we now have a source of homomorphic images of an Ω-algebra A: quotient algebras A/E, where E is a congruence relation on A. But perhaps there are homomorphic images of A that have nothing to do with quotient algebras. Not so, as we are about to prove. The first step is to show that associated with every morphism there is a natural congruence relation.

Lemma 1.   Let $\phi: A \to A'$ be a morphism of Ω-algebras. Then the kernel relation E_ϕ,

$$a E_\phi b \quad \Leftrightarrow \quad \phi(a) = \phi(b),$$

is a congruence relation on A.

Proof.   From set theory we know that E_ϕ is an equivalence relation; we now show that E_ϕ has the substitution property. For any $\omega \in \Omega$, we have

$$a_i E_\phi b_i \quad (i = 1, \ldots, n)$$

$\Rightarrow \phi(a_i) = \phi(b_i)$ defn. of E_ϕ

$\Rightarrow \omega(\phi(a_1), \ldots, \phi(a_n)) = \omega(\phi(b_1), \ldots, \phi(b_n))$

$\Rightarrow \phi(\omega(a_1, \ldots, a_n)) = \phi(\omega(b_1, \ldots, b_n))$ ϕ is a morphism

$\Rightarrow \omega(a_1, \ldots, a_n) E_\phi \omega(b_1, \ldots, b_n)$ defn. of E_ϕ. $\square$

We now show that every homomorphic image of an Ω-algebra is given, up to isomorphism, by a quotient algebra.

Theorem 2 (*Universal Isomorphism Theorem*). Let A and A' be Ω-algebras, with A' a homomorphic image of A under $\phi: A \to A'$. Then $A' \cong A/E_\phi$, where $\psi: [a] \mapsto \phi(a)$ is an isomorphism $A/E_\phi \to A'$. The situation is summarized by the following mapping diagram:

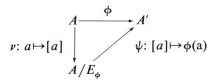

$$v: a \mapsto [a] \qquad\qquad \psi: [a] \mapsto \phi(a)$$

Proof. Since E_ϕ is a congruence relation (Lemma 1), A/E_ϕ is a quotient algebra (Theorem 1). We now show that $\psi: [a] \mapsto \phi(a)$ is an isomorphism $A/E_\phi \to A'$. That ψ is well-defined and bijective follows from our Decomposition Theorem for Functions (Theorem 1 of Section I.4.4); that ψ is a morphism is the point at issue here, which we now verify.

For any $\omega \in \Omega$ and $[a_1], \ldots, [a_n] \in A/E_\phi$,

$$\psi(\omega([a_1], \ldots, [a_n]))$$

$$\begin{aligned}
&= \psi([\omega(a_1, \ldots, a_n)]) && \text{defn. of } \omega \text{ on } A/E_\phi \\
&= \phi(\omega(a_1, \ldots, a_n)) && \text{defn. of } \psi \\
&= \omega(\phi(a_1), \ldots, \phi(a_n)) && \phi \text{ is a morphism} \\
&= \omega(\psi([a_1]), \ldots, \psi([a_n])) && \text{defn. of } \psi,
\end{aligned}$$

and so ψ is a morphism, hence an isomorphism, as claimed. $\square$

Example 1. The map $a \mapsto r_m(a)$ is an epimorphism of rings $\mathbf{Z} \to \mathbf{Z}_m$ (Example 2 of Section III.4.1). The kernel relation E_{r_m} satisfies

$$\begin{aligned}
a\, E_{r_m}\, b &\Leftrightarrow r_m(a) = r_m(b) && \text{defn. of } E_{r_m} \\
&\Leftrightarrow a \equiv_m b && \text{Lemma 1(c) of Section II.3.1,}
\end{aligned}$$

so that $E_{r_m} = \equiv_m$. By Lemma 1, $\equiv_m$ is a congruence relation on $\mathbf{Z}$ (as already explicitly shown in Example 1 of Section 1.1). Thus by Theorem 1, $\mathbf{Z}/\equiv_m$ is a quotient algebra of $\mathbf{Z}$, with operations given by

$$\begin{aligned}
[a] + [b] &= [a + b], \\
-[a] &= [-a], \\
0 &= [0], \\
[a][b] &= [ab], \\
1 &= [1].
\end{aligned}$$

By Theorem 2, $\psi: [a] \mapsto r_m(a)$ is an isomorphism $\mathbf{Z}/\equiv_m \to \mathbf{Z}_m$, in accordance with the following mapping diagram:

$$v: a \mapsto [a] \begin{array}{ccc} \mathbf{Z} & \xrightarrow{\;r_m\;} & \mathbf{Z}_m \\ \Big\downarrow & \nearrow & \\ \mathbf{Z}/\equiv_m & & \psi: [a] \mapsto r_m(a) \end{array} \qquad \Box$$

* * *

We now drop down a level of abstraction and see what the Universal Isomorphism Theorem gives us for the case of rings.

2.   QUOTIENT RINGS

The theory of quotient rings, indeed of rings generally, revolves around the beautiful and aptly named concept of an *ideal*.

2.1   Ideals and Quotient Rings

Let R be a ring. From our Universal Isomorphism Theorem, we know that all homomorphic images of R are given up to isomorphism by quotient rings of the form R/E where E is a congruence relation on R. Our goal now is to characterize the congruence relations, or equivalently the s.p. partitions, of any ring R. The all-important definition is the following:

Definition.   An *ideal* I in a ring R is a nonempty subset of R that satisfies

1. $a, b \in I \Rightarrow a - b \in I$;
2. $a \in I, r \in R \Rightarrow ar, ra \in I$.

Postulate 1 makes I an additive subgroup of R (Proposition 1 of Section III.3.3). It follows from group theory (Propositions 2 and 3 of Section III.3.3) that the system of cosets of I in R ("$R \bmod I$"),

$$R/I = \{a + I \mid a \in R\},$$

is a partition of R, where

$$a + I = b + I \Leftrightarrow a - b \in I,$$

or, in terms of the induced equivalence relation E_I,

$$a E_I b \Leftrightarrow a - b \in I.$$

We emphasize that R/I and R/E_I are one and the same: the coset $a + I$ is the equivalence class $[a]_{E_I}$.

Notation.   For $a E_I b$ we also use the notation

$$a \equiv b \pmod I$$

("a is congruent to $b \bmod I$").

An important class of ideals (for us the most important) are those of

Example 1   (*Principal ideals*).   If R is a commutative ring and $m \in R$, then the set of all ring multiples of m,

$$(m) = \{ km \,|\, k \in R \}$$

is (i) an ideal and (ii) the smallest ideal of R containing m. [Proof of (i): For every km, $lm \in (m)$ and $r \in R$, (m) contains $km - lm = (k - l)m$ and $r(km) = (rk)m$. Hence (m) is an ideal. Proof of (ii): Any ideal of R containing m has to contain all multiples of m (postulate 2 for ideals).] We call (m) the *principal ideal generated by* m, *principal* because it is generated by a single element.

We note that $x E_{(m)} y \Leftrightarrow x - y \in (m)$, so that the equivalence relation $E_{(m)}$ induced by (m) is our congruence mod m relation, $\equiv_m$, of Example 1, Section 1.1. For $a \equiv_m b$ we also write $a \equiv b \,(\mathrm{mod}\, m)$ rather than $a \equiv b \,(\mathrm{mod}\, (m))$. $\square$

Example 2.   Let $a_1, a_2, \ldots, a_n \in R$, R a commutative ring. It is easily verified that

$$(a_1, \ldots, a_n) = \left\{ \sum_{i=1}^{n} r_i a_i \,\middle|\, r_i \in R \right\}$$

is an ideal, moreover the smallest ideal of R that contains $a_1, \ldots, a_n$ (Exercise 1). Accordingly, we refer to $(a_1, \ldots, a_n)$ as the *ideal generated by* $a_1, \ldots, a_n$. Of course $(a_1, \ldots, a_n)$ is a principal ideal if and only if there is some $b \in R$ such that $(a_1, \ldots, a_n) = (b)$. $\square$

* * *

In Example 1, the equivalence relation $E_{(m)}$ induced by the principal ideal (m) has turned out to be a congruence relation, namely $\equiv_m$. We now establish the far-reaching result that the equivalence relation induced by *any* ideal is a congruence relation.

Theorem 1.   Let R be a ring and I an ideal in R. Then E_I is a congruence relation on R.

Proof.   Appealing only to the fact that I is an additive subgroup of R, one easily verifies (do it) that E_I has the substitution property with respect to $+$ and $-$. To complete the proof we must show that E_I has the substitution property with respect to multiplication, which we now do in detail.

$$a E_I c \text{ and } b E_I d$$

$$\Rightarrow a = c + k, \quad b = d + l \qquad\qquad (k, l \in I)$$

$(*)$ $\qquad\qquad \Rightarrow ab = cd + (cl + kd + kl).$

Now cl, kd, kl are all in I by postulate 2 for an ideal; hence $cl + kd + kl$ is in I, since I is an additive subgroup of R by postulate 1. From $(*)$, $ab \, E_I \, cd$ as required. $\square$

Theorem 2 (*Quotient rings*).   Let I be an ideal in a ring R. Then

(a) The collection $R/I = \{a + I \mid a \in R\}$ of additive cosets of I in R is a quotient algebra ("quotient ring") under the operations

$$(a + I) + (b + I) = (a + b) + I,$$
$$-(a + I) = -a + I,$$
$$0_{R/I} = 0 + I = I,$$
$$(a + I)(b + I) = ab + I,$$
$$1_{R/I} = 1 + I;$$

(b) R/I is a homomorphic image of R under the natural map ν: $a \mapsto a + I$;
(c) R/I is a ring (commutative if R is).

Proof.   By Theorem 1, E_I is a congruence relation on R. Since R/I is nothing more than R/E_I, where the coset $a + I$ is the equivalence class $[a]_{E_I}$, we obtain parts (a) and (b) of the theorem as an immediate consequence (no further proof required) of our Universal Quotient Algebra Theorem (Theorem 1 of Section 1.2).

Part (c) follows from our result that any homomorphic image of a (commutative) ring is a (commutative) ring (Theorem 2 of Section IV.1.3). □

The quotient ring $Z/(m)$.   Since (m) is an ideal in $\mathbf{Z}$, we conclude by Theorem 2 that $\mathbf{Z}/(m) = \{a + (m) \mid a \in \mathbf{Z}\}$ under the coset operations of Theorem 2(a) is a ring ("quotient ring of integers mod m")—a ring, moreover, that is a homomorphic image of $\mathbf{Z}$ under the map $a \mapsto a + (m)$.

As noted in Example 1,

$$a + (m) = b + (m) \quad \Leftrightarrow \quad a \equiv_m b.$$

Thus $\mathbf{Z}/(m)$ and $\mathbf{Z}/\equiv_m$ are one and the same; the coset $a + (m) \in \mathbf{Z}/(m)$ is the equivalence class $[a] \in \mathbf{Z}/\equiv_m$. Theorem 1 of Section II.3.1 becomes

Theorem 3.

(a) For any $a + (m) \in \mathbf{Z}/(m)$, $a + (m) = r_m(a) + (m)$.
(b) For any $a, b \in \mathbf{Z}_m$, $a \neq b \Rightarrow a + (m) \neq b + (m)$.

Thus each coset $a + (m)$ has a unique representative in $\mathbf{Z}_m$, namely $r_m(a)$.

Example 3.   Let $m = 8$. Then the eight distinct elements of $\mathbf{Z}/(8)$ are $a + (8)$ for $0 \leq a < 8$. In $\mathbf{Z}/(8)$ we have, for example,

$$[3 + (8)] + [7 + (8)] = 10 + (8) = 2 + (8),$$
$$[3 + (8)][7 + (8)] = 21 + (8) = 5 + (8),$$
$$-[3 + (8)] = -3 + (8) = 5 + (8). \quad □$$

The quotient ring $R[x]/(m(x))$.   Let $m(x)$ be a monic polynomial in $R[x]$, R commutative. (To avoid trivialities we shall always assume $\deg m(x) > 0$.) Then $(m(x))$ is an ideal in $R[x]$, and from Theorem 2 we conclude that $R[x]/(m(x))$ = $\{a(x) + (m(x))|\, a(x) \in R[x]\}$ is a ring ("quotient ring of polynomials mod $m(x)$"), a homomorphic image of $R[x]$ under the map $a(x) \mapsto a(x) + (m(x))$.

Proceeding as in the previous example, we have that $R[x]/(m(x))$ and $R[x]/\equiv_{m(x)}$ are one and the same; the coset $a(x) + (m(x)) \in R[x]/(m(x))$ is the equivalence class $[a(x)] \in R[x]/\equiv_{m(x)}$. Theorem 2 of Section IV.3.2 becomes

Theorem 4.

(a) For any $a(x) + (m(x)) \in R[x]/(m(x))$,
$$a(x) + (m(x)) = r_{m(x)}(a(x)) + (m(x)).$$

(b) If $\deg a(x)$, $\deg b(x)$ are $< \deg m(x)$, then
$$a(x) \neq b(x) \;\Rightarrow\; a(x) + (m(x)) \neq b(x) + (m(x)).$$

Thus each coset $a(x) + (m(x))$ has a unique representative having degree $<$ $\deg m(x)$, namely $r_{m(x)}(a(x))$.

Corollary.   If $|R| = k$ and $\deg m(x) = n$, then $|R[x]/(m(x))| = k^n$.

Proof.   Immediate from Theorem 4. $\square$

We now wish to identify the ground ring R with a certain "obvious" subring of $R[x]/(m(x))$. To this end, consider the map $a \mapsto a + (m(x))$ $(a \in R)$. It is trivially verified to be a monomorphism $R \to R[x]/(m(x))$. Thus $R[x]/(m(x))$ includes a subring $\{a + (m(x))|\, a \in R\}$ isomorphic with R. *We henceforth agree to identify* $a + (m(x)) \in R[x]/(m(x))$ *with* $a \in R$ *and accordingly to regard* $R[x]/(m(x))$ *as an extension ring of* R [or, equivalently, R as a subring of $R[x]/(m(x))$].

Lemma 1.   Let $a(x) = a_n x^n + a_{n-1} x^{n-1} + \cdots + a_0 \in R[x]$. Then in $R[x]/I$, where I is some ideal $(m(x))$, we have
$$a(x + I) = a(x) + I.$$

Proof.   As agreed, we identify each coefficient $a_i \in R$ with $a_i + I \in R[x]/I$. We then have
$$a(x + I) = (a_n + I)(x + I)^n + (a_{n-1} + I)(x + I)^{n-1} + \cdots + (a_0 + I)$$
$$= (a_n x^n + I) + (a_{n-1} x^{n-1} + I) + \cdots + (a_0 + I)$$
$$\text{(defn. of multiplication in } R[x]/I)$$
$$= (a_n x^n + a_{n-1} x^{n-1} + \cdots + a_0) + I$$
$$\text{(defn. of addition in } R[x]/I)$$
$$= a(x) + I. \;\square$$

The following theorem exploits Lemma 1 to provide what is essentially a convenient notation for $R[x]/(m(x))$, wherein the rather cumbersome coset notation, $a(x) + (m(x))$, is replaced by a polynomial notation.

Theorem 5. In $R[x]$, let $I = (m(x))$ where $m(x) = x^n + $ l.d.t. Let $\alpha = x + I$. Then

(a) $R[x]/I = R[\alpha]$;
(b) $m(\alpha) = 0$;
(c) Each $\beta \in R[\alpha]$ can be uniquely expressed in the form

$$a_0 + a_1\alpha + \cdots + a_{n-1}\alpha^{n-1} \qquad (a_i \in R).$$

Proof. (a) Recall that $R[\alpha]$, the ring generated by $R \cup \{\alpha\}$, consists of the values $a(\alpha)$ of all polynomials $a(x) \in R[x]$ (Proposition 3 of Section IV.1.3). We thus have

$$\begin{aligned} R[\alpha] &= \{a(\alpha)|\, a(x) \in R[x]\} \\ &= \{a(x+I)|\, a(x) \in R[x]\} \\ &= \{a(x) + I |\, a(x) \in R[x]\} \qquad \text{Lemma 1} \\ &= R[x]/I. \end{aligned}$$

(b) $\begin{aligned}[t] m(\alpha) &= m(x+I) \\ &= m(x) + I \qquad \text{Lemma 1} \\ &= 0 + I \qquad\quad m(x) \in I \\ &= 0 \qquad\qquad\ \text{agreed identification.} \end{aligned}$

(c) Consider any $\beta = a(\alpha) \in R[\alpha]$. From Theorem 4(a), $a(x) + I = r(x) + I$ where $r(x) = r_{m(x)}(a(x))$. Hence by Lemma 1, $a(x + I) = r(x + I)$; i.e., $a(\alpha) = r(\alpha)$. Thus each $\beta \in R[x]$ can be expressed in the form $a_0 + a_1\alpha + \cdots + a_{n-1}\alpha^{n-1}$.

As for uniqueness, suppose that we have

$$a_0 + a_1\alpha + \cdots + a_{n-1}\alpha^{n-1} = b_0 + b_1\alpha + \cdots + b_{n-1}\alpha^{n-1}.$$

Then $a(x + I) = b(x + I)$, which is to say that $a(x) + I = b(x) + I$. But $\deg a(x)$, $\deg b(x)$ are $< n$, whence $a(x) = b(x)$ by Theorem 4(b), which gives $a_0 = b_0$, $a_1 = b_1, \ldots, a_{n-1} = b_{n-1}$, as required. $\square$

In the proof of Theorem 5(c) we established the useful

Corollary. For any $a(x) \in R[x]$,

$$a(\alpha) = r(\alpha) \quad \text{where } r(x) = r_{m(x)}(a(x)).$$

Consequently, operations on polynomial expressions $a(\alpha) \in R[\alpha]$ can be carried out $\mod m(\alpha)$ as if α were an indeterminate. (But note that α is not an indeterminate; it is the coset $x + I$.)

Example 4. Let $m(x) = x^3 + 1 \in \mathbf{Z}[x]$. Letting $\alpha = x + (x^3 + 1)$, the elements of

$Z[x]/(x^3 + 1)$ can be uniquely represented in the form

$$a_0 + a_1\alpha + a_2\alpha^2 \quad (a_i \in Z),$$

where $\alpha^3 + 1 = 0$. For example, if $a(\alpha) = 1 + 3\alpha^2$, $b(\alpha) = 4 - 6\alpha$, then

$$a(\alpha) + b(\alpha) = \; 5 - 6\alpha + 3\alpha^2,$$
$$a(\alpha)b(\alpha) = \; 4 - 6\alpha + 12\alpha^2 - 18\alpha^3$$
$$= 22 - 6\alpha + 12\alpha^2 \quad (\text{using } \alpha^3 = -1). \; \square$$

The quotient ring $R[[x]]/(x^m)$. Let $R[[x]]$ be a power series ring and m a fixed positive integer. We then take the ideal (x^m) and form the quotient ring $R[[x]]/(x^m)$ ("the quotient ring of power series mod x^m").
The congruence relation $\equiv_{x^m}$ induced by (x^m) satisfies

$$a(x) \equiv_{x^m} b(x) \quad \Leftrightarrow \quad a(x) - b(x) \in (x^m)$$
$$\Leftrightarrow \quad a_i = b_i \quad (i = 0, \dots, m-1).$$

Hence $a(x), b(x) \in R[[x]]$ are congruent mod x^m whenever they agree in their first m terms.
We introduce the *trunction* mod x^m map T_m,

$$T_m: \sum_{i=0}^{\infty} a_i x^i \mapsto \sum_{i=0}^{m-1} a_i x^i.$$

Then analogous to Theorems 3 and 4 we have

Theorem 6.

(a) For any $a(x) + (x^m) \in R[[x]]/(x^m)$,

$$a(x) + (x^m) = T_m(a(x)) + (x^m).$$

(b) If $\deg a(x)$, $\deg b(x)$ are $< m$, then

$$a(x) \neq b(x) \quad \Rightarrow \quad a(x) + (x^m) \neq b(x) + (x^m).$$

Proof. Easy. $\square$

Thus each coset $a(x) + (x^m)$ has a unique polynomial representative having degree $< m$, namely $T_m(a(x))$.

2.2 Isomorphism Theorem for Rings

Taking stock, we now have a source of homomorphic images of a ring R, namely quotient rings R/I where I is an ideal in R. But are all homomorphic images of a ring of this form? As an important preliminary step to establishing that the answer is yes, we show that associated with every homomorphism of rings is a natural ideal.

Definition.   Let $\phi: R \to R'$ be a morphism of rings. Then the (*ring-theoretic*) *kernel* of ϕ, ker ϕ or K_ϕ, is defined by

$$K_\phi = \{a \in R \mid \phi(a) = 0\}.$$

Remark.   We distinguish between the above (ring-theoretic) kernel K_ϕ and the (set-theoretic) kernel relation E_ϕ—the former is a subset of R, the latter is a relation on R. We shall see in the next theorem that the two notions of kernel are intimately related, making the twofold use of the term "kernel" quite appropriate.

Theorem 1.   Let $\phi: R \to R'$ be a morphism of rings. Then

(a) K_ϕ is an ideal of R;
(b) $E_\phi = E_{K_\phi}$. (In words: The kernel relation of ϕ $[a E_\phi b \Leftrightarrow \phi(a) = \phi(b)]$ is the equivalence relation induced by K_ϕ regarded as an additive subgroup of R $[a E_{K_\phi} b \Leftrightarrow a - b \in K_\phi]$.)

Proof.   (a) Since the inverse image of a subalgebra is a subalgebra (Theorem 2 of Section III.4.2), it follows that $K_\phi = \phi^{-1}(\{0\})$ is a subgroup of the additive group of R; postulate 1 for an ideal thus holds. As for postulate 2, if $a \in K_\phi$, $r \in R$, then

$$\phi(ar) = \phi(a)\phi(r) = 0\phi(r) = 0,$$

$$\phi(ra) = \phi(r)\phi(a) = \phi(r)0 = 0,$$

so that $ar, ra \in K_\phi$.
    (b) Since K_ϕ is an ideal of R, hence an additive subgroup of R, we have the basic equivalence from group theory

$(*)$ $a E_{K_\phi} b \Leftrightarrow a - b \in K_\phi$.

We then have

$$a E_\phi b \Leftrightarrow \phi(a) = \phi(b) \qquad \text{defn. of } E_\phi$$
$$\Leftrightarrow \phi(a - b) = 0 \qquad \phi \text{ is a morphism of rings}$$
$$\Leftrightarrow a - b \in K_\phi \qquad \text{defn. of } K_\phi$$
$$\Leftrightarrow a E_{K_\phi} b \qquad \text{by } (*). \ \square$$

Now for the specialization of our Universal Isomorphism Theorem (Theorem 2 of Section 1.2) to rings.

Theorem 2 (*Ring Isomorphism Theorem*).   Let R and R' be rings, with R' a homomorphic image of R under $\phi: R \to R'$. Let $K = K_\phi$, the kernel of ϕ. Then $R' \cong R/K$, where $\psi: a + K \mapsto \phi(a)$ is an isomorphism $R/K \to R'$. The situation is summarized by the following mapping diagram:

$$
\begin{array}{ccc}
R & \xrightarrow{\ \phi\ } & R' \\
{\scriptstyle v:\,a\mapsto a+K}\Big\downarrow & \diagup & {\scriptstyle \psi:\,a+K\mapsto\phi(a)} \\
R/K & &
\end{array}
$$

Proof. Since K is an ideal (Theorem 1(a)), R/K is a quotient ring (Theorem 2 of Section 2.1). We now show that $\psi: a + K \mapsto \phi(a)$ is an isomorphism $R/K \to R'$.

By our Universal Isomorphism Theorem, the map $\psi: [a]_{E_\phi} \mapsto \phi(a)$ is an isomorphism $R/E_\phi \to R'$, where E_ϕ is the (set-theoretic) kernel relation of ϕ. But by Theorem 1(b), E_ϕ is nothing more than $E_K\,(=E_{K_*})$, the equivalence relation induced by the (ring-theoretic) kernel of ϕ. Hence $R/E_\phi = R/E_K$.

Finally, since K is an ideal in R (Theorem 1(a)), R/E_K can be written as R/K, where the equivalence class $[a]_{E_K} = [a]_{E_\phi}$ is the coset $a + K$. Thus the isomorphism

$$\psi: [a]_{E_\phi} \mapsto \phi(a) \quad \text{from } R/E_\phi \text{ to } R'$$

is in fact the isomorphism

$$\psi: a + K \mapsto \phi(a) \quad \text{from } R/K \text{ to } R'. \ \square$$

Example 1 (*Isomorphism between* $\mathbf{Z}/(m)$ *and* $\mathbf{Z}_m$). The kernel of the remainder mod m epimorphism $a \mapsto r_m(a)$ from $\mathbf{Z}$ to $\mathbf{Z}_m$ is (m), for we have

$$a \in \ker r_m \Leftrightarrow r_m(a) = 0$$

$$\Leftrightarrow a \in (m).$$

Hence by the Ring Isomorphism Theorem, $\mathbf{Z}_m$ is isomorphic with the quotient ring $\mathbf{Z}/(m)$; the map ψ in the mapping diagram below is an isomorphism.

$$
\begin{array}{ccc}
\mathbf{Z} & \xrightarrow{\ r_m\ } & \mathbf{Z}_m \\
{\scriptstyle v:\,a\mapsto a+(m)}\Big\downarrow & \diagup & {\scriptstyle \psi:\,a+(m)\mapsto r_m(a)} \\
\mathbf{Z}/(m) & &
\end{array}
$$

$\square$

Example 2 (*Another construction of modular polynomial rings*). Let $m(x)$ be a monic polynomial in $R[x]$, R a commutative ring. On $R[x]_{m(x)} = \{a(x) \in R[x] | \deg a(x) < \deg m(x)\}$ we define $+, -, 0, 1$ as on $R[x]$, and $\odot$ by

$$a(x) \odot b(x) = r_{m(x)}(a(x)b(x)) \qquad (\text{"multiplication mod } m(x)\text{"}).$$

Using Lemma 1 of Section IV.3.2, it is straightforward to show that $r_{m(x)}: a(x) \mapsto r_{m(x)}(a(x))$ is an epimorphism from $[R[x]; +, -, 0, \cdot, 1]$ to $[R[x]_{m(x)}; +, -, 0, \odot, 1]$. Hence $R[x]_{m(x)}$, like $R[x]$, is a commutative ring (Theorem 2 of Section IV.1.3)—"the ring of polynomials mod $m(x)$."

Proceeding as in Example 1, we note that the kernel of $r_{m(x)}$ is $(m(x))$, for we have

$$a(x) \in \ker r_{m(x)} \Leftrightarrow r_{m(x)}(a(x)) = 0$$

$$\Leftrightarrow a(x) \in (m(x)).$$

By the Ring Isomorphism Theorem, the map

$$\psi: a(x) + (m(x)) \mapsto r_{m(x)}(a(x))$$

is an isomorphism $R[x]/(m(x)) \to R[x]_{m(x)}$. $\square$

Example 3 (*Construction of truncated power series rings*). Let $R[[x]]$ be a power series ring and m a fixed integer > 0. We define the set

$$R[[x]]_m = \{a(x) \in R[[x]] \| \deg a(x) < m\}$$

of all "truncated" power series having degree $< m$. On $R[[x]]_m$ we define $+, -, 0, 1$ as on $R[[x]]$, and $\odot$ using the truncation $\mod x^m$ operator T_m by

$$a(x) \odot b(x) = T_m(a(x)b(x)) \quad (\text{"multiplication } \mod x^m\text{"}).$$

The claim now is that T_m is an epimorphism from $[R[[x]]; +, -, 0, \cdot, 1]$ to $[R[[x]]_m; +, -, 0, \odot, 1]$.

Proof of claim: We check that T_m preserves multiplication (the other operations are trivially preserved). For $a(x), b(x) \in R[[x]]$, we can write

$$\left. \begin{array}{l} a(x) = T_m(a(x)) + x^m s(x) \\ b(x) = T_m(b(x)) + x^m t(x) \end{array} \right\} s(x), t(x) \in R[[x]].$$

Then

$$a(x)b(x) = T_m(a(x))T_m(b(x)) + x^m u(x)$$

for some $u(x) \in R[[x]]$, so that

$$T_m(a(x)b(x)) = T_m(T_m(a(x))T_m(b(x))) \qquad \text{defn. of } T_m$$

$$= T_m(a(x)) \odot T_m(b(x)) \qquad \text{defn. of } \odot.$$

Also, T_m is trivially surjective $R[[x]] \to R[[x]]_m$, completing the proof of the claim.

The kernel of T_m is as follows:

$$a(x) \in \ker T_m \Leftrightarrow T_m(a(x)) = 0$$

$$\Leftrightarrow a_i = 0 \quad \text{for } i = 0, 1, \ldots, m-1$$

$$\Leftrightarrow a(x) \in (x^m).$$

By the Ring Isomorphism Theorem, the map

$$\psi: a(x) + (x^m) \mapsto T_m(a(x))$$

is an isomorphism $R[[x]]/(x^m) \to R[[x]]_m$. $\square$

*   *   *

Recapitulating the major results of these last two sections:

1. If I is an ideal in a ring R, then R/I is a homomorphic image of R (Theorem 2 of Section 2.1); and conversely,
2. If R' is a homomorphic image of R, then R' is isomorphic to a quotient ring R/I, where I is an ideal in R (namely, the kernel of the homomorphism) (Theorem 2 of this section).

In the last three examples, we have shown that the rings $\mathbf{Z}_m$ of integers mod m, $R[x]_{m(x)}$ of polynomials mod $m(x)$, and $R[[x]]_m$ of (truncated) power series mod x^m are all isomorphic to quotient rings—$\mathbf{Z}/(m)$, $R[x]/(m(x))$, and $R[[x]]/(x^m)$, respectively.

3. FURTHER THEORY OF IDEALS

In developing this further theory of ideals, we have two related goals in mind: (1) to acquire information about the kinds of abstractions that particular rings of interest can have, and (2) to acquire information that will enable us to infer properties of rings R' of the form $R' \cong R/I$ from properties of the ideal I. This latter information might be used either to analyze a given ring R' or to synthesize a ring R' with desirable properties.

3.1 Principal Ideal Domains

A *principal ideal domain* (PID) is an integral domain in which every ideal is principal. Thus, if D is a PID and I is an ideal in D, then $I = (a)$ for some $a \in D$.

To assure that we are not engaged in trivialities, we show that not every integral domain is a PID.

Example 1. In $F[x, y]$ we claim that the ideal

$$(x, y) = \{xa(x, y) + yb(x, y) \mid a(x, y), b(x, y) \in F[x, y]\}$$

of all polynomials having zero constant term is not a principal ideal.

Proof of claim: Suppose to the contrary that $(x, y) = (g(x, y))$ for some $g(x, y) \in F[x, y]$. Since $x = x \cdot 1 + y \cdot 0$ and $y = x \cdot 0 + y \cdot 1$ are both in $(x, y) = (g(x, y))$, we have

$$\text{(i) } x = g(x, y)a(x, y), \qquad \text{(ii) } y = g(x, y)b(x, y)$$

for some $a(x, y), b(x, y) \in F[x, y]$. From (i), $g(x, y) \neq 0$ has degree zero in y; from (ii), $g(x, y)$ has degree zero in x. Hence $g(x, y) = g$ is a nonzero constant in F. But $g \in (g)$ is not in (x, y), which contradicts $(x, y) = (g)$ and forces the desired conclusion that no such polynomial $g(x, y)$ exists. □

We now show that the three most important integral domains for us—$\mathbf{Z}$, $F[x]$, and $F[[x]]$—are all PIDs. We can kill two of these birds with one stone.

Theorem 1.   A Euclidean domain is a PID.

Proof.   Let D be a Euclidean domain with degree function d, and let I be an ideal in D. If $I = \{0\}$, then I is the principal ideal (0). So assume $I \neq \{0\}$. Let $m \neq 0$ be an element of I having minimum degree. We claim that $I = (m)$.

Proof of the claim: $(m) \subseteq I$, trivially; it is the other direction that is interesting. For any $a \in I$, we divide by m to obtain

$$a = mq + r, \qquad d(r) < d(m) \quad \text{or} \quad r = 0.$$

Since a, m are in I, it follows from the postulates for an ideal that $r = a - mq$ is in I. But then the above possibility $d(r) < d(m)$ is excluded by the choice of m as a minimum degree element of I. Hence $r = 0$, giving $a = mq \in (m)$. Thus $I \subseteq (m)$, which completes the proof of the claim $I = (m)$. $\square$

In an integral domain, the generator of a principal ideal is determined only up to associates:

$$(a) = (b) \Leftrightarrow a \sim b$$

(Exercise 2 of Section 2). Thus in $\mathbf{Z}$ we may choose the generator m of a (necessarily) principal ideal to be ≥ 0; in $F[x]$ we may choose the generator to be monic. We then obtain

Corollary 1. $\mathbf{Z}$ is a PID in which every ideal has the form (m) where $m \geq 0$.

Corollary 2. $F[x]$ is a PID in which every ideal $\neq (0)$ has the form $(m(x))$ where $m(x)$ is monic.

Theorem 2. $F[[x]]$ is a PID.

Proof.   Let I be an ideal in $F[[x]]$. (We proceed much as in the proof of Theorem 1.) If $I = \{0\}$, then I is the principal ideal (0). So assume $I \neq \{0\}$. Let m be the minimum order of any element of I. We claim that $I = (x^m)$.

Proof of claim: Any $a(x) \in I$ has order $\geq m$, hence has the form

$$a(x) = a_n x^n + a_{n+1} x^{n+1} + \cdots \qquad (n \geq m)$$
$$= x^m \left(a_n x^{n-m} + a_{n+1} x^{n-m+1} + \cdots \right),$$

so that $a(x) \in (x^m)$, giving $I \subseteq (x^m)$.

As for the other (more interesting) direction, let $b(x) \neq 0$ in I have minimum order m. Then

$$b(x) = b_m x^m + b_{m+1} x^{m+1} + \cdots \qquad (b_m \neq 0)$$
$$= x^m \left(b_m + b_{m+1} x^{m+1} + \cdots \right),$$

so that $b(x) = x^m c(x)$ where $c(x)$ is invertible in $F[[x]]$ (since its constant term b_m is nonzero). Hence

$$x^m = b(x) c(x)^{-1} \in I,$$

which gives $(x^m) \subseteq I$ and completes the proof of the claim $I = (x^m)$. $\square$

Corollary. $F[[x]]$ is a PID in which every ideal $\neq (0)$ is of the form (x^m) for $m \geq 0$.

Theorems 1 and 2 yield remarkably complete information about $\mathbf{Z}$, $F[x]$, and $F[[x]]$ regarding their possible abstractions. Since the ideals in $\mathbf{Z}$ are all of the form (m) for $m \geq 0$, the only homomorphic images of $\mathbf{Z}$, excluding $\mathbf{Z}$ itself, are the rings $\mathbf{Z}_m$ of integers mod m; similarly, the only homomorphic images of $F[x]$ are the rings $F[x]_{m(x)}$ of polynomials mod $m(x)$. Theorem 2 tells us that the only way to abstract $F[[x]]$ is to truncate each power series after some fixed number of terms; it is not possible, for example, to take every second, third, or nth term and obtain a homomorphic image of $F[[x]]$.

Simple rings. Any ring (commutative or otherwise) has two so-called *improper* ideals, the zero ideal $(0) = \{0\}$ and the unit ideal $(1) = R$, with corresponding homomorphic images

$$R/(0) \cong R, \qquad R/(1) \cong \{0\}.$$

We refer to $R/(0)$ and $R/(1)$ [and to any homomorphic images R' that are isomorphic to either of $R/(0)$ or $R/(1)$] as *improper* homomorphic images.

A ring whose only ideals are the improper ones (0) and (1) is called *simple*. By the Ring Isomorphism Theorem, any homomorphic image of a ring R is isomorphic to R/I where I is an ideal in R. Thus a simple ring does not have any proper homomorphic images, i.e., cannot be abstracted (hence the term "simple").

More generally, a universal algebra A is simple if there are no congruence relations on A (except for the trivial ones that lump all of A together in one equivalence class or put each element into its own equivalence class). The problem of cataloging simple algebras is, in general, a most difficult one. But for commutative rings we have a very sharp result:

Theorem 3. Let R be a nontrivial commutative ring. Then

$$R \text{ is simple} \Leftrightarrow R \text{ is a field.}$$

Proof. ($\Leftarrow$) Let $I \neq (0)$ be an ideal in R, a field. (We must show that $I = R$.) Since $I \neq (0)$, some $x \neq 0$ is in I. Then $x^{-1}x = 1 \in I$, so that $r \cdot 1 \in I$ for all $r \in R$. Thus $I = R$.

($\Rightarrow$) Let R be simple, and let $x \in R$ be nonzero. (We must show that x^{-1} exists in R.) Consider the principal ideal (x). Since $x \in (x)$, $(x) \neq (0)$. But R is simple. Therefore $(x) = R$, which means, in particular, that $1 \in (x)$. Thus $1 = ax$ for some $a \in R$, making $a = x^{-1}$. $\square$

Thus fields are PIDs of an especially restricted sort: their only ideals are the improper ones (0) and (1); fields, moreover, are characterized by this property.

Since proper ideals and proper homomorphic images go hand in hand, we conclude the

Corollary.   Let R be a nontrivial ring. Then

$$R \text{ is a field} \Leftrightarrow R \text{ has no proper homomorphic images.}$$

Thus we have the result that fields are precisely those rings that are unabstractable. What this means is that elements of any field are so individualistic as to make their coalescence into an equivalence class of a congruence relation impossible.

g.c.d.'s in PIDs.   Our final result about PIDs is that like Euclidean domains they have the g.c.d. property.

Theorem 4.   Let D be a PID and $a,\ b \in D$. Then a, b have a g.c.d. g expressible in the form

$$g = sa + tb \qquad (s, t \in D).$$

Proof.   (Cf. the proof of Theorem 1 of Section IV.4.2.) For $a, b \in D$, consider the ideal

$$(a, b) = \{sa + tb \mid s, t \in D\}.$$

Since D is a PID, it follows that $(a, b) = (g)$ for some $g \in D$, whence $g = sa + tb$ for some $s,\ t \in D$. The claim now is that g is a g.c.d. of a, b.

Proof of claim: Since $a = 1a + 0b$ and $b = 0a + 1b$ are both in $(a, b) = (g)$, it immediately follows that $g \mid a$ and $g \mid b$, which is to say that g is a common divisor of a, b. If $c \mid a$ and $c \mid b$, then $c \mid sa + tb$ trivially. Thus $g = sa + tb$ is a *greatest* common divisor of a and b. $\square$

3.2   Unital Subrings; Prime Subfields

The Ring Isomorphism Theorem combined with the PID property of $\mathbf{Z}$ allows us to sharply characterize the unital subring of any ring R,

$$[1] = \{n \cdot 1 \mid n \in \mathbf{Z}\}.$$

(Recall that the unital subring, the subring generated by 1, is the smallest subring of R—Example 3 of Section IV.1.2.)

Theorem 1.   Let R be a ring. Then

(a) $[1] \cong \mathbf{Z}$ if char $R = 0$;
(b) $[1] \cong \mathbf{Z}_m$ if char $R = m$.

Proof.   Consider the *unital morphism* $\phi \colon \mathbf{Z} \to R$, defined by

$$\phi(n) = n \cdot 1.$$

ϕ is clearly an *epi*morphism $\mathbf{Z} \to [1]$; we examine its kernel.

(a) char $R = 0$. Then $n \cdot 1 \neq 0$ for any $n \neq 0$, which means that $\ker \phi = (0)$. By the Ring Isomorphism Theorem,

$$[1] \cong \mathbf{Z}/(0) \cong \mathbf{Z}.$$

(b) char $R = m$. Then $m \cdot 1 = 0$, but $m' \cdot 1 \neq 0$ for $0 < m' < m$. This is to say that m is the least positive integer in $\ker \phi$. But $\ker \phi$ is an ideal $\neq (0)$ in $\mathbf{Z}$, hence must be a principal ideal (k) for $k > 0$ (Corollary 1 to Theorem 1 of Section 3.1). Since k is the least positive integer of (k), it immediately follows that $k = m$. By the Ring Isomorphism Theorem,

$$[1] \cong \mathbf{Z}/(m) \cong \mathbf{Z}_m. \quad \square$$

Prime subfields. Let E be a field. A subring F of E is called a *subfield* of E if F is itself a field; E is called an *extension field* of F. The smallest subfield (with respect to inclusion) of a field E is called the *prime subfield* of E.

As for the uniqueness of the prime subfield, let P_1 and P_2 both be prime subfields of a field E. Then $P_1 \subseteq P_2$ (since P_1 is a prime subfield) and $P_2 \subseteq P_1$ (since P_2 is a prime subfield), giving $P_1 = P_2$. As for the existence of a prime subfield of any field E, consider $\cap \mathcal{F}$ where $\mathcal{F}$ is the collection of all subfields of E. Clearly $\cap \mathcal{F}$ is a subfield of E, since each $F \in \mathcal{F}$ is a subfield of E. Also $\cap \mathcal{F} \subseteq F$ for all subfields F of E. Hence $\cap \mathcal{F}$ is a prime subfield of E.

The following theorem sharply characterizes prime subfields.

Theorem 2. Let E be a field and P the prime subfield of E. Then

(a) $P \cong \mathbf{Z}_p$ if char $E = p$;
(b) $P \cong \mathbf{Q}$ if char $E = 0$.

Proof. (a) char $E = p$. We first examine the unital subring, $[1]$, of E. By Theorem 1, $[1]$ is isomorphic to $\mathbf{Z}_p$. Since E is a field, p must be prime (Theorem 6 of Section IV.2.1). Hence $\mathbf{Z}_p$ is a field (Theorem 3 of Section IV.2.1), which in turn means that $[1]$ ($\cong \mathbf{Z}_p$) is a field. But $[1]$ is the smallest *subring* of E, hence *a fortiori* must be the smallest subfield of E, i.e., the prime subfield of E. Thus we have $P = [1] \cong \mathbf{Z}_p$, as required.

(b) char $E = 0$. In this case, $[1]$ is isomorphic to $\mathbf{Z}$, again by Theorem 1. Hence $[1]$ is now an integral domain but not a field. Now any subring, hence any subfield, of E includes $[1]$, so that the prime subfield of E must be the smallest subfield of E that includes $[1]$. But the smallest subfield of an integral domain is its field of quotients (Theorem 1, Part 5, of Section IV.2.2). Thus we have

$$P = Q([1])$$
$$\cong Q(\mathbf{Z}) \qquad \text{Proposition 2 of Sec. IV.2.2}$$
$$= \mathbf{Q}. \quad \square$$

Remark. Part (a) of the above theorem accounts for the terminology "prime" subfield. Part (b), however, shows that the terminology isn't entirely appropriate. (We would prefer the term "unital subfield," but no one seems to use it, so we won't either.)

3.3 Prime and Maximal Ideals

Throughout this section, R denotes a commutative ring.

Our Ring Isomorphism Theorem tells us in general terms about the homomorphic images of a ring R: they are all quotient rings, each having the form R/I where I is an ideal in R.

Our goal now is to acquire more specialized information about these quotient rings R/I. In particular we ask, for the case of a commutative ring R, what kinds of ideals I make R/I (1) an integral domain, (2) a field? The answers center around the following two definitions.

Definition. An ideal P in R is *prime* if

$$ab \in P \quad \Rightarrow \quad a \in P \text{ or } b \in P.$$

Definition. An ideal M in R is *maximal* if

$$M \subset I \subset R \quad \text{for no ideal } I \text{ in } R.$$

In terms of a diagram:

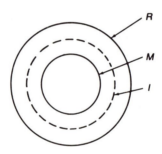

(M is maximal if no ideal of the form I exists in R.)

Example 1. The ideal (6) is not prime in **Z**, since $3 \times 2 \in (6)$ whereas $3 \notin (6)$, $2 \notin (6)$. The ideal (7), on the other hand, is prime: if $ab \in (7)$, then $7 \mid ab$, whence $7 \mid a$ or $7 \mid b$ (since 7 is prime), so that $a \in (7)$ or $b \in (7)$. ☐

Example 2. The ideal (x) is *not* maximal in **Z**$[x]$, since $(x) \subset (x, 2) \subseteq$ **Z**$[x]$. (Cf. Exercise 1.) We shall shortly see that $(x, 2)$ is maximal. ☐

The importance of our definitions of prime and maximal ideal is wrapped up in the following two theorems.

Theorem 1. R/P is an integral domain $\Leftrightarrow P$ is prime.

Theorem 2. R/M is a field $\Leftrightarrow M$ is maximal.

Proof of Theorem 1.

R/P is an integral domain

$\Leftrightarrow$ $(a + P)(b + P) = P \Rightarrow a + P = P$ or $b + P = P$

(no zerodivisors; recall that $0_{R/P} = P$)

$\Leftrightarrow$ $ab + P = P \Rightarrow a + P = P$ or $b + P = P$

(defn. of multiplication on R/P)

$\Leftrightarrow$ $ab \in P \Rightarrow a \in P$ or $b \in P$

(criterion for coset equality:
$x + P = y + P \Leftrightarrow x - y \in P$)

$\Leftrightarrow$ P is a prime ideal. $\square$

Proof of Theorem 2.   ($\Leftarrow$) Let M be a maximal ideal in R. We must show that R/M is a field. Since R/M is a commutative ring, we need only show that any nonzero element of R/M has an inverse.

So for any $a + M \in R/M$, $a + M \neq M$ ($= 0_{R/M}$), consider

$$(a, M) = \{sa + m \mid s \in R, m \in M\}.$$

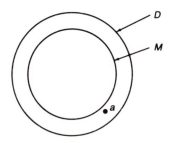

It is trivially verified that (a, M) is an ideal of R (indeed, as the notation suggests, (a, M) is the ideal generated by a and M—the smallest ideal of R that includes $\{a\} \cup M$).

Now clearly

$$M \subseteq (a, M) \subseteq R.$$

Since $a \notin M$ (otherwise $a + M = M$), it follows that $M \neq (a, M)$. Hence $(a, M) = R$ because M is a maximal ideal. Thus $1 \in R$ is in (a, M), so that

$$1 = sa + m \quad \text{for some } s \in R, m \in M.$$

We then have

$$1 + M = (sa + m) + M$$
$$= sa + M$$
$$= (s + M)(a + M) \qquad \text{defn. of multiplication on } R/M,$$

which gives $(a + M)^{-1} = s + M$.

($\Rightarrow$) Suppose R/M is a field, and assume to the contrary that there exists an ideal I such that

$$M \subset I \subset R.$$

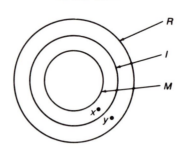

Then there exists $x \in I - M$, $y \in R - I$, as illustrated.

Now the map $\phi \colon R/M \to R/I$ defined by

$$\phi \colon a + M \mapsto a + I$$

is easily verified to be a morphism (check it out). We note two facts about the kernel of ϕ (which of course we know to be an ideal—Theorem 1(a) of Section 2.2):

1. $\ker \phi \neq \{0_{R/M}\}$: Since $\phi(x + M) = x + I = I = 0_{R/I}$, there is a nonzero element of R/M, namely $x + M \neq M$, that is in the kernel.
2. $\ker \phi \neq R/M$: Since $\phi(y + M) = y + I \neq I$, there is an element of R/M, namely $y + M$, that is not in the kernel.

We have thus shown that R/M has a *proper* ideal, namely the kernel of ϕ. But R/M is a field, hence has no proper ideals (Theorem 3 of Section 3.1). This contradiction forces the desired conclusion that no such ideal I exists, i.e., that M is maximal. $\square$

The following corollary, though not at all obvious from the definitions of prime and maximal ideal, is a trivial consequence of Theorems 1 and 2.

Corollary.   A maximal ideal is a prime ideal.

Proof.   A field is an integral domain. $\square$

Example 3.   Let us return to Example 2 and show that the $(x, 2)$ is a maximal ideal in $\mathbf{Z}[x]$. If we can show that $\mathbf{Z}[x]/(x, 2)$ is a field, then we need only appeal to Theorem 2 to conclude that $(x, 2)$ is maximal. We now propose to show that $\mathbf{Z}[x]/(x, 2)$ is a field via our Ring Isomorphism Theorem. The

"0-evaluation mod 2" map

$$\phi: a(x) \mapsto r_2(a(0))$$

is readily shown to be a morphism (indeed ϕ is the composition of two familiar morphisms: 0-evaluation, $a(x) \mapsto a(0)$, and reduction mod 2, $b \mapsto r_2(b)$). Clearly ϕ is an *epi*morphism $\mathbf{Z}[x] \to \mathbf{Z}_2$. Thus by the Ring Isomorphism Theorem we have an isomorphism

$$\mathbf{Z}[x]/\ker \phi \cong \mathbf{Z}_2.$$

Now $\mathbf{Z}_2$ is a field; hence so is $\mathbf{Z}[x]/\ker \phi$. Thus if we can show that $\ker \phi = (x, 2)$, then we are done. We leave that verification as a straightforward exercise for the reader. (Or better, see Exercise 4 for any easy generalization of this example.) $\square$

We now investigate the connection between prime elements and prime ideals.

Theorem 3. In any integral domain,

$$(p) \text{ prime} \Rightarrow p \text{ is prime.}$$

Proof. Exercise 6. $\square$

Corollary. In an integral domain D,

$$m \text{ composite} \Rightarrow D/(m) \text{ is a ring with zerodivisors.}$$

Proof. If m is composite, then (m) is not a prime ideal by Theorem 3, whence $D/(m)$ is a ring with zerodivisors by Theorem 1. $\square$

Interestingly enough, the converse to Theorem 3 does not hold, at least in an arbitrary integral domain. (The explanation is to be found in the "nonstandard" divisibility phenomenon exhibited in Exercise 12(b) of Section IV.4.) But in PIDs we get not only the converse of Theorem 3 but more:

Theorem 4. In a PID (hence in a Euclidean domain),

$$p \text{ prime} \Rightarrow (p) \text{ is maximal.}$$

Proof. Let D be a PID and $p \in D$ a prime. Let I be an ideal in D such that $(p) \subseteq I \subseteq D$. (To conclude that (p) is maximal, we must show that $I = (p)$ or $I = D$.)

Since D is a PID, we can express I as (q) for some $q \in D$. Thus we have

$$(p) \subseteq (q) \subseteq D.$$

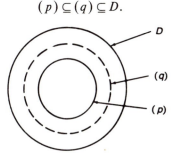

Now $p \in (p)$, so that $p \in (q)$, giving $p = xq$ for some $x \in D$. Since p is prime, x or q must be a unit.

CASE 1. x a unit. Then $q = x^{-1}p \in (p)$, so that $(q) \subseteq (p)$, whence $(q) = (p)$.

CASE 2. q a unit. Then $q^{-1}q = 1 \in (q)$, giving $(q) = D$. $\square$

Corollary.   In a PID,

$$p \text{ prime} \Rightarrow D/(p) \text{ is a field.}$$

Proof.   If p is prime, then (p) is maximal by Theorem 4, whence $D/(p)$ is a field by Theorem 2. $\square$

Thus, for the Euclidean domains (hence PIDs) $\mathbf{Z}$ and $F[x]$, we conclude from the corollaries to Theorems 3 and 4 that

1. $\mathbf{Z}/(p) \cong \mathbf{Z}_p$ is a field $\Leftrightarrow$ p is prime;
2. $F[x]/(m(x)) \cong F[x]_{m(x)}$ is a field $\Leftrightarrow$ $m(x)$ is irreducible over F.

Example 4. $x^2 - 2 = (x + \sqrt{2})(x - \sqrt{2})$ is reducible over $\mathbf{R}$. Hence $\mathbf{R}[x]/(x^2 - 2)$ is not a field, indeed is not even an integral domain (Corollary to Theorem 3). Let us verify this explicitly.

Let $\alpha = x + (x^2 - 2)$. Then $\mathbf{R}[x]/(x^2 - 2) = \mathbf{R}[\alpha]$ is given by

$$\mathbf{R}[\alpha] = \{a + b\alpha \mid a, b \in \mathbf{R}\},$$

where $\alpha^2 - 2 = 0$ (Theorem 5 of Section 2.1). We then have

$$(\sqrt{2} + \alpha)(\sqrt{2} - \alpha) = 2 - \alpha^2 = 0,$$

so that $\sqrt{2} + \alpha$ and $\sqrt{2} - \alpha$ are zerodivisors. $\square$

Example 5. $x^2 - 2$ is irreducible over $\mathbf{Q}$ (appealing to the irrationality of $\sqrt{2}$). Hence $\mathbf{Q}[x]/(x^2 - 2)$ is a field. Let us develop a formula for inverses in this field.

With $\alpha = x + (x^2 - 2)$, $\mathbf{Q}[x]/(x^2 - 2) = \mathbf{Q}[\alpha]$ is given by

$$\mathbf{Q}[\alpha] = \{a + b\alpha \mid a, b \in \mathbf{Q}\},$$

where $\alpha^2 - 2 = 0$. Employing the axioms of a field in $\mathbf{Q}[\alpha]$, we can write

$$\frac{1}{a + b\alpha} = \frac{1}{a + b\alpha} \cdot \frac{a - b\alpha}{a - b\alpha}$$

$$= \frac{a - b\alpha}{a^2 - 2b^2} \quad (\text{using } \alpha^2 = 2)$$

$$= \frac{a}{a^2 - 2b^2} + \frac{-b}{a^2 - 2b^2}\alpha \in \mathbf{Q}[\alpha]. \square$$

Remark.   Unfortunately this method of computing inverses (essentially the high school algebra process of "rationalizing the denominator") breaks down if the degree of the modulus polynomial is > 2. However, we shall eventually (Chapter VII) derive an algorithm for computing inverses in the general case $F[x]/(m(x))$.

Example 6. Let $m(x) = x^2 + x + 1 \in \mathbb{Z}_2[x]$. Since $m(0), m(1) \neq 0$, $m(x)$ has no linear factors, hence is irreducible over $\mathbb{Z}_2$. Thus $\mathbb{Z}_2[x]/(x^2 + x + 1)$ is a field. With $\alpha = x + (x^2 + x + 1)$, $\mathbb{Z}_2[x]/(x^2 + x + 1) = \mathbb{Z}_2[\alpha]$ is given by

$$\mathbb{Z}_2[\alpha] = \{0, 1, \alpha, 1 + \alpha\},$$

where $\alpha^2 + \alpha + 1 = 0$. Addition and multiplication tables for $\mathbb{Z}_2[\alpha]$ are as follows:

+	0	1	α	$1 + \alpha$
0	0	1	α	$1 + \alpha$
1	1	0	$1 + \alpha$	α
α	α	$1 + \alpha$	0	1
$1 + \alpha$	$1 + \alpha$	α	1	0

$\cdot$	1	α	$1 + \alpha$
1	1	α	$1 + \alpha$
α	α	$1 + \alpha$	1
$1 + \alpha$	$1 + \alpha$	1	α

We see that $\alpha^{-1} = 1 + \alpha$ in $\mathbb{Z}_2[\alpha]$. Thus we have constructed a field with 4 elements. (This is seen to be the field of Example 5 of Section IV.2.1.) ☐

EXERCISES

Section

1. Show that any congruence relation on $\mathbb{Z}$ with respect to addition is also a congruence relation with respect to multiplication. What about the converse?

2. (*Lattice-theoretic properties of congruence relations.*)
   (a) Let $\mathcal{E}_A$ be the set of equivalence relations on a set A. Prove that $\mathcal{E}_A$ under the partial order of inclusion is a lattice in which
   $$E_1 \wedge E_2 = E_1 \cap E_2,$$
   $$E_1 \vee E_2 = (E_1 \cup E_2)_{\text{tr}}.$$
   [Recall from Exercise 7 of Section II.2: $(E_1 \cup E_2)_{\text{tr}}$ is the transitive closure of $E_1 \cup E_2$, given by $\bigcup_{i=1}^{\infty}(E_1 \cup E_2)^i$.] Why is $E_1 \vee E_2$ not equal to $E_1 \cup E_2$?
   (b) Now let $\mathcal{C}_A$ be the set of all congruence relations on an Ω-algebra A. Prove that $\mathcal{C}_A$ is a sublattice of $\mathcal{E}_A$.

3. (*Application of Exercise 2 to automata*—no background in automaton theory required!) Consider the automaton specified by the following table:

states S	next-state δ_0	δ_1	output λ
a	f	d	0
b	e	a	0
c	b	d	1
d	b	c	1
e	b	a	0
f	a	e	1

Here we have the set $S = \{a, b, c, d, e, f\}$ of states together with next state functions $\delta_0, \delta_1: S \to S$. $\delta_0(s)$ is the next state of the automaton when its current state is s and it receives an input of 0; $\delta_1(s)$ applies when it receives an input of 1. Here we have $\delta_0(a) = f$, $\delta_1(a) = d$, etc. For the time being, we ignore the output section (λ) of the automaton.

(a) Determine the lattice Π_{sp} of all s.p. partitions of S (the substitution property should be with respect to both next-state functions). [*Hint:* There are six.]

(b) For each s.p. partition π, draw up a table for the corresponding quotient automaton (still ignoring the output section).

We now bring the output section of the automaton into the picture. $\lambda(s)$ designates the output of the machine when the automaton passes into state s. (In our model $\lambda(s)$ is either 0 or 1, and is independent of the state from which the transition is made.)

Definition. A partition π on the state set S of an automaton is *output consistent* if for all $s, t \in S$,

$$s \, E_\pi \, t \Rightarrow \lambda(s) = \lambda(t).$$

(c) Show that in general the set of all output consistent s.p. partitions of S is a sublattice of Π_{sp}. Identify this sublattice of Π_{sp} obtained in (a). [*Remark:* The quotient automaton determined by the greatest (coarsest) output consistent s.p. partition is called the *reduced* automaton corresponding to the given automaton A, so-called because it is the automaton with the fewest number of states having the same input-output characteristics as A.] (Note 1.)

Section 2

1. Verify the claim of Example 2 of Section 2.1, that $\{\sum_{i=1}^{n} a_i r_i \mid r_i \in R\}$ is the smallest ideal containing $a_1, \dots, a_n \in R$, R a commutative ring.

2. In an integral domain, prove: $(a) = (b) \Leftrightarrow a \sim b$.

3. For R a commutative ring, show that the set of all polynomials in $R[x, y]$ with zero constant term is an ideal.

4. For R a commutative ring, establish the following isomorphisms:
   (a) $R[x, y]/(x, y) \cong R$ [*Hint:* Consider the evaluation map $a(x, y) \mapsto a(0, 0)$];
   (b) $R[x, y]/(x + y) \cong R[x]$;
   (c) $R[x]/(x^2 + 1) \cong R[x]/(x^2 + 2x + 2)$ [*Hint:* $x^2 + 2x + 2 = (x + 1)^2 + 1$].

5. (*Lattice-theoretic properties of ideals.*)
   (a) Show that the set of all ideals of a ring R, partially ordered by inclusion, is a lattice in which

$$I \wedge J = I \cap J,$$
$$I \vee J = I + J = \{a + b \mid a \in I, b \in J\}.$$

   (b) In $\mathbf{Z}$, what is $(4) \wedge (6)$? $(4) \vee (6)$? Generalize to $(a) \wedge (b)$, $(a) \vee (b)$ in an arbitrary Euclidean domain.

Problems 6–8 deal with *quotient groups*. Let $[G; \cdot, ^{-1}, 1]$ be a group.

6.   **Definition.** A subgroup $N \leq G$ is a *normal* subgroup of G if $gN = Ng$ (i.e., if every left coset is also a right coset).

(a) Show that if N is a normal subgroup of G, then G/N is an s.p. partition of G. Conclude that G/N under the operations

$$(gN)(hN) = ghN,$$

$$(gN)^{-1} = g^{-1}N,$$

$$1_{G/N} = N,$$

is a homomorphic image of G, hence a group ("quotient group"). (Cf. Theorem 2 of Section 2.1.)

(b) Let $\phi: G \to G'$ be a morphism of groups. Show that the (*group-theoretic*) *kernel*, defined by

$$K_\phi = \{a \in G | \phi(a) = 1\},$$

is a normal subgroup of G. (Cf. Theorem 1 of Section 2.2.)

(c) (*Isomorphism Theorem for Groups.*) Let $\phi: G \to G'$ be an epimorphism of groups. Then $G' \cong G/K_\phi$, where $\psi: gK_\phi \mapsto \phi(g)$ is an isomorphism.

(Cf. Theorem 2 of Section 2.2.)

7.   Let f be the permutation on $\{1, 2, 3\}$ defined by $1 \mapsto 1$, $2 \mapsto 3$, $3 \mapsto 2$. Show that $T = \{1_3, f\}$ is a non-normal subgroup of the symmetric group S_3. Exhibit the cosets of S_3 and verify directly that S_3/T is not an s.p. partition of S_3.

8.   Use the map $x \mapsto e^{2\pi i x}$ (and any of its familiar properties) to show that the multiplicative group of the complex numbers having absolute value 1 is isomorphic to the quotient group $\mathbf{R}/\mathbf{Z}$ of the additive group of reals by the additive group of integers.

Section 3

1.   Show that the ideal $(x, 2)$ in $\mathbf{Z}[x]$ is not principal (hence that $\mathbf{Z}[x]$ is not a PID).

2.   (*Simplicity of the matrix ring $M_n(F)$.*) Show that the ring $M_n(F)$ of all $n \times n$ matrices over a field F has no proper ideals. [*Hint:* Let $I \neq \{0\}$ contain a nonzero matrix $A = (a_{ij})$ with $a_{rs} \neq 0$. Now consider the product $a_{rs}^{-1} E_{ir} A E_{si}$ $(1 \leq i \leq n)$, where E_{ij} is the matrix with 1 in the (i,j) position, 0 elsewhere.]

3.   Show that the ideal (x) is
      (i) prime but not maximal in $\mathbf{Z}[x]$;
      (ii) maximal in $\mathbf{Q}[x]$.

4.   For m a positive integer, establish the isomorphism $\mathbf{Z}[x]/(x, m) \cong \mathbf{Z}_m$. Conclude that (x, m) is maximal if and only if m is a prime.

5.   (a) Let F be a field and let α be a fixed element of F. In the ring F^F of all functions $F \to F$, show that $M_\alpha = \{f \in F^F | f(\alpha) = 0\}$ is a maximal ideal.

(b) When is the coset $f + M_\alpha$ invertible in F^F/M_α? For the case $F = \mathbf{R}$, $\alpha = 0$, is the coset containing (i) e^x, (ii) $\sin x$ invertible?

6. Prove Theorem 3 of Section 3.3.

7. (a) Verify that $\mathbf{Q}[x]/(x^2 - 3)$ is a field. What is the inverse of the coset containing $2 + 5x$?

   (b) Verify that $\mathbf{Z}_7[x]/(x^2 - 3)$ is a field. What is the inverse of the coset containing $2 + 5x$?

8. Prove: If F is an ordered field, then $F[x]/(x^2 + 1)$ is a field but not an ordered field.

9. Which of the following are fields:
   (a) $\mathbf{Z}_2[x]/(x^4 + x^3 + x^2 + x + 1)$;
   (b) $\mathbf{Z}_2[x]/(x^4 + x^3 + x + 1)$;
   (c) $\mathbf{Z}_2[x]/(x^4 + x + 1)$;
   (d) $\mathbf{Z}_5[x]/(x^5 + x^3 + x^2 + x + 1)$?

10. Construct a field with 125 elements.

NOTES

Section 1

1. For an extensive development of a lattice-theoretic approach to automata, see J. Hartmanis and R. E. Stearns, *Algebraic Structure Theory of Sequential Machines* (Englewood Cliffs, N.J.: Prentice-Hall, 1966).

ELEMENTS OF FIELD THEORY

The main goal of this chapter is to construct and analyze *finite* fields, a class of fields that is of interest not only for pure algebra but also for a variety of applications of algebra, in particular for algebraic computing.

Although we focus our attention on those aspects of field theory that are germane to our interest in finite fields, the first segment of our development (extension fields) applies with equal force to infinite fields, in particular to the familiar number fields **Q** and **R**. Indeed, the entire chapter deals with some of the most elegant and useful ideas of all algebra.

* * *

Throughout this chapter F denotes a field, the "ground field," and E denotes an extension field of F (E ≥ F).

1. EXTENSION FIELDS

1.1 The Root Adjunction Problem

Consider the polynomial $x^3 - 2$ first as a polynomial over **Q**. None of its roots are rational, hence it has no linear factors over **Q** (Factor Theorem), hence it is irreducible over **Q** (Proposition 6 of Section IV.4.1). However, when we extend the coefficient field from $\mathbf{Q}$ to $\mathbf{R}$, the situation changes. $x^3 - 2$ has the real root $\sqrt[3]{2}$, hence the linear factor $x - \sqrt[3]{2}$, hence is reducible over $\mathbf{R}$.

This little example suggests the following *root adjunction problem:*

> *Given a field F and an irreducible polynomial m(x) over F, extend F to a field E in which m(x) has a root.*

The following constructive solution to this problem, due to Kronecker (1823–1891), makes effective use of our theory of quotient rings.

Theorem 1. Let $m(x)$ be an irreducible polynomial over F. Then $E = F[x]/(m(x))$ is an extension of F in which $m(x)$ has a root.

Proof. (We have done most of the work already.) Since $m(x)$ is irreducible over F, $F[x]/(m(x))$ is a field (Corollary to Theorem 4, Section V.3.3). Moreover, $F[x]/(m(x))$ is an extension field of F in accordance with our agreed identification of $a + (m(x)) \in F[x]/(m(x))$ with $a \in F$ (see the discussion following the corollary to Theorem 4, Section V.2.1). Finally, $\alpha = x + (m(x))$ is a root of $m(x)$ in $F[x]/(m(x))$ (Theorem 5(b) of Section V.2.1). $\square$

Corollary. Let $m(x)$ be an irreducible polynomial over a field F. Then $E = F[x]/(m(x))$ is an extension over which $m(x)$ has a linear factor (hence over which $m(x)$ is reducible).

Proof. Factor Theorem. $\square$

Examples 5 and 6 of Section V.3.3 provide instances of the above theorem. Another example is

Example 1. We have already noted that $x^3 - 2$ is irreducible over $\mathbf{Q}$. The quotient ring $\mathbf{Q}[x]/(x^3 - 2)$ is an extension field of $\mathbf{Q}$ in which $x^3 - 2$ has a root, namely $\alpha = x + (x^3 - 2)$.

We can represent $\mathbf{Q}[x]/(x^3 - 2)$ as

$$\mathbf{Q}[\alpha] = \left\{ a_0 + a_1\alpha + a_2\alpha^2 \mid a_i \in \mathbf{Q} \right\}$$

where $\alpha^3 - 2 = 0$ (Theorem 5 of Section V.2.1). By dividing $x^3 - 2$ by the linear factor $x - \alpha$, we can provide an explicit factorization of $x^3 - 2$:

$$
\require{enclose}
\begin{array}{r}
x^2 + \alpha x + \alpha^2 \\
x - \alpha \enclose{longdiv}{x^3 - 2 } \\
\underline{x^3 - \alpha x^2 } \\
\alpha x^2 - 2 \\
\underline{\alpha x^2 - \alpha^2 x } \\
\alpha^2 x - 2 \\
\underline{\alpha^2 x - 2} \qquad \text{(using } \alpha^3 = 2\text{)} \\
0
\end{array}
$$

Thus over $\mathbf{Q}[\alpha]$, $x^3 - 2$ has the factorization

$$x^3 - 2 = (x - \alpha)(x^2 + \alpha x + \alpha^2),$$

in contrast to its irreducibility over $\mathbf{Q}$. $\square$

Theorem 2. Let $f(x)$ be any polynomial over F having degree ≥ 1, where $f(x)$ need not be irreducible. Then F can be extended to a field E in which $f(x)$ has a root (hence a linear factor).

Proof. Let $g(x)$ be an irreducible (over F) factor of $f(x)$, so that $f(x) = g(x)h(x)$. By Theorem 1, $E = F[x]/(g(x))$ is an extension of F in which $g(\alpha) = 0$. Hence in E, $f(\alpha) = g(\alpha)h(\alpha) = 0$. $\square$

We know by Theorem 2 of Section IV.3.3 that a polynomial $f(x)$ of degree n over a field F has at most n roots in F or in any extension of F (where a root α of multiplicity k counts as k roots $\alpha_1, \alpha_2, \ldots, \alpha_k$ each equal to α). We now ask, does there exist an extension of F in which $f(x)$ has its full complement of n roots?

Theorem 3. Let $f(x)$ be a polynomial of degree $n \geq 1$ over a field F. Then F can be extended to a field E in which $f(x)$ has n roots.

Proof. By induction on $n = \deg f(x)$.

Basis $(n = 1)$. If $f(x) = ax + b \in F[x]$ has degree 1, then $f(x)$ has a single root $-b/a$ in F.

Induction. Assume the statement of the theorem for all polynomials of degree n over a field, and let $f(x) \in F[x]$ have degree $n + 1$. By Theorem 2, $f(x)$ has a root α in an extension E of F. By the Factor Theorem, $f(x)$ can be factored over E as

$$(*) \qquad\qquad f(x) = (x - \alpha)g(x).$$

Now $g(x) \in E[x]$ has degree n. By the induction hypothesis, E can be extended to a field E' in which $g(x)$ has n roots $\beta_1, \beta_2, \ldots, \beta_n$. But from $(*)$ any root of $g(x)$ is a root of $f(x)$. Thus E' is an extension of F in which $f(x)$ has $n + 1$ roots, namely $\alpha, \beta_1, \beta_2, \ldots, \beta_n$. $\square$

Definition. A *root field over* F of a polynomial $f(x) \in F[x]$ is a field $E \geq F$ that satisfies

1. All the roots of $f(x)$ lie in E;
2. E is a smallest extension of F that satisfies 1 in that no proper subfield of E is an extension of F that contains all the roots of $f(x)$.

Theorem 4. Any polynomial $f(x) \in F[x]$ has a root field over F.

Proof. By Theorem 3, there exists a field $E \geq F$ that contains all the roots of $f(x)$. Then

$$\cap \{E' | F \leq E' \leq E \quad \text{and} \quad E' \text{ contains all roots of } f(x)\}$$

is evidently an extension field of F that satisfies 1 and 2. $\square$

See Exercise 3 for an important factorization property of $f(x)$ over its root field.

1.2 Analysis of Simple Extension Fields

Our viewpoint now changes from the synthesis of extension fields to the analysis of extension fields.

We assume that a ground field F and an extension field $E \geq F$ are both given. We now wish to ascertain the structure of the subfield of E generated by F and a single element $\alpha \in E$. We denote this extension field of F—the smallest extension field of F that contains α—by $F(\alpha)$ and refer to it as "F adjoined by α." The general setup is as follows:

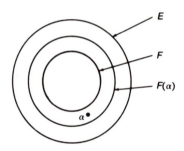

$F(\alpha)$ is called a *simple* extension (of F) in that a single element is adjoined to F. One might adjoin several elements $\alpha_1, \alpha_2, \ldots, \alpha_r \in E$ to F in forming the *multiple*, or *iterated*, extension $F(\alpha_1, \ldots, \alpha_r)$ of F—the smallest extension field of F that contains $\alpha_1, \ldots, \alpha_r$.

Definition. Let $\alpha \in E \geq F$. α is called *algebraic over* F if α satisfies a polynomial equation over F: $f(\alpha) = 0$ for some $f(x) \in F[x]$; otherwise α is *transcendental over* F. $F(\alpha)$ is called a (simple) *algebraic* or *transcendental* extension of F according to whether α is algebraic or transcendental over F.

Example 1. Take $F = \mathbf{Q}$, $E = \mathbf{C}$. This puts us in the classical setting for the investigation of field extensions. When we refer to a complex number as being "algebraic" or "transcendental" we mean algebraic or transcendental over $\mathbf{Q}$.

Some examples:

(a) $\sqrt[3]{2}$ is algebraic, since $\sqrt[3]{2}$ is a root of $x^3 - 2 \in \mathbf{Q}[x]$;
(b) $i = \sqrt{-1}$ is algebraic, since i is a root of $x^2 + 1 \in \mathbf{Q}[x]$;

(c) $\sqrt[3]{2 + \sqrt{5}}$ is algebraic, for if $x = \sqrt[3]{2 + \sqrt{5}}$, then $x^3 = 2 + \sqrt{5}$, or $(x^3 - 2)^2 = 5$, or $x^6 - 4x^3 - 1 = 0$.

(d) e and π are both transcendental—familiar facts though quite difficult to prove (Note 1). $\square$

Remark. In dealing with field extensions, the ground field F must be clearly specified. For example, e and π are both trivially algebraic over **R**.

Let $\alpha \in E$ be algebraic over F. We define the *minimum polynomial of α over F*, denoted $m_\alpha(x)$, as the monic polynomial of smallest degree in $F[x]$ having α as a root.

Proposition 1 (*Properties of the minimum polynomial*). Let $\alpha \in E$ have minimum polynomial $m_\alpha(x)$ over F.

(a) $m_\alpha(x)$ is unique;
(b) $m_\alpha(x)$ is irreducible over F;
(c) For any $f(x) \in F[x]$,

$$f(\alpha) = 0 \Leftrightarrow m_\alpha(x) \mid f(x).$$

Proof. (a) Suppose to the contrary that α has two distinct minimum polynomials

$$m(x) = x^n + m_{n-1}x^{n-1} + \cdots + m_0,$$
$$\overline{m}(x) = x^n + \overline{m}_{n-1}x^{n-1} + \cdots + \overline{m}_0.$$

Then α is a root of $m(x) - \overline{m}(x)$, a nonzero polynomial of degree $< n$ which can be made monic by dividing through by the leading coefficient. But this contradicts the smallest-degree property of the minimum polynomial, forcing the desired conclusion $m(x) = \overline{m}(x)$.

(b) Suppose to the contrary that $m_\alpha(x)$ is reducible over F: $m_\alpha(x) = g(x)h(x)$ where $\deg g(x)$, $\deg h(x) \geq 1$. Then

$$m_\alpha(\alpha) = 0 = g(\alpha)h(\alpha),$$

so that $g(\alpha) = 0$ or $h(\alpha) = 0$. This makes α the root of a polynomial in $F[x]$ (either $g(x)$ or $h(x)$) that (i) has degree $< \deg m_\alpha(x)$ and (ii) can be made monic by dividing through by the leading coefficient. But this contradicts the smallest-degree property of $m_\alpha(x)$, forcing the conclusion that $m_\alpha(x)$ is irreducible over F.

(c) If $f(x) = m_\alpha(x)g(x)$, then $f(\alpha) = m_\alpha(\alpha)g(\alpha) = 0$.

In the other (more interesting) direction, suppose that $f(\alpha) = 0$. Dividing $f(x)$ by $m_\alpha(x)$ gives

$$f(x) = m_\alpha(x)q(x) + r(x), \qquad \deg r(x) < \deg m_\alpha(x),$$

so that

$$r(\alpha) = f(\alpha) - m_\alpha(\alpha)q(\alpha) = 0.$$

If $r(x) \neq 0$, then $r(x)$ can be made monic by dividing through by the leading coefficient and we will have found a monic polynomial of degree $< \deg m_\alpha(x)$ having α as a root. This contradiction to the smallest-degree property of $m_\alpha(x)$ forces $r(x) = 0$, so that $m_\alpha(x) | f(x)$ as required. $\square$

Proposition 2 (*A useful characterization of the minimum polynomial*). If $\alpha \in E$ is a root of a monic irreducible polynomial $m(x)$ over F, then $m(x)$ is the minimum polynomial of α over F.

Proof. Let the minimum polynomial of α over F be $m_\alpha(x)$. By Proposition 1(c), $m_\alpha(x) | m(x)$. But $m(x)$ is irreducible. Hence $m(x) = c m_\alpha(x)$ $(c \in F)$. Since $m(x)$ and $m_\alpha(x)$ are both monic, $c = 1$. $\square$

Proposition 2 can be restated thus: *an irreducible polynomial over a field F is the minimum polynomial over F of any of its roots.*

Now for the main result of this section, which tells us what an algebraic extension $F(\alpha)$ looks like up to isomorphism.

Theorem 3.   Let $\alpha \in E$ be algebraic over $F \leq E$ with minimum polynomial $m_\alpha(x)$ over F. Then

$$F(\alpha) \cong F[x]/(m_\alpha(x)).$$

Proof.   We have here a glorious opportunity to apply our Ring Isomorphism Theorem.

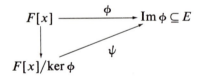

If we can find a morphism $\phi: F[x] \to E$ whose image is $F(\alpha)$ and whose kernel is $(m_\alpha(x))$, then we are done.

The natural candidate for ϕ is (what else) the evaluation morphism

$$\phi_\alpha: f(x) \mapsto f(\alpha).$$

The claim now is that (i) $\ker \phi_\alpha = (m_\alpha(x))$ and (ii) $\operatorname{Im} \phi_\alpha = F(\alpha)$.
   Proof of (i): We have

$$\ker \phi_\alpha = \{ f(x) \in F[x] | f(\alpha) = 0 \}$$

i.e., the kernel is the ideal of all polynomials in $F[x]$ having α as a root. Now any ideal of $F[x]$ is principal, whose generator may be taken as the monic polynomial having smallest degree. For the case of $\ker \phi_\alpha$, this smallest degree monic polynomial is (by definition) $m_\alpha(x)$, the minimum polynomial of α over F. Thus $\ker \phi_\alpha = (m_\alpha(x))$, as required.
   Proof of (ii): Any $f(\alpha) \in \operatorname{Im} \phi_\alpha$ must be in $F(\alpha)$ by closure, using only that $F(\alpha)$ is a ring containing $F \cup \{\alpha\}$. Hence $\operatorname{Im} \phi_\alpha \subseteq F(\alpha)$.

As for the other (more interesting) direction, by the Ring Isomorphism Theorem we have

$$\text{Im } \phi_\alpha \cong F[x]/\ker \phi = F[x]/(m_\alpha(x)).$$

Now $m_\alpha(x)$ is irreducible over F (Proposition 1(b)), which means that $F[x]/(m_\alpha(x))$, hence $\text{Im } \phi_\alpha$, is a field. Thus $\text{Im } \phi_\alpha$ is a subfield of E that contains each $a \in F$ (because $\phi_\alpha(a) = a$) and α (because $\phi_\alpha(x) = \alpha$). But then $F(\alpha) \subseteq \text{Im } \phi_\alpha$, since $F(\alpha)$, by definition, is the *smallest* subfield of E that includes $F \cup \{\alpha\}$. Thus $\text{Im } \phi_\alpha = F(\alpha)$ as required.

From the Ring Isomorphism Theorem we conclude that

$$\psi: a(x) + (m(x)) \mapsto a(\alpha)$$

is an isomorphism $F[x]/(m_\alpha(x)) \to F(\alpha)$. $\square$

Corollary (*Representation of $F(\alpha)$*). Let $\alpha \in E$ have minimum polynomial $m_\alpha(x) = x^n + \text{l.d.t.}$ over F. Then each $\beta \in F(\alpha)$ can be uniquely represented in the form

$$(*) \qquad \beta = a_0 + a_1 \alpha + \cdots + a_{n-1} \alpha^{n-1} \qquad (a_i \in F).$$

Proof. We shall exploit the isomorphism

$$\psi: a(x) + (m_\alpha(x)) \mapsto a(\alpha)$$

from $F[x]/(m_\alpha(x))$ to $F(\alpha)$.

   (i) Each coset of $F[x]/(m_\alpha(x))$ can be expressed in the form $a(x) + (m_\alpha(x))$ with $\deg a(x) < n$ (Theorem 4(a) of Section V.2.1). Since ψ is surjective, it follows that each $\beta = a(\alpha) \in F(\alpha)$ can be expressed in the form $(*)$.

   (ii) Let $a(x), b(x)$ have degree $< n$, $a(x) \neq b(x)$. Then $a(x) + (m_\alpha(x)) \neq b(x) + (m_\alpha(x))$ (Theorem 4(b) of Section V.2.1), and since ψ is injective, $a(\alpha) \neq b(\alpha)$. Thus each $\beta \in F(\alpha)$ has a *unique* representation of the form $(*)$. $\square$

Remark. Since $m_\alpha(\alpha) = 0$ in $F(\alpha)$, arithmetic in $F(\alpha)$ is essentially polynomial arithmetic modulo $m_\alpha(\alpha)$, regarding α as an indeterminate (which it isn't!).

Example 2. Let us examine the structure of $Q(\sqrt[3]{2})$, the rationals adjoined by $\sqrt[3]{2} \in R$. (We shall use the easily shown irrationality of $\sqrt[3]{2}$.) The claim now is that the minimum polynomial of $\sqrt[3]{2}$ over Q is $x^3 - 2$.

   First, $\sqrt[3]{2}$ is a root of $x^3 - 2$. Second, $x^3 - 2$ is irreducible over Q, for if it were not, it would have a linear factor over Q, hence a rational root in Q, which would contradict the irrationality of $\sqrt[3]{2}$. By Propostion 2, $x^3 - 2$ is the minimum polynomial of $\sqrt[3]{2}$.

   By Theorem 3, we have the isomorphism

$$Q(\sqrt[3]{2}) \cong Q[x]/(x^3 - 2).$$

And by the corollary to Theorem 3, each $\beta \in \mathbf{Q}(\sqrt[3]{2})$ can be uniquely expressed in the form

$$(*) \qquad\qquad a + b\sqrt[3]{2} + c\sqrt[3]{4} \qquad (a, b, c \in \mathbf{Q}).$$

In particular, since $\mathbf{Q}(\sqrt[3]{2})$ is a field, each $\beta \neq 0$ in this field has a reciprocal of the form $(*)$; i.e.,

$$\frac{1}{a + b\sqrt[3]{2} + c\sqrt[3]{4}} = d + e\sqrt[3]{2} + f\sqrt[3]{4}$$

for some $d, e, f \in \mathbf{Q}$, a fact that would not at all be obvious without Theorem 3. (In Chapter VII we shall develop an algorithm for computing such inverses in algebraic extension fields.) $\square$

As the companion to Theorem 3 we have

Theorem 4. Let $\alpha \in E$ be transcendental over $F \leq E$. Then

$$F(\alpha) \cong F(x)$$

(where $F(x)$ as usual is the field of rational functions in an indeterminate x).

Proof. As in Theorem 3, consider the evaluation morphism $\phi_\alpha \colon f(x) \mapsto f(\alpha)$. We let the reader complete the proof (Exercise 7). $\square$

Although we are not especially interested in transcendental extensions, the reader should note that they are quite different from algebraic extensions (stemming, of course, from the fact that an algebraic element satisfies a polynomial equation over its ground field whereas a transcendental element does not). In this regard note that a simple algebraic extension of a finite field is finite, whereas a simple transcendental extension of a finite field is infinite.

$$*\quad *\quad *$$

We have focussed our attention on *simple* (principally algebraic) extensions. The theory of *iterated* extensions, where several elements (perhaps even infinitely many) are adjoined to the ground field, is more involved. Yet simple extensions continue to play an important role even for the case of iterated extensions. We state without proof the following interesting result: If the ground field F has characteristic zero or is finite, then any iterated extension $F(\alpha_1, \ldots, \alpha_r)$ of F by finitely many algebraic elements over F is in fact a *simple* algebraic extension $F(\beta)$ of F (Note 2). Instances of this phenomenon are to be found in Exercises 4 and 5.

2. THE MULTIPLICATIVE GROUP OF A FINITE FIELD

The most prominent feature of finite fields concerns their multiplicative groups —they are cyclic. This section is devoted to establishing that result and exploring some of its consequences.

2.1   Cyclic Property of Finite Fields

We shall establish our main result about the structure of finite fields—that the multiplicative group is cyclic—via a theorem about abelian groups. We require the following

> **Lemma 1.**   Let a and b be elements of an abelian group, and let $o(a) = m$ and $o(b) = n$ where $(m, n) = 1$. Then $o(ab) = mn$.

Proof.   Let $r = o(ab)$. Now $(ab)^{mn} = (a^m)^n (b^n)^m = 1 \cdot 1 = 1$, so that $r \leq mn$ (whether or not $(m, n) = 1$).

As for the other direction ($mn \leq r$), we first note that $sm + tn = 1$ for integers s and t, since $(m, n) = 1$. We then have

$$a^r = (a^{sm+tn})^r$$
$$= (a^m)^{sr}(a^n)^{tr}$$
$$= (a^n)^{tr}$$
$$= (a^n b^n)^{tr} \qquad (b^n = 1)$$
$$= ((ab)^r)^{nt}$$
$$= 1,$$

so that $m | r$. Symmetrically, $n | r$. But $(m, n) = 1$ means that $mn | r$ (Lemma 5(b) of Section IV.4.2). Thus $mn \leq r$, so that $o(ab) = mn$. $\square$

> **Theorem 1.**   Let G be a finite abelian group and let m be the order of a maximal order element of G. Then the order of any element of G divides m.

Proof.   Let $a \in G$ have maximal order m, and let $b \in G$ have order n. Assume to the contrary that $n \nmid m$. Then there must be some prime p in the prime power factorizations of m and n that occurs with greater exponent in n than in m (otherwise n would divide m). Thus m and n can be expressed in the form

$$m = p^e q, \qquad n = p^f r$$

where $f > e \geq 0$, $p \nmid q$, $p \nmid r$.

Now $o(a^{p^e}) = q$ and $o(b^r) = p^f$. Since $(p^f, q) = 1$, $o(a^{p^e}b^r) = p^f q$ by Lemma 1. But $p^f q > m$, in contradiction to m being the maximal order of elements in G. Hence the desired conclusion $n | m$. $\square$

> **Theorem 2**   (*Main result*).   Let E be a finite field. Then E^*, the multiplicative group of E, is cyclic.

Proof.   Let $|E| = r$, and let $\alpha \in E^*$ have maximal order m. (Our task is to show that $m = r - 1$.)

One direction is easy. By Lagrange's Theorem, the order of any element of E^* divides the order of E^*. Hence $m | (r - 1)$, giving $m \leq r - 1$.

For the other direction $(m \geq r - 1)$, we examine the roots of $x^m - 1$ in E^*. For any $b \in E^*$, $o(b) = n$, we have $n \mid m$ by Theorem 1, so that $m = kn$. Then $b^m = (b^n)^k = 1$, making every one of the $r - 1$ elements of E^* a root of $x^m - 1$. But $x^m - 1$ has at most m distinct roots (Theorem 2 of Section IV.3.3), whence $m \geq r - 1$. $\square$

We call any generator of the cyclic group E^* a *primitive element* of E.

We have just proved that any finite field has a primitive element. But group theory tells us more:

Theorem 3. Let $|E| = r$. Then E has $\phi(r - 1)$ primitive elements ($\phi = $ Euler's phi function). In particular, if $\alpha \in F^*$ is primitive, then α^i is primitive whenever $(i, r - 1) = 1$.

Proof.   Theorem 4 of Section III.3.4. $\square$

Thus, if we can find one primitive element we can easily get all of them. The problem in practice is to find just one. It is not a simple matter to (efficiently) determine a primitive element of even the modular fields $\mathbf{Z}_p$ for large values of p (we shall examine the problem and come up with a much better than brute force solution in Chapter IX). For the present, we test out Theorems 2 and 3 on a small finite field.

Example 1.   In $\mathbf{Z}_7$, 3 is primitive. Check: $3^1 = 3$, $3^2 = 2$, $3^3 = 6$, $3^4 = 4$, $3^5 = 5$, $3^6 = 1$. We draw up a table of the orders of elements in $\mathbf{Z}_7^*$.

a	$o(a)$
1	1
2	3
3	6
4	3
5	6
6	2

In accordance with Theorem 3, $\mathbf{Z}_7^*$ has $\phi(6) = 2$ primitive elements, 3 and 5. $\square$

2.2   Finite Fields as Algebraic Extensions

We now bring our general theory of field extensions to bear on our study of finite fields. Our overall goal is to describe the structure of any finite extension field E in terms of a given ground field $F \leq E$. As we shall see, the cyclic property has a dominant influence on the structure of finite fields.

Lemma 1.   Let E be a finite field, $|E| = r$. Then any $\beta \in E^*$ is a root of $x^{r-1} - 1$.

Proof.   By Lagrange's Theorem, the order k of any $\beta \in E^*$ divides $|E^*| = r - 1$. Hence $r - 1 = km$ for some $m \in \mathbf{P}$, so that $\beta^{r-1} = (\beta^k)^m = 1$. $\square$

Theorem 1. Let F be a finite field and E a finite extension field of F. Then any $\beta \in E$ is algebraic over F.

Proof. Let $|E| = r$. By Lemma 1 any $\beta \in E^*$ is a root of $x^{r-1} - 1 \in F[x]$. Also 0 is the root of $x \in F[x]$. Thus, each $\beta \in E$ is algebraic over F. $\square$

The next theorem accounts for our preoccupation with simple algebraic extensions.

Theorem 2. Let F be a finite field and E any finite extension field of F. Then E is a simple algebraic extension of F.

Proof. Since E is a finite field, it contains a primitive element α. Since the powers of α exhaust E^*, it follows that F adjoined by α exhausts all of E; i.e., $E = F(\alpha)$. Thus E is a *simple* extension of F, whence by Theorem 1 is a simple *algebraic* extension of F. $\square$

We are now in a position to garner a result that yields crucial information about the structure of finite fields.

Theorem 3. Let F be a finite field and E a finite extension field of F. Then

$$E \cong F[x]/(m_\alpha(x))$$

where $m_\alpha(x)$ is the minimum polynomial over F of a primitive element $\alpha \in E$.

Proof. From Theorem 2, E is the simple algebraic extension $F(\alpha)$ of F. By Theorem 3 of Section 1.2,

$$F(\alpha) \cong F[x]/(m_\alpha(x)). \quad \square$$

By the corollary to Theorem 3 of Section 1.2, each $\beta \in E$ can be uniquely expressed in the form

$$\beta = a_0 + a_1\alpha + \cdots + a_{n-1}\alpha^{n-1} \qquad (a_i \in F)$$

where $n = \deg m_\alpha(x)$. This gives the following

Corollary. Let F be a finite field, $|F| = q$, and let E be a finite extension of F. Then $|E| = q^n$ for some positive integer n.

We now make a distinguished choice for the ground field F. According to Theorem 2 of Section V.3.2, a field E of characteristic p has a unique prime (smallest) subfield isomorphic with $\mathbf{Z}_p$, namely its unital subring of "integers":

$$[1] = \{0, 1, 2 \cdot 1, \ldots, (p-1) \cdot 1\}.$$

Making the obvious identification of $a \cdot 1 \in E$ with $a \in \mathbf{Z}_p$ ($0 \leq a < p$), we henceforth regard $\mathbf{Z}_p$ *itself as the unique prime subfield of any field E having characteristic p.*

Choosing the ground field F in Theorem 3 and its corollary to be $\mathbf{Z}_p$ gives the following important

Theorem 4. Any finite field has p^n elements for some prime p (the characteristic of the field) and positive integer n (the degree of the minimum polynomial over $\mathbf{Z}_p$ of a primitive element of the field).

We now introduce the widely used notation $GF(q)$ ("Galois field of order q") to denote any finite field with q elements. (By Theorem 4, q must be a prime power.) This notation, which honors the discoverer of finite fields, Evariste Galois (Note 1), anticipates a result that we shall eventually prove: any two fields with the same number of elements are isomorphic.

We now present a detailed example of a finite field: $GF(2^4)$. This example serves not only to illustrate what we already know about finite fields but also to motivate what we still would like to find out.

Example 1 (*Analysis of* $GF(2^4)$). The polynomial $x^4 + x + 1$ is irreducible over $\mathbf{Z}_2$ [$= GF(2)$], which can be verified by exhaustion (check that $x^4 + x + 1$ has no linear or quadratic factors over $\mathbf{Z}_2$).

Now $x^4 + x + 1$ has a root α in an extension of $\mathbf{Z}_2$, namely in $\mathbf{Z}_2[x]/(x^4 + x + 1)$ (Theorem 1 of Section 1.1). Moreover, $x^4 + x + 1$ is the minimum polynomial of α over $\mathbf{Z}_2$ (Proposition 2 of Section 1.2). By the corollary to Theorem 3 of Section 1.2, each element of $\mathbf{Z}_2(\alpha)$ can be uniquely represented in the form

$$a_0 + a_1\alpha + a_2\alpha^2 + a_3\alpha^3 \qquad (a_i \in \mathbf{Z}_2).$$

Thus $\mathbf{Z}_2(\alpha)$ is $GF(2^4)$, an extension field of $\mathbf{Z}_2$ with 16 elements.

In Fig. 1 we have drawn up a table of elements of $\mathbf{Z}_2(\alpha)$ together with their minimum polynomials over $\mathbf{Z}_2$.

As a sample check, let us verify that α^5 has minimum polynomial $x^2 + x + 1$ over $\mathbf{Z}_2$, as claimed. We note that $x^2 + x + 1$ is irreducible over $\mathbf{Z}_2$, since it has no roots in $\mathbf{Z}_2$. Also,

$$
\begin{aligned}
(\alpha^5)^2 &+ \alpha^5 + 1 \\
&= \alpha^{10} + \alpha^5 + 1 \\
&= (1 + \alpha + \alpha^2) + (\alpha + \alpha^2) + 1 \quad \text{(from Fig. 1)} \\
&= 0,
\end{aligned}
$$

so that $x^2 + x + 1$ is indeed the minimum polynomial of α^5. The other minimum polynomials could be similarly checked.

Now for some important observations and questions that are raised by this example.

From Fig. 1 we note that the powers α^i exhaust $GF(2^4)^*$. This is because we chose an irreducible polynomial $x^4 + x + 1$ that has a *primitive* element of

Elements $\beta = a_0 + a_1\alpha + a_2\alpha^2 + a_3\alpha^3$ of $\mathbf{Z}_2(\alpha)$
where $\alpha^4 + \alpha + 1 = 0$

β	a_0	a_1	a_2	a_3	minimum polynomial over $\mathbf{Z}_2$
0	0	0	0	0	x
1	1	0	0	0	$x + 1$
α	0	1	0	0	$x^4 + x + 1$
α^2	0	0	1	0	$x^4 + x + 1$
α^3	0	0	0	1	$x^4 + x^3 + x^2 + x + 1$
$1 + \alpha = \alpha^4$	1	1	0	0	$x^4 + x + 1$
$\alpha + \alpha^2 = \alpha^5$	0	1	1	0	$x^2 + x + 1$
$\alpha^2 + \alpha^3 = \alpha^6$	0	0	1	1	$x^4 + x^3 + x^2 + x + 1$
$1 + \alpha + \alpha^3 = \alpha^7$	1	1	0	1	$x^4 + x^3 + 1$
etc. α^8	1	0	1	0	$x^4 + x + 1$
α^9	0	1	0	1	$x^4 + x^3 + x^2 + x + 1$
α^{10}	1	1	1	0	$x^2 + x + 1$
α^{11}	0	1	1	1	$x^4 + x^3 + 1$
α^{12}	1	1	1	1	$x^4 + x^3 + x^2 + x + 1$
α^{13}	1	0	1	1	$x^4 + x^3 + 1$
α^{14}	1	0	0	1	$x^4 + x^3 + 1$
$1 = \alpha^{15}$	1	0	0	0	$x + 1$

Fig. 1   A representation of GF(2^4).

GF(2^4) for a root. Such an irreducible polynomial is called a *primitive poly-nomial*. In general, an irreducible polynomial of degree n over GF(q) is called primitive if it has a primitive element of GF(q^n) for a root.

The point to be made is that not every irreducible polynomial is primitive. For example, $x^4 + x^3 + x^2 + x + 1$ can readily be shown to be irreducible over $\mathbf{Z}_2$. Thus each element of GF(2^4) $\simeq \mathbf{Z}_2[x]/(x^4 + x^3 + x^2 + x + 1)$ can be represented as $a_0 + a_1\beta + a_2\beta^2 + a_3\beta^3$ ($a_i \in \mathbf{Z}_2$) where $\beta^4 + \beta^3 + \beta^2 + \beta + 1 = 0$. But β is not primitive. A simple check shows that β has order 5 in GF(2^4).

Several questions come to the fore:

Are $\mathbf{Z}_2[x]/(x^4 + x + 1)$ and $\mathbf{Z}_2[x]/(x^4 + x^3 + x^2 + x + 1)$ isomorphic? More generally, is GF(p^n) unique up to isomorphism?

Does GF(p^n) exist for every prime p and positive integer n?

Are all the roots of a primitive polynomial primitive? (Note that for the primitive polynomial $x^4 + x + 1$ of our example, it is true. From Fig. 1, $x^4 + x + 1$ has roots $\alpha, \alpha^2, \alpha^4, \alpha^8$, each of which is primitive by Theorem 3 of Section 2.1.)

Are all the roots of a minimum, or irreducible, polynomial simple? (Again, for our example it is true. As a case in point, the irreducible polynomial $x^4 + x^3 + x^2 + x + 1$ has distinct roots $\alpha^3, \alpha^6, \alpha^9, \alpha^{12}$; hence these roots must be simple by Theorem 2 of Section IV.3.3. Similarly for the other minimum polynomials displayed in Fig. 1.)

How can we compute minimum polynomials (short of exhaustive testing)? The remainder of our study of finite fields is devoted to answering these questions. $\square$

3. UNIQUENESS AND EXISTENCE OF FINITE FIELDS

3.1 Uniqueness of GF(p^n)

On the way to proving the uniqueness of GF(p^n) we shall obtain some useful information about irreducible polynomials over finite fields. We shall require the following

Lemma 1.   Let $a_1(x), a_2(x), \ldots, a_L(x) \in F[x]$ be distinct irreducible factors of $f(x) \in F[x]$. Then

$$\prod_{l=1}^{L} a_l(x) \mid f(x).$$

Proof.   Lemma 5(b) of Section IV.4.2 + induction. $\square$

Let GF(q) be a given ground field and GF(q^n) its extension (q as always denotes a prime power). We now ascertain the irreducible factorization of the polynomial $x^{q^n-1} - 1$ (i) over GF(q^n) and (ii) over GF(q).

For notational simplicity, we let $Q = q^n - 1$ in what follows.

Proposition 1.   Let GF(q^n) have distinct elements $\alpha_0 = 0, \alpha_1, \ldots, \alpha_Q$. Then over GF($q^n$), $x^Q - 1$ has the irreducible factorization

$$x^Q - 1 = \prod_{i=1}^{Q} (x - \alpha_i).$$

Proof.   Each α_i is a root of $x^Q - 1$ (Lemma 1 of Section 2.2); hence by the Factor Theorem, $x^Q - 1$ has a linear factor $x - \alpha_i$ for each $\alpha_i \in$ GF(q^n)*. Thus by Lemma 1,

$$x^Q - 1 = c(x) \prod_{i=1}^{Q} (x - \alpha_i).$$

In order for the degrees and leading coefficients of both sides to match, we must have $c(x) = 1$. $\square$

Proposition 2.   Let $m_1(x), \ldots, m_L(x)$ be the distinct minimum polynomials over GF(q) of the elements of GF(q^n)*. Then over GF(q), $x^Q - 1$ has the irreducible factorization

$$x^Q - 1 = \prod_{l=1}^{L} m_l(x).$$

Proof. Each $m_l(x)$ is the minimum polynomial over $GF(q)$ of some $\alpha \in GF(q^n)^*$, which α is also a root of $x^Q - 1$. Hence $m_l(x) \mid x^Q - 1$ by Proposition 1(c) of Section 1.2. Then, since each $m_l(x)$ is irreducible, $\prod_{l=1}^{L} m_l(x) \mid x^Q - 1$ by Lemma 1, so that

$$(*) \qquad\qquad x^Q - 1 = c(x) \prod_{l=1}^{L} m_l(x).$$

From $(*)$, $\deg[\prod_l m_l(x)] \leq Q$. But each of the Q elements of $GF(q^n)^*$ is a root of some $m_l(x)$, hence a root of $\prod_l m_l(x)$. Thus $\deg[\prod_l m_l(x)] \geq Q$, making $\deg[\prod_l m_l(x)] = Q$ and forcing $c(x)$ to be a constant. But this constant must be unity for the righthand side of $(*)$ to be monic. $\square$

Comparing the factorization of $x^Q - 1$ over $GF(q^n)$ and over $GF(q)$, we conclude the

Corollary. Let $\alpha \in GF(q^n)$ have minimum polynomial $m_\alpha(x)$ over $GF(q)$. Then $m_\alpha(x)$ has distinct roots, all of which lie in $GF(q^n)$.

Example 1. Using the information about $GF(2^4)$ given in Fig. 1 of Section 2.2, we can apply Propositions 1 and 2 to immediately write down the irreducible factorizations of $x^{15} - 1 = x^{2^4-1} - 1$ (i) over $GF(2^4)$ and (ii) over $GF(2)$.
(i) Over $GF(2^4)$,

$$x^{15} - 1 = \prod_{i=0}^{14} (x - \alpha^i)$$

where α is a primitive element of $GF(2^4)$, e.g., a root of $x^4 + x + 1$.
(ii) Over $GF(2)$,

$$x^{15} - 1 = (x + 1)(x^2 + x + 1)(x^4 + x + 1)$$
$$\times (x^4 + x^3 + 1)(x^4 + x^3 + x^2 + x + 1). \quad \square$$

Now for the main result of this section: any two finite fields with the same number of elements are isomorphic. We have already shown that any finite field has p^n elements for some prime p and positive integer n (Theorem 4 of Section 2.2). The proof which follows makes interesting use of unique factorization in a Euclidean domain.

Theorem 3 (*Uniqueness of* $GF(p^n)$). Any two fields with p^n elements are isomorphic.

Proof. If F is a field with p^n elements, then F has characteristic p (Exercise 2), hence is an extension field of $\mathbf{Z}_p$. By Proposition 2, $x^{p^n-1} - 1$ has the irreducible factorization over $\mathbf{Z}_p$

$$(*) \qquad\qquad x^{p^n-1} - 1 = \prod_{l=1}^{L} m_l(x)$$

where the $m_l(x)$'s are the distinct minimum polynomials over $\mathbf{Z}_p$ of the elements of F^*.

Choosing α to be a primitive element of F, we have by Theorem 3 of Section 1.2

$$F = \mathbf{Z}_p(\alpha) \cong \mathbf{Z}_p[x]/(m_\alpha(x)).$$

Now suppose G is also a field with p^n elements. By exactly the same reasoning as above, $x^{p^n-1} - 1$ has the irreducible factorization over $\mathbf{Z}_p$

$$(**) \qquad\qquad x^{p^n-1} - 1 = \prod_{l=1}^{\bar{L}} \bar{m}_l(x)$$

where the $\bar{m}_l(x)$'s are the distinct minimum polynomials over $\mathbf{Z}_p$ of the elements of G^*.

But $(*)$ and $(**)$ yield two factorizations of $x^{p^n-1} - 1$ into irreducible monic polynomials over $\mathbf{Z}_p$. By unique factorization in the Euclidean domain $\mathbf{Z}_p[x]$, the family of $m_l(x)$'s must coincide with the family of $\bar{m}_l(x)$'s. In particular, $m_\alpha(x)$ must equal $m_\beta(x)$ for some $\beta \in G$, whence by Theorem 3 of Section 1.2,

$$\mathbf{Z}_p(\beta) \cong \mathbf{Z}_p[x]/(m_\alpha(x)).$$

Since $\mathbf{Z}_p[x]/(m_\alpha(x)) \cong F$ has p^n elements, so does the subfield $\mathbf{Z}_p(\beta)$ of G. Hence $\mathbf{Z}_p(\beta)$ must be all G. Combining the above results,

$$F = \mathbf{Z}_p(\alpha) \cong \mathbf{Z}_p[x]/(m_\alpha(x)) \cong \mathbf{Z}_p(\beta) = G,$$

giving $F \cong G$. $\square$

3.2   Existence of GF(p^n)

We require a preliminary result concerning pth powers in a domain of characteristic p.

Lemma 1.   In a domain D of characteristic $p \neq 0$ (hence p is prime),

$$(a+b)^p = a^p + b^p.$$

Proof.   By the binomial theorem (Exercise 6 of Section IV.1) we have

$$(*) \qquad (a+b)^p = a^p + \binom{p}{1}a^{p-1}b + \cdots + \binom{p}{p-1}ab^{p-1} + b^p.$$

We now claim that for $1 \leq k \leq p-1$, the binomial coefficient

$$\binom{p}{k} = \frac{p(p-1)(p-2)\cdots(p-k+1)}{k!}$$

has p as a factor. First we note that $k!$ divides $p(p-1)\cdots(p-k+1)$, since $\binom{p}{k}$ is an integer. But $(k!,p) = 1$, since p is prime. Hence $k!$ divides $(p-1)\cdots(p-k+1)$ by Lemma 5(a) of Section IV.4.2, giving $\binom{p}{k} = mp$ for some integer m. But for any x in a domain of characteristic p, $mp \cdot x = m \cdot (p \cdot x) = m \cdot 0 = 0$. Thus every term on the righthand side of $(*)$ vanishes except a^p and b^p. $\square$

Corollary. $(a + b)^{p^n} = a^{p^n} + b^{p^n}$.

Proof. Lemma 1 + induction on n. $\square$

Now for the main result.

Theorem 1. For every prime p and positive integer n there is a field with p^n elements.

Proof. Consider the polynomial $f(x) = x^{p^n} - x \in \mathbf{Z}_p[x]$. From Theorem 4 of Section 1.1, $f(x)$ has a root field: an extension E of $\mathbf{Z}_p$ that contains all the roots of $f(x)$ and such that no proper subfield of E also contains all the roots of $f(x)$. We now show that the root field of $f(x)$ yields a field with p^n elements.

First, by Theorem 3 of Section 1.1, $f(x) = x^{p^n} - x$ has p^n roots, not necessarily distinct, in its root field. But

$$f'(x) = p^n x^{p^n - 1} - 1 = -1 \neq 0,$$

so by Theorem 4 of Section IV.3.3, the p^n roots of $f(x)$ *are* distinct.

Next we claim that the roots of $f(x) = x^{p^n} - x$ form a field. If r_1, r_2 are roots of $f(x)$, then

$$\begin{aligned}
f(r_1 + r_2) &= (r_1 + r_2)^{p^n} - (r_1 + r_2) \\
&= r_1^{p^n} + r_2^{p^n} - (r_1 + r_2) \qquad \text{Corollary to Lemma 1} \\
&= r_1 + r_2 - (r_1 + r_2) \qquad \text{r_1, r_2 are roots of $x^{p^n} - x$} \\
&= 0,
\end{aligned}$$

so that the roots of $f(x)$ are closed under $+$. It is even easier to check that the roots are closed under the other field operations as well.

Thus the p^n roots of $f(x)$ themselves form a field, clearly the smallest field containing the roots of $f(x)$. We have therefore constructed a field with p^n elements: the root field of $x^{p^n} - x$ over $\mathbf{Z}_p$. (Note 1.) $\square$

Other important properties of finite fields are established in the exercises.

EXERCISES

Section 1

1. Show that the following numbers are algebraic over $\mathbf{Q}$: (a) $\sqrt{2}\,i$, (b) $\sqrt[5]{1 + \sqrt{3}}$, (c) $\sqrt{2} + \sqrt{3}$, (d) $\sqrt{2} + \sqrt[3]{2}$.

2. Let E be an extension field of F, $\alpha \in E$. Show that $F[\alpha] = F(\alpha)$. (*Recall:* $F[\alpha]$ is the smallest extension *ring* of F that contains α; $F(\alpha)$ is the smallest extension *field* of F that contains α.)

3. Let E be a root field for $f(x) \in F[x]$. Show that in E, $f(x)$ "splits" into linear factors,

$$f(x) = c \prod_i (x - \alpha_i) \qquad (c \in F).$$

(Because of this result, a root field is also called a "splitting field.")

4. Show that for every positive integer n, a root field for the polynomial $x^n - 1$ over $\mathbf{Q}$ can be obtained as a simple algebraic extension of $\mathbf{Q}$. (Use your knowledge of complex numbers!)

5. Show that $\mathbf{Q}(\sqrt{2}, \sqrt{3}) = \mathbf{Q}(\sqrt{2} + \sqrt{3})$.

6. Establish the isomorphisms:
    (a) $\mathbf{C} \cong \mathbf{R}[x]/(x^2 + 1)$;
    (b) $\mathbf{Q}(\sqrt{5}) \cong \mathbf{Q}[x]/(x^2 - 5)$.

7. Prove Theorem 4 of Section 1.2.

8. Is $F(x) = F(x^2, x^3)$? Is $F(x) = F(x^4, x^6)$?

Sections 2 and 3

1. Find all primitive elements of $\mathbf{Z}_{11}$.

2. Show that if F is a finite field with p^n elements, then F has characteristic p.

3. (a) Prove the uniqueness of pth roots in $GF(p^n)$.
    (b) Derive a formula for computing the pth root of any $\alpha^i \in GF(p^n)^*$, where α is a primitive element of $GF(p^n)$. [*Hint:* Exploit $\alpha^{p^n} = \alpha$.]
    (c) Compute:
        (i) $\sqrt{\alpha^{19}}$ where α is a primitive element of $GF(2^5)$.
        (ii) $\sqrt[3]{\alpha^5}$ where α is a primitive element of $GF(3^4)$.

4. (*Subfields of* $GF(q^n)$.)
    (a) (*Preliminary result.*)   Prove: For positive integers m, n, and q

$$m \mid n \Rightarrow (x^{q^m - 1} - 1) \mid (x^{q^n - 1} - 1).$$

(b) Prove that $GF(q^m)$ is a subfield of $GF(q^n)$ if and only if $m \mid n$. [*Hint for "if" part:* Argue that each root of $x^{q^m - 1} - 1$ is a root of $x^{q^n - 1} - 1$.]

5. Prove: If $f(x)$ is a polynomial over $GF(q)$, then

$$f(x)^{q^k} = f(x^{q^k}) \qquad (k = 1, 2, 3, \dots).$$

Conclude that if α is a root of $f(x)$, then $\alpha^q, \alpha^{q^2}, \alpha^{q^3}, \dots$ are also roots of $f(x)$.

6. (*Roots of irreducible polynomials over* $GF(q)$.)   Let $m(x)$ be a monic irreducible polynomial of degree n over $GF(q)$.
    (a) Argue that $m(x)$ has a root, call it α, in $GF(q^n)$ and that $GF(q^n)$ is the smallest extension of $GF(q)$ that contains α.
    (b) (*Key result for computing minimum polynomials.*)   Prove that all the roots of $m(x)$ are given by $\alpha, \alpha^q, \dots, \alpha^{q^{n-1}}$, and that these roots are distinct. [*Hint for proving distinctness:* assume otherwise and show that α would then lie in $GF(q^r)$ with $r < n$, contradicting (a).] Conclude that an irreducible polynomial over a finite field has only simple roots.

    (c) Prove that the roots of $m(x)$ all have the same multiplicative order e in $GF(q^n)$ where e is the least positive integer such that $m(x) | x^e - 1$. (Assume $m(x) \neq x$, which is the minimum polynomial of 0.)

7. (*Application of Exercise* 6(b).)

    (a) Let α be a root of the primitive polynomial $x^4 + x + 1$ over $GF(2)$. Compute the minimum polynomial over $GF(2)$ of (i) α^7, (ii) α^{10}. (Cf. Example 1 of Section 2.2.)

    (b) Let α be a primitive element of $GF(3^4)$. What are the roots of the minimum polynomial of α^2 (i) over $GF(3)$, (ii) over $GF(3^2)$?

8. Let α be a root of $x^2 + 2x + 2 \in GF(3)[x]$.

    (a) Show that α is a primitive element of $GF(3^2)$.

    (b) Determine the minimum polynomial over $GF(3)$ of each element of $GF(3^2)$.

    (c) Factor $x^9 - x$ over $GF(3)$.

9. Establish an explicit isomorphism between $\mathbf{Z}_2[x]/(x^4 + x + 1)$ and $\mathbf{Z}_2[x]/(x^4 + x^3 + x^2 + x + 1)$.

NOTES

Section 1

1. The demonstrations that e and π are transcendental—by Hermite in 1873 (e) and by Lindemann in 1822 (π)—were landmark achievements of nineteenth century mathematics. See Chapter 5, Section 2, of Herstein (1964) for a proof of the transcendence of e and an idea of the kind of the mathematics involved in proving transcendentality.

    When we refer to the familiar elementary functions $\log x$, e^x, $\sin x$, etc., as *transcendental* functions, much the same idea is at work as with transcendental numbers. A function $\mathbf{R} \rightarrow \mathbf{R}$ is said to be a *transcendental* function if it is transcendental over the ring $P_{\mathbf{R}}$ of all polynomial functions $\mathbf{R} \rightarrow \mathbf{R}$. Thus $y = f(x) \in \mathbf{R}^{\mathbf{R}}$ is transcendental if it satisfies no polynomial of the form $a(x, y) = \Sigma_i a_i(x) y^i$, $a_i(x) \in \mathbf{R}[x]$; otherwise $f(x)$ is called an *algebraic* function.

    For example, $y = \sqrt[3]{x^2 + 1}$ is clearly an algebraic function, for we have $y^3 - x^2 - 1 = 0$. On the other hand, just as with transcendental numbers, it is not so easy to demonstrate the transcendentality of functions, even the elementary ones. See R. W. Hamming [*Amer. Math. Monthly* **77** (1970), 294–297] for an interesting article on this problem.

2. See Section 9.1 of Goldstein (1973).

Section 2

1. The crowning achievement of Evariste Galois (1811–1832) was his "Galois theory" of equations in which he established necessary and sufficient conditions for a polynomial equation to be solvable in exact terms, i.e., by rational operations and root extraction. His brilliant theory, entirely group-theoretic in nature, marked the beginning of group theory (the term "group" is due to Galois) and as such was an early landmark in the development of modern algebra.

It is remarkable that Galois, whose life was so short and so miserable, could have accomplished so much. For a brief but readable account of Galois' life and work, see Chapter 16 of E. Kramer, *The Nature and Growth of Modern Mathematics* (New York: Hawthorn Books, 1970).

Section 3

1. For a quite different approach to proving the existence of $GF(p^n)$ (combinatorial rather than algebraic) see R. C. Mullin [*Amer. Math. Monthly* **71** (1964), 901–902].

ARITHMETIC IN EUCLIDEAN DOMAINS

Integers and polynomials are the basic objects of algebraic computing. Thus we are interested in computational aspects of Euclidean domains.

In Section 1 our viewpoint is domain-specific, where we examine from a complexity viewpoint the familiar algorithms for polynomial and integer arithmetic. In the last two sections, our viewpoint is more abstract. In the setting of an arbitrary Euclidean domain, we derive, analyze, and apply what is undoubtedly the most venerated algorithm of computing: Euclid's algorithm for computing greatest common divisors.

1. COMPLEXITY OF INTEGER AND POLYNOMIAL ARITHMETIC

The classical pencil-and-paper algorithms for adding, subtracting, multiplying, and dividing integers and polynomials are ingrained in the consciousness of every mathematically literate person. Our goal here is to analyze the computational complexity of these algorithms, specifically to establish bounds on the computing time required by these algorithms as a function of the size of the inputs. Of course what we have in mind when carrying out such an analysis is the implementation and large scale application of these algorithms on a computer.

John D. Lipson, Elements of Algebra and Algebraic Computing. ISBN 0-201-04115-4.

In analyzing the complexity of algorithms, it is convenient, and usually sufficiently informative, to specify computing times only up to order of magnitude, i.e., up to a constant factor. The "big oh" notation lets us do this: If f is a real-valued function of positive integers $m, n, \ldots$, we write $O(f(m, n, \ldots))$ ("order $f(m, n, \ldots)$") to stand for a quantity which is bounded by $C|f(m, n, \ldots)|$, where C is a positive real constant independent of $m, n, \ldots$. For example, if $T(m, n) = (n + 1)^2(2m - 1)$, then $T(m, n) = O(n^2 m)$, since

$$(n + 1)^2(2m - 1) = 2n^2 m (1 + 1/n)^2 (1 - 1/2m) \leq 8n^2 m.$$

If $T(m, n)$ above is the time to execute an algorithm whose inputs are characterized by two independent size parameters m and n, then the $O(n^2 m)$ bound tells us that the computation time grows quadratically with n and linearly with m. Thus doubling n will result in a roughly fourfold increase in computing time, whereas doubling m will roughly only double the computing time. It is this kind of broad pattern of computational behavior that the big oh notation nicely reveals.

Occasionally we may wish to retain the constant of the highest order terms, in particular when comparing algorithms with the same order of magnitude time bounds. The big oh notation can be used here as well to suppress lower order (uninteresting) information. For example, for $T(m, n)$ above we have

$$T(m, n) = 2n^2 m + O(mn).$$

$$* \quad * \quad *$$

Remark.   We assume familiarity with the "school algorithms" for polynomial and integer arithmetic. Accordingly, our derivations of these algorithms are only as detailed as is required to obtain computing time bounds.

1.1   Polynomial Arithmetic

Let $a(x) = \sum_{i=0}^{m} a_i x^i$ and $b(x) = \sum_{i=0}^{n} b_i x^i$ be polynomials over a field F. We are interested in deriving bounds, as functions of the degrees of $a(x)$ and $b(x)$, on the times required to compute $a(x) \pm b(x)$, $a(x)b(x)$, and $q(x), r(x)$ satisfying the Division Property: $a(x) = b(x)q(x) + r(x)$, $\deg r(x) < n$.

We assume that operations over the given field F can be carried out in time bounded by a constant. For example F might be a fixed precision floating point implementation of the real field as to be found on any general purpose computer. Or, more germane to our pursuits, F might be a finite field $\mathbf{Z}_p$ where p is a prime that fits within a computer word—a "wordsize" prime. In either case we have a field in which elementary operations can be computed in constant $[O(1)]$ time.

Polynomial addition.   Let $a(x) = \sum_{i=0}^{m} a_i x^i$, $b(x) = \sum_{i=0}^{n} b_i x^i$, and suppose that $m \geq n$. Then

$$a(x) + b(x) = a_m x^m + \cdots + a_{n+1} x^{n+1} + (a_n + b_n)x^n + \cdots + (a_0 + b_0).$$

The number of additions is $n + 1$. But if we make the conservative assumption that our polynomial addition algorithm has to at least access the coefficients $a_m, \ldots, a_{n+1}$, then the computation time for polynomial addition will grow with m rather than with n. A symmetric argument holds for the case $n \geq m$, and in either case we can state

Proposition 1. Let $A(m, n)$ be the time required to add (or subtract) two polynomials having degrees m and n. Then $A(m, n) = O(\max(m, n))$.

Polynomial multiplication. Let $a(x) = \sum_{i=0}^{m} a_i x^i$, $b(x) = \sum_{i=0}^{n} b_i x^i$ have degrees m and n. Then $a(x)b(x)$ has degree $m + n$. The amount of work required to compute the $m + n + 1$ coefficients of $a(x)b(x)$ is clearly indicated by the pencil-and-paper scheme used to collect like terms:

$$
\begin{array}{cccccccc}
 & & \times & \times & \times & \times & \times & & (m = 4) \\
 & & & & \times & \times & \times & & (n = 2) \\
\hline
 & & \times & \times & \times & \times & \times & & \\
 & \times & \times & \times & \times & \times & & & \\
\times & \times & \times & \times & \times & & & & \\
\hline
\times & \times & \times & \times & \times & \times & \times & &
\end{array}
$$

Each coefficient of $a(x)$ is multiplied by each coefficient of $b(x)$. Hence there are $(m + 1)(n + 1)$ coefficient multiplications [and slightly fewer coefficient additions—$(m + 1)(n + 1) - (m + n - 1)$ to be exact]. We therefore conclude

Proposition 2. Let $M(m, n)$ be the time required to multiply two polynomials having degrees m and n. Then $M(m, n) = O(mn)$.

Remark. The argument leading up to Proposition 2 might lead one to conjecture that *any* algorithm for multiplying $a(x)$ by $b(x)$ must at the very least multiply every coefficient of $a(x)$ by every coefficient of $b(x)$. This would imply that every polynomial multiplication algorithm requires at least $O(mn)$ time and would in turn allow us to conclude the optimality (up to a constant) of the classical algorithm. However, the conjecture is false! We shall eventually (Chapter IX) derive an algorithm that runs in time much faster than $O(mn)$, at least in the important case $m \approx n$.

Polynomial division (with remainder). Here we divide $a(x)$ (degree m) by $b(x)$ (degree n) to obtain the unique quotient $q(x)$ and remainder $r(x)$ satisfying the Division Property:

$$a(x) = b(x)q(x) + r(x), \qquad \deg r(x) < n.$$

If $m < n$ then $q(x) = 0$ and $r(x) = a(x)$, with no need for any computation. We therefore restrict our attention to the interesting case $m \geq n$.

Let us examine the layout of the first step of polynomial "long division" (the classical division algorithm):

$$
\begin{array}{r}
q_{m-n}x^{m-n} \\
\hline
b_n x^n + \cdots + b_0 \overline{\smash{\big)}\, a_m x^m + a_{m-1}x^{m-1} + \cdots + a_{m-n}x^{m-n} + \cdots + a_0} \\
c_{m-1}x^{m-1} + \cdots + c_{m-n}x^{m-n} \\
\hline
d_{m-1}x^{m-1} + \cdots + d_{m-n}x^{m-n}
\end{array}
$$

Each coefficient of $b(x)$ (except the leading one) is multiplied by $q_{m-n} = b_n^{-1}a_m$ to compute the c_i's. This requires $n+1$ multiplications (counting the one to compute $q_{m-n} = b_n^{-1}a_m$) and then n subtractions to compute the d_i's. This step is repeated for each of the $m-n+1$ terms of $q(x)$, giving an overall total of $(n+1)(m-n+1)$ multiplications and $n(m-n+1)$ subtractions (plus one division to compute b_n^{-1}). We therefore conclude

Proposition 3. Let $D(m,n)$ be the time required to divide a degree m polynomial by a degree n polynomial $(m \geq n)$ to produce a quotient of degree $m-n$ and a remainder of degree $< n$. Then $D(m,n) = O(n(m-n))$.

Corollary. The time to divide $a(x)$ by $b(x)$, computing $q(x)$ and $r(x)$ satisfying the Division Property

$$ a(x) = b(x)q(x) + r(x), \qquad \deg r(x) < \deg b(x), $$

is essentially the time required to compute the product $b(x)q(x)$.

Addition and multiplication of multivariate polynomials. An r-variate polynomial $a(x_1, \ldots, x_r)$ over a field F can be regarded as a polynomial $a(x_1, \ldots, x_r) = \sum_i a_i x_r^i$ in one variable x_r whose coefficients a_i are polynomials in $x_1, \ldots, x_{r-1}$; i.e., $a_i = a_i(x_1, \ldots, x_{r-1})$. In essence, we regard r-variate polynomials as members of $F[x_1, \ldots, x_{r-1}][x_r]$. (We have already encountered the case $r = 2$ in Example 5 of Section IV.1.1.)

The formulas for addition and multiplication,

$$ c(x) = a(x) + b(x) \iff c_k = a_k + b_k, $$

$$ c(x) = a(x)b(x) \iff c_k = \sum_{i+j=k} a_i b_j, $$

lead naturally to recursive extensions of the classical univariate addition and multiplication algorithms. If the coefficients a_k, b_k above are polynomials in $r > 0$ variables, then the indicated coefficient additions and multiplications are carried out by recursion; if $r = 0$, then these coefficient operations are operations in the ground field F.

For simplicity, let us assume that $a(x_1, \ldots, x_r)$ and $b(x_1, \ldots, x_r)$ have degrees m and n respectively in each of their r variables. We note that a polynomial $a(x_1, \ldots, x_r)$ of degree m in each of r variables has $(m+1)^r$ terms when expanded into the form

$$a(x_1, \ldots, x_r) = \sum_{e \in N^r} a_e x_1^{e_1} x_2^{e_2} \cdots x_r^{e_r}.$$

Addition. By the same argument leading to Proposition 1, the time to compute $a(x_1, \ldots, x_r) \pm b(x_1, \ldots, x_r)$ will grow no faster than the maximum number of terms in either of $a(x_1, \ldots, x_r)$ or $b(x_1, \ldots, x_r)$. Thus we have

Proposition 4. Let $A(m, n, r)$ be the time to add (or subtract) two polynomials having degrees m and n in each of the same r variables. Then $A(m, n, r) = O(\max(m, n)^r)$.

Multiplication. Again, by the same argument leading to Proposition 2, the time to compute $a(x_1, \ldots, x_r)b(x_1, \ldots, x_r)$ will essentially be the time to multiply each of the $(m + 1)^r$ terms of $a(x_1, \ldots, x_r)$ by each of the $(n + 1)^r$ terms of $b(x_1, \ldots, x_r)$. We conclude

Proposition 5. Let $M(m, n, r)$ be the time to multiply two polynomials having degrees m and n in the same r variables. Then $M(m, n, r) = O((mn)^r)$.

We refer the reader to Note 1 for a discussion of the computer representation of univariate and multivariate polynomials.

1.2 Integer Arithmetic

In this section dealing with the representation and manipulation of integers, we restrict our attention to unsigned integers. The accommodation of signs causes no essential difficulties.

Our standard decimal representation of integers is intimately connected with polynomials; the m-digit decimal integer $a_{m-1} \cdots a_1 a_0$ represents the quantity

$$a_{m-1} 10^{m-1} + \cdots + a_1 10 + a_0,$$

i.e., the value of the polynomial $\sum_{i=0}^{m-1} a_i x^i$ at $x = 10$. (Example: $347 = 3 \times 10^2 + 4 \times 10 + 7$.)

This polynomial (positional) representation generalizes to any integral base $B \geq 2$. The m-digit base B number $(a_{m-1} \cdots a_1 a_0)_B$ (where each a_i is a base B digit, $0 \leq a_i < B$) represents the quantity

$$a_{m-1} B^{m-1} + \cdots + a_1 B + a_0,$$

i.e., the value of $a(x) = \sum_{i=0}^{m-1} a_i x^i$ at $x = B$. [Example: $(347)_8 = 3 \times 8^2 + 4 \times 8 + 7 = 231$ decimal.]

In a computer setting these ideas about the base B representation of integers find application in the following way. The computer's hardware can only perform operations on integers whose size does not exceed the wordsize W of the machine. To accommodate integers having size $> W$, we must provide for the representation and manipulation of *multiple-precision* integers—integers of the form

$$a = (a_{m-1} \cdots a_1 a_0)_B$$

where the base B is chosen $\leq W$. The base B digits of such an integer a are then stored in m storage locations, and operations on these digits can be performed by the machine's hardware. This puts us in a position to implement the classical algorithms for adding, subtracting, multiplying, and dividing integers, only our digits are base B ($B \gg 10$) rather than decimal, thus providing our computer with a *multiple-precision integer arithmetic* capability.

For a fixed base B, the number of digits m of an integer $a = a_{m-1} \cdots a_1 a_0$ [$= (a_{m-1} \cdots a_1 a_0)_B$] is called the *precision*, or *length*, of a. If an integer a has length m, then clearly $a < B^m$; conversely, if $a < B^m$, then the length of a can be taken to be $\leq m$.

A particularly convenient choice for the base B is a power of 10, the largest power of 10 that is $\leq W$. The power of 10 choice neatly solves the problem of converting integers from decimal to base B on input and from base B to decimal on output. To illustrate, suppose that our computer has a 9 bit (= binary digit) word, with one bit taken up for the sign. This gives $W = 2^8 - 1 = 127$, and we choose $B = 100$. Then, for example, the decimal integer 34709 is converted on input to the three digit base 100 number 03 47 09—a trivial conversion. On output the conversion back to decimal is equally trivial.

We now turn to the problem of establishing bounds on the time required to add, subtract, multiply, and divide base B (unsigned) integers. The complicating issue that we are confronted with in the integer case (which we did not have in the polynomial case) is the problem of analyzing the effects of carry propagation. The major thrust of our analysis is to show that carries do not become too large even when manipulating integers of arbitrary size.

Addition. Let a and b be base B (positive) integers of length m and n respectively, say $m \geq n$. Let $c = a + b$. Since $a, b < B^m$, it follows that $c < 2B^m$. The length of c, therefore, is $\leq m + 1$. Thus we can cast the addition process in the familiar format

$$a_{m-1} \cdots a_1 a_0$$
$$\underline{+ \ b_{m-1} \cdots b_1 b_0} \qquad (b_{m-1}, b_{m-2}, \ldots, b_n = 0)$$
$$c_m c_{m-1} \cdots c_1 c_0$$

The computation of the c_i's is carried out according to the algorithm of Fig. 1, which we have expressed in an Algol-like language (Note 2).

We assume that our language admits the representation of "small" integers ($< B^2$) and their manipulation by operations $+$, $-$, $\times$, div, and mod. The operations div and mod yield the quotient and remainder in accordance with the integer Division Property: if $a \operatorname{div} b = q$ and $a \operatorname{mod} b = r$, then $a = bq + r$ with $0 \leq r < |b|$.

```
            begin
1.              γ := 0;
2.              for   i := 0 until m − 1 do
                begin
3.                  t := a_i + b_i + γ;
4.                  c_i := t mod B;
5.                  γ := t div B
                end;
6.              c_m := γ
            end
```

Fig. 1   Base B addition algorithm.

Let γ_j be the value of the variable γ after j iterations of the **for** loop $(0 \leq j \leq m)$. γ_j $(j \geq 1)$ is then the carry from the computation of digit c_{j-1} (line 4), and $\gamma_0 = 0$ (line 1). We now claim that each γ_j is either 0 or 1. Proof: $\gamma_0 = 0$ by line 1. Inductively, if $\gamma_j \leq 1$ $(0 \leq j < m)$, then

$$\gamma_{j+1} = (a_j + b_j + \gamma_j) \operatorname{div} B$$
$$\leq [2(B-1) + 1]/B$$
$$= (2B - 1)/B < 2,$$

which establishes the claim.

Hence each c_i is computed by two additions of base B digits (line 3) followed by a division by B of a quantity $t \leq 2B - 1$ (lines 4 and 5). Thus each of the $m + 1$ digits of c can be computed in time bounded by a constant (i.e., a quantity independent of m or n), from which we conclude

Proposition 1. Let $A(m, n)$ be the time required to add an m precision integer to an n precision integer. Then $A(m, n) = O(\max(m, n))$.

We leave it as an easy exercise to verify that the bound of Proposition 1 applies to the subtraction of an m precision integer from an n precision integer (Exercise 2).

Multiplication.   Let a and b be base B integers of length m and n, and let $c = ab$. Then $c < B^{m+n}$, so that the length of c is $\leq m + n$. In symbolic terms we have

$$a_{m-1} \cdots a_1 a_0$$
$$\times b_{n-1} \cdots b_1 b_0$$
$$\overline{c_{m+n-1} \cdots\cdots c_1 c_0}$$

Let $b^{(k)} = b_{k-1} \cdots b_0$ $(0 < k \leq n)$, $b^{(0)} = 0$, and let $c^{(k)}$ be the partial product

$$ab^{(k)} = c^{(k)}_{m+k-1} \cdots c^{(k)}_1 c^{(k)}_0 \qquad (0 < k \leq n).$$

---

begin
    for $i := 0$ **until** $m - 1$ **do** $c_i^{(0)} := 0$;
    for $k := 0$ **until** $n - 1$ **do**
        β: compute digits $c_i = c_i^{(k+1)}$ of $ab^{(k+1)}$ from
            digits $c_i = c_i^{(k)}$ of $ab^{(k)}$
end
$\{ab = ab^{(n)} = c_{m+n-1} \cdots c_1 c_0\}$

Fig. 2a High-level base B multiplication algorithm.

The complete product $c = ab = ab^{(n)}$ is formed by computing in turn the partial products $ab^{(k)}$ $(k = 1, \dots, n)$ as in Fig. 2a.

To refine statement β (Fig. 2a), we note that

$$c^{(k+1)} = ab^{(k+1)}$$
$$= a\big(b_k B^k + b^{(k)}\big)$$
$(*)$
$$= ab_k B^k + c^{(k)}.$$

From $(*)$ we see that the digits $c_0, c_1, \dots, c_{k-1}$ of $c^{(k+1)}$ are unaltered from those of $c^{(k)}$ and that the remaining digits c_{k+i} of $c^{(k+1)}$ $(i = 0, 1, \dots, m - 1)$ can be computed according to the algorithm of Fig. 2b, which yields the desired refinement of β.

Now for the analysis of our base B multiplication algorithm. In Fig. 2b, for fixed k, let γ_j be the value of γ after j iterations of the **for** loop $(0 \le j \le m)$. γ_j $(j \ge 1)$ is then the carry from the computation of digit c_{k+j-1} (line 4), where $\gamma_0 = 0$ (line 0). The claim now is that each $\gamma_j < B$. Proof: $\gamma_0 = 0$ by line 1.

       β: **begin**
1. $\gamma := 0$;
2.   **for** $i := 0$ **until** $m - 1$ **do**
             begin
3. $t := a_i \times b_k + c_{k+i} + \gamma$;
4. $c_{k+i} := t \bmod B$;
5. $\gamma := t \operatorname{div} B$
             end;
            $c_{k+m} := \gamma$
       end

Fig. 2b Computation of $c^{(k+1)} = ab^{(k+1)}$ from $c^{(k)} = ab^{(k)}$.

Inductively, if $\gamma_j < B$ $(0 \le j < m)$, then

$$\gamma_{j+1} = (a_j \times b_k + c_{k+j} + \gamma_j) \,\text{div}\, B$$
$$\le \left[(B-1)^2 + 2(B-1) \right]/B$$
$$= (B^2 - 1)/B < B,$$

which establishes the claim.

Hence each c_{k+i} $(0 \le i < m)$ is computed by one multiplication and two additions (line 3) and a division (lines 4 and 5) of quantities all bounded by $(B-1)^2 + 2(B-1) = B^2 - 1$—a constant. Thus the time to compute the m digits c_{k+i}, $0 \le i < m$, of $ab^{(k+1)}$ from the digits c_i of $ab^{(k)}$ is $O(m)$.

The algorithm of Fig. 2b is executed for $k = 0, 1, \ldots, n-1$ (Fig. 2a). Thus the execution time to compute the complete product $ab = ab^{(n)}$ is $O(mn)$, which we record as

Proposition 2. Let $M(m, n)$ be the time required to multiply an m precision integer by an n precision integer. Then $M(m, n) = O(mn)$.

We note that the above algorithm for integer multiplication differs from the usual pencil-and-paper method. In the modified algorithm an addition of a partial product occurs immediately after it is formed, whereas in the pencil-and-paper method all partial products are first computed before they are added. We illustrate:

Pencil-and-paper	Modified (product digits underlined)
7 4 7 8	7 4 7 8
4 6 2 5	4 6 2 5
3 7 3 9 0	3 7 3 9 0
1 4 9 5 6	1 4 9 5 6
4 4 8 6 8	1 8 6 9 5
2 9 9 1 2	4 4 8 6 8
3 4 5 8 5 7 5 0	4 6 7 3 7
	2 9 9 1 2
	3 4 5 8 5

The pencil-and-paper algorithm has the obvious disadvantage that storage is required for all the partial products—storage that grows as $m \times n$. The modified algorithm, on the other hand, requires no such auxiliary storage.

Division. Let a and b be base B integers having length m and n. Here we divide a by b to obtain q and r satisfying the Division Property:

$$a = bq + r, \qquad 0 \le r < b.$$

If $a < b$, then no computation is required: $q = 0$ and $r = a$ satisfy the Division Property. So we assume that $a \geq b$ (hence that $m \geq n$).

First we bound the length ℓ of the quotient q. Since $B^{m-1} \leq a < B^m$ and $B^{n-1} \leq b < B^n$, it follows that

$$q \leq a/b < B^m/B^{n-1} = B^{m-n+1};$$

hence $\ell \leq m - n + 1$.

As we did with polynomial division, let us examine the first step of integer long division, which yields the high order digit of the quotient.

$$
\begin{array}{r}
q_{m-n} \\
b_{n-1} \cdots b_0 \overline{)a_m \cdots a_{m-n} \cdots a_1 a_0} \\
c_m \cdots c_{m-n} \\
\hline
d_m \cdots d_{m-n}
\end{array}
$$

q_{m-n} must be chosen to satisfy

$$0 \leq (d_m \cdots d_{m-n})_B < (b_{n-1} \cdots b_0)_B.$$

The c_i's above are obtained by multiplying a base B digit (q_{m-n}) by a length n integer (b), and the d_i's are obtained by subtracting two integers of length $\leq n + 1$. Hence the time to perform the first stage of integer long division is $O(n)$. (This time bound holds even if the proper choice of q_{m-n} is determined by trial and error—there are only B choices!)

The work involved in carrying out this first stage is repeated for each of the $m - n + 1$ digits of q. Thus we conclude

Proposition 3.   Let $D(m, n)$ be the time required to divide an m precision integer by an n precision integer. Then $D(m, n) = O(n(m - n))$.

Corollary.   The time to divide a by b, computing q and r satisfying the Division Property

$$a = bq + r, \qquad 0 \leq r < b,$$

is essentially the time required to compute the product bq.

Remark.   The fact that the $O(n(m - n))$ time bound is insensitive to the particular trial and error method used to determine the successive digits of the quotient should serve as a warning that a lot of computational sins may be obscured by a big oh analysis of an algorithm. Of course any decent implementation of integer long division will take care to determine a good approximation to the correct quotient digit based on a trial division of the current dividend by the first (very) few leading digits of the divisor (Note 3).

2.   COMPUTATION OF GREATEST COMMON DIVISORS

One of our major results about Euclidean domains is that a Euclidean domain D is a g.c.d. domain: any $a, b \in D$ with a, b not both zero have a g.c.d. g, expressible in the form $g = sa + tb$ (Theorem 1 of Section IV.4.2). Our goal now is to derive Euclid's algorithm for computing g and then to derive Euclid's so-called extended algorithm for computing s and t as well.

Our derivations take place in the setting of an abstract Euclidean domain. We assume that our (Algol-like) language allows for the manipulation of values from an arbitrary Euclidean domain D with degree function d. In particular we assume that our language provides a *Division Algorithm* in the form of two operations "div" and "mod" which return, respectively, a preferred quotient and remainder in accordance with the Division Property of a Euclidean domain: If $a, b \in D$, $b \neq 0$, then $a \operatorname{div} b = q$ and $a \bmod b = r$ satisfy $a = bq + r$, with $d(r) < d(b)$ or $r = 0$. We shall also assume that distinct values in the range of the mod m function are inequivalent, so that

$$(*) \qquad\qquad a \equiv a' \pmod{b} \quad \Rightarrow \quad a \bmod b = a' \bmod b.$$

This condition eliminates, for example, the likes of a mod function for **Z** that returns 2 for $8 \bmod 3$ and -1 for $11 \bmod 3$.

For the Euclidean domain **Z**, we define $a \operatorname{div} b$ and $a \bmod b$ to return, respectively, the unique quotient q and remainder r that satisfy $a = bq + r$, $0 \le r < |b|$. And for the Euclidean domain $F[x]$, we define $a(x) \operatorname{div} b(x)$ and $a(x) \bmod b(x)$ to return, respectively, the unique quotient $q(x)$ and remainder $r(x)$ that satisfy $a(x) = b(x)q(x) + r(x)$, $\deg r(x) < \deg b(x)$. We note that these mod functions for **Z** and $F[x]$ both satisfy the above requirement $(*)$.

Note. Our derivation will make extensive use of the ideas presented in the Appendix (The Invariant Relation Theorem) of this chapter. At this point the reader should peruse the important ideas presented there.

2.1 Derivation of Euclid's Algorithm

Our derivation of Euclid's algorithm shall exploit

Proposition 1 *(Useful properties of the gcd)*. In a Euclidean domain D,

(a) $\gcd(a, 0) \sim a$;
(b) If $b \neq 0$, $\gcd(a, b) = \gcd(b, a \bmod b)$.

Note. Recall that $\gcd(a, b)$ denotes the unique preferred g.c.d. of a and b (e.g., $\gcd(a, b) > 0$ in **Z**; $\gcd(a(x), b(x))$ is monic in $F[x]$). Since any associate of a g.c.d. of a and b is a g.c.d. of a and b (Proposition 4 of Section IV.4.1), $\gcd(a, b) \sim c$ means that c is a g.c.d. of a and b.

Proof of Proposition 1. (a) By a straightforward appeal to the definition of g.c.d., a is a g.c.d. of a and 0.

(b) If $b \neq 0$, then $a \bmod b = r$ is defined and satisfies the Division Property $a = bq + r$. Now any common divisor c of a and b is also a common divisor of b and $a - bq = r$; conversely, any common divisor of b and r is also a common divisor of $bq + r = a$ and b. Since a, b and b, r have the same common divisors, they must have the same *greatest* common divisors. Hence $\gcd(a, b) = \gcd(b, r)$.

$\square$

Now for our derivation of Euclid's algorithm, which affords an important example of the application of the Invariant Relation (IR) Theorem to algorithm development.

The key idea behind Euclid's algorithm for computing a g.c.d. of a, b is the manipulation of two auxiliary variables $a0, a1$ in a loop

$$\ell: \textbf{while } a1 \neq 0 \textbf{ do } \beta$$

in such a way that

$$p: \gcd(a, b) = \gcd(a0, a1)$$

is maintained as a loop invariant of ℓ.

With p so maintained, on termination of ℓ we can conclude p and $\neg \gamma$ (IR Theorem), where γ is the **while** condition "$a1 \neq 0$." This gives

$$\gcd(a, b) = \gcd(a0, a1) \qquad \text{by } p$$

$$= \gcd(a0, 0) \qquad a1 = 0 \, (\neg \gamma)$$

$$\sim a0 \qquad \text{Proposition 1(a),}$$

so that $a0$ can be returned as a g.c.d. of a and b.

Using just the above considerations, we can give a high level (= partially developed) version of Euclid's algorithm, which we present in the form of a procedure (Note 1).

> **procedure** GCD(a, b);
> **begin**
> 1.　　initialize $a0, a1$ so that p holds;
> 2.　　**while** $a1 \neq 0$ **do**
> 3.　　β: redefine $a0$ by $a0'$, $a1$ by $a1'$ so that p is maintained
> 　　　　and measurable work is done;
> 4.　　**return** $a0$
> **end**.

To complete the development of GCD, we must refine lines 1 and 3.

Line 1 (initialization): Set $a0$ to a, $a1$ to b; then p holds trivially.

Line 3 ($\{p \text{ and } \gamma\} \, \beta \, \{p\}$): Before execution of the loop body β, we can assume

$$(*) \qquad p \text{ and } \gamma: \quad \gcd(a, b) = \gcd(a0, a1) \text{ and } a1 \neq 0.$$

Let β redefine the variables $a0, a1$ by the values $a0', a1'$. We now exploit Proposition 1(b). If we let

$$a0' = a1, \qquad a1' = a0 \bmod a1$$

(noting that $a0 \bmod a1$ is defined since $a1 \neq 0$), then $\gcd(a0, a1) = \gcd(a0', a1')$. From $(*)$ we then have $\gcd(a, b) = \gcd(a0', a1')$. The invariance of p is thus maintained by β.

Moreover, measurable work is done in executing β. We have $d(a1') < d(a1)$, so that execution of β decreases the degree of $a1$ by at least one. Since either $0 \le d(a1) \le d(b)$ or $a1 = 0$, β is executed at most $d(b) + 1$ times before $a1 = 0$.

We have now derived and simultaneously proved the correctness, including termination, of the following algorithm, Euclid's algorithm, for computing g.c.d.'s in a Euclidean domain.

Note. For convenience we shall manipulate $a0$ and $a1$ as an ordered pair $(a0, a1)$. In an ordered pair assignment $(x, y) := (u, v)$, the right-hand side expressions u and v are both computed before being assigned to the left-hand side variables x and y. Thus $(x, y) := (u, v)$ is equivalent to the sequence of scalar assignments

$$\text{temp} := u; \quad y := v; \quad x := \text{temp}.$$

For example, $(x, y) := (y, x)$ interchanges the values of the variables x and y.

Algorithm 1 (*Euclid's algorithm over a Euclidean domain D*).

Input. $a, b \in D$, not both zero.
Output. A g.c.d. of a, b.

```
procedure GCD(a, b);
begin
        (a0, a1) := (a, b);
        {Loop invariant: gcd(a, b) = gcd(a0, a1)}
        while a1 ≠ 0 do
                (a0, a1) := (a1, a0 mod a1);
        return a0
end □
```

This concludes our "modern" derivation of Euclid's algorithm, an algorithm that is over two thousand years old.

Example 1. Let us execute GCD over $\mathbf{Z}$ with $a = 228$, $b = 612$. The following values for $a0, a1$ are computed:

iteration	$a0$	$a1$
0	228	612
1	612	228
2	228	156
3	156	72
4	72	12
5	12	0

GCD returns $12 = \gcd(228, 612)$.

Note that $GCD(a, b)$ works properly if $a < b$ (as was the case above) but that one iteration can always be saved by invoking $GCD(a, b)$ with $a \ge b$. □

Example 2. We execute GCD over $\mathbf{Q}[x]$ with $a(x) = x^3 - 2$, $b(x) = 2x^2 - 3$.

iteration	$a0$	$a1$
0	$x^3 - 2$	$2x^2 - 3$
1	$2x^2 - 3$	$\frac{3}{2}x - 2$
2	$\frac{3}{2}x - 2$	$\frac{5}{9}$
3	$\frac{5}{9}$	0

Here GCD returns $\frac{5}{9} \sim 1 = \gcd(x^3 - 2, 2x^2 - 3)$. Thus $x^3 - 2$ and $2x^2 - 3$ are relatively prime. (Note that GCD does not in general return the preferred monic g.c.d.) $\square$

Example 3. With $a(x)$ and $b(x)$ as in Example 2, we execute GCD this time over $\mathbf{Z}_5[x]$.

iteration	$a0$	$a1$
0	$x^3 + 3$	$2x^2 + 2$
1	$2x^2 + 2$	$4x + 3$
2	$4x + 3$	0

Here GCD returns $4x + 3 \sim x + 2 = \gcd(x^3 + 3, 2x^2 + 2)$. (Thus $x^3 - 2$ and $2x^2 - 3$ are not relatively prime over $\mathbf{Z}_5$—cf. Example 2.) $\square$

2.2 Analysis of Euclid's Algorithm over Z and F[x]

With $a_0 = a$, $a_1 = b$, Euclid's algorithm $\mathrm{GCD}(a, b)$ computes a *remainder sequence* $a_2, a_3, \ldots, a_{n+1} = 0$ according to

$$a_2 = a_0 - a_1 q_1, \qquad d(a_2) < d(a_1),$$

$$a_3 = a_1 - a_2 q_2, \qquad d(a_3) < d(a_2),$$

$$\vdots$$

$$a_{i+1} = a_{i-1} - a_i q_i, \qquad d(a_{i+1}) < d(a_i),$$

$$\vdots$$

$$a_{n+1} = a_{n-1} - a_n q_n = 0.$$

$a_{i+1} = a_{i-1} \bmod a_i$ is the value of $a1$ after the ith iteration; a_n, the last nonzero remainder, is the returned g.c.d. of a, b.

The main complexity issue concerning Euclid's algorithm is the number of iterations—the number n in the above scheme—that are required to compute a g.c.d. of two elements a and b. Our argument that the **while** loop in GCD terminates gives us the following bound:

Proposition 1. Let D be an arbitrary Euclidean domain, and let $a, b \in D$ satisfy $N \geq d(a) \geq d(b)$. Then Euclid's algorithm requires at most $N + 1 = O(N)$ iterations.

The integer case. A glance at Example 1 suggests that the $O(N)$ bound given by Proposition 1 for the number of iterations required by Euclid's algorithm to compute a g.c.d. of integers $0 \leq a$, $b \leq N$ is unduly pessimistic. We now show that in the integer case a much smaller bound applies.

For convenience we assume that a and b are nonnegative integers. (There is no loss of generality, since $\gcd(a, b) = \gcd(|a|, |b|)$.) With $a_0 = a$, $a_1 = b$, Euclid's algorithm computes an integer remainder sequence $a_2, a_3, \ldots, a_{n+1} = 0$ according to

$$a_2 = a_0 - a_1 q_1, \qquad a_2 < a_1,$$

$$\vdots$$

$$(*) \qquad\qquad a_{i+1} = a_{i-1} - a_i q_i, \qquad a_{i+1} < a_i,$$

$$\vdots$$

$$a_{n+1} = a_{n-1} - a_n q_n = 0.$$

Here $a_n = g = \gcd(a, b)$. We also note that since $a_1 > a_2 > \cdots > a_n > a_{n+1} = 0$, the quotients $q_i = a_{i+1} \text{ div } a_i$ must satisfy $q_i \geq 1$ for $1 \leq i \leq n - 1$, and $q_n \geq 2$.

From $(*)$ we have

$$a_{i-1} = a_i q_i + a_{i+1} \qquad (1 \leq i \leq n - 1)$$

$$> a_{i+1} q_i + a_{i+1}$$

$$= a_{i+1}(q_i + 1).$$

Hence

$$\prod_{i=1}^{n-1} a_{i-1} > \prod_{i=1}^{n-1} a_{i+1}(q_i + 1).$$

Cancelling $\prod_{i=2}^{n-2} a_i$ from both sides gives

$$a_0 a_1 > a_{n-1} a_n \prod_{i=1}^{n-1} (q_i + 1),$$

$$= a_n^2 q_n \prod_{i=1}^{n-1} (q_i + 1).$$

Substituting $a_0 = a$, $a_1 = b$, $a_n = g$, and using $q_i \geq 1$ for $1 \leq i \leq n - 1$ and $q_n \geq 2$, we obtain

$$ab > g^2 2^n \geq 2^n.$$

If $a \geq b$, then $a^2 \geq ab > 2^n$, or $n < 2 \log a$, and we have proved the following

Theorem 2. Let $N \geq a \geq b \geq 0$. Then Euclid's algorithm computes $\gcd(a, b)$ in $< 2 \log N = O(\log N)$ iterations. (Note 2.)

The polynomial case. Let $a(x), b(x) \in F[x]$ with $N \geq \deg a(x) \geq \deg b(x)$. The bound of Proposition 1—$N + 1$ iterations for Euclid's algorithm to compute a g.c.d. of $a(x), b(x)$—is actually tightest possible. But it hardly tells the whole story, since successive iterations will involve divisions of smaller and smaller degree polynomials. A much more realistic reflection of the time to compute polynomial g.c.d.'s is given by the time required to carry out the coefficient operations. This we now derive.

With $a_0 = a(x)$, $a_1 = b(x)$, Euclid's algorithm $\mathrm{GCD}(a(x), b(x))$ computes a polynomial remainder sequence $a_2, a_3, \ldots, a_{N+1} = 0$ according to

$$a_2 = a_0 - a_1 q_1, \qquad \deg a_2 < \deg a_1,$$

$$\vdots =$$

$(*)$
$$a_{i+1} = a_{i-1} - a_i q_i, \qquad \deg a_{i+1} < \deg a_i,$$

$$\vdots =$$

$$a_{N+1} = a_{N-1} - a_N q_N = 0.$$

The ith iteration requires the division of a_{i-1} by a_i in order to obtain the remainder a_{i+1}. Using ordinary polynomial division, the time T_i to perform the division at the ith iteration is essentially the time to compute the product $a_i q_i$ (Corollary to Proposition 3, Section 1.1). Thus

$$T_i \leq c(\deg a_i)(\deg q_i)$$

for some constant c. Hence a bound on the total time T to perform Euclid's algorithm is given by

$$T = \sum_{i=1}^{N} T_i \leq c \sum_{i=1}^{N} (\deg a_i)(\deg q_i)$$

$$\leq cN \left(\sum_{i=1}^{N} \deg q_i \right).$$

Now from $(*)$ we have

$$\deg a_{i-1} = \deg a_i + \deg q_i, \qquad 1 \leq i \leq N.$$

Hence

$$\sum_{i=1}^{N} \deg q_i = \sum_{i=1}^{N} (\deg a_{i-1} - \deg a_i)$$

$$= \deg a_0 - \deg a_N$$

$$\leq \deg a_0 \leq N.$$

Thus $T \leq cN^2$, and we have proved

Theorem 3. For $a(x), b(x) \in F[x]$, let $N \geq \deg a(x) \geq \deg b(x)$. Then Euclid's algorithm computes a g.c.d. of $a(x), b(x)$ in time $T = O(N^2)$, assuming that coefficient operations take constant $[O(1)]$ time.

2.3 Euclid's Extended Algorithm

In a Euclidean domain D, a g.c.d. g of $a, b \in D$ is expressible in the form

$$g = sa + tb.$$

Our goal now is to extend Euclid's algorithm (procedure GCD of Algorithm 1, Section 2.1) to compute s and t as well as g.

As with our derivation of the ordinary (unextended) Euclid's algorithm, our main tool is the loop invariant concept. We introduce two pairs of variables $(s0, s1), (t0, t1)$ into procedure GCD and manipulate them to maintain the invariance of

$$p: a0 = s0 \times a + t0 \times b, \quad \text{and}$$
$$a1 = s1 \times a + t1 \times b.$$

A high level version of

$$\text{EUCLID}(a, b, g, s, t),$$

which for inputs a, b yields g, s, t such that g is a g.c.d. of a, b and $g = sa + tb$, is given by

procedure EUCLID(a, b, g, s, t);
begin
 $(a0, a1) := (a, b)$;
α. $(s0, s1) := (s0', s1')$;
 $(t0, t1) := (t0', t1')$;
 {Loop invariant: $a0 = s0 \times a + t0 \times b$,
         $a1 = s1 \times a + t1 \times b$}
 while $a1 \neq 0$ **do**
  $(a0, a1) := (a1, a0 \bmod a1)$;
β. $(s0, s1) := (s0', s1')$;
  $(t0, t1) := (t0', t1')$
 end;
 $g := a0$;
 $s := s0$;
 $t := t0$
end.

Since $(a0, a1)$ is manipulated just as in GCD, the value of $a0$ on termination of EUCLID's **while** loop is a g.c.d. of a, b. By the invariance of p, we have

$$a0 = s0 \times a + t0 \times b.$$

Hence on returning from EUCLID, g, s, and t satisfy

$$g = s \times a + t \times b,$$

where g is a g.c.d. of a, b. Thus our partially developed EUCLID works correctly.

To complete the derivation of EUCLID, we determine values $s0', s1', t0', t1'$ at α and β such that the invariance of p is maintained at both places.

At α (initialization): on entry to the **while** loop, we have $(a0, a1) = (a, b)$, $(s0, s1) = (s0', s1')$, $(t0, t1) = (t0', t1')$. Since

$$a = a0 = 1 \times a + 0 \times b,$$
$$b = b0 = 0 \times a + 1 \times b,$$

p will obviously hold if we let

$$s0' = 1, \qquad t0' = 0, \qquad s1' = 0, \qquad t1' = 1.$$

At β: Here we assume that p holds at the top of the **while** loop, i.e.,

$$(*)\qquad \begin{aligned} a0 &= s0 \times a + t0 \times b, \quad \text{and} \\ a1 &= s1 \times a + t1 \times b. \end{aligned}$$

After execution of the loop body we have $(a0, a1) = (a0', a1')$, $(s0, s1) = (s0', s1')$, $(t0, t1) = (t0', t1')$, where

$$(**)\qquad \begin{aligned} a0' &= a1, \\ a1' &= a0 \bmod a1 \\ &= a0 - a1 \times q, \quad \text{where } q = a0 \operatorname{div} a1. \end{aligned}$$

Now let us determine $s0', s1', t0', t1'$ so that p holds for the primed values. We have

$$\begin{aligned} a0' &= a1 &&\text{from } (**) \\ &= s1 \times a + t1 \times b &&\text{from } (*), \end{aligned}$$

and

$$\begin{aligned} a1' &= a0 - a1 \times q &&\text{from } (**) \\ &= s0 \times a + t0 \times b \\ &\quad -(s1 \times a + t1 \times b)q &&\text{from } (*) \\ &= (s0 - s1 \times q)a + (t0 - t1 \times q)b. \end{aligned}$$

Hence p is maintained if we let

$$s0' = s1, \qquad\qquad t0' = t1,$$
$$s1' = s0 - s1 \times q, \qquad t1' = t0 - t1 \times q.$$

We have thus derived and simultaneously proved the correctness of

Algorithm 1   (*Euclid's Extended Algorithm over a Euclidean domain D*).

Input. $a, b \in D$, not both zero.
Output. g, s, t such that g is a g.c.d. of a, b and $g = sa + tb$.

```
procedure EUCLID(a, b, g, s, t);
begin
        (a0, a1) := (a, b);
        (s0, s1) := (1, 0);
        (t0, t1) := (0, 1);
        {Loop invariant:  a0 = s0×a + t0×b,
                          a1 = s1×a + t1×b}
        while a1 ≠ 0 do
        begin
                q := a0 div a1;
                (a0, a1) := (a1, a0 − a1×q);
                (s0, s1) := (s1, s0 − s1×q);
                (t0, t1) := (t1, t0 − t1×q)
        end;
        g := a0;
        s := s0;
        t := t0
end  □
```

Of course we could replace the assignment for $(a0, a1)$ in Euclid's **while** loop by $(a0, a1) := (a1, a0 \bmod a1)$ as in procedure GCD of Section 2.1. The **while** loop of EUCLID as written is intended to emphasize the complete symmetry between the manipulations of $(a0, a1)$ and those of $(s0, s1)$ and $(t0, t1)$.

Example 1. Let us execute EUCLID over **Z** with $a = 228$, $b = 612$, thus "extending" Example 1 of Section 2.1. The following values are computed:

iteration	q	a0	a1	s0	s1	t0	t1
0	—	228	612	1	0	0	1
1	0	612	228	0	1	1	0
2	2	228	156	1	− 2	0	1
3	1	156	72	− 2	3	1	− 1
4	2	72	12	3	− 8	− 1	3
5	6	12	0	− 8	51	3	− 19

EUCLID returns $g = \gcd(228, 612) = 12$, $s = -8$, and $t = 3$. Check: $12 = -8 \times 228 + 3 \times 612$. □

Example 2. We extend Example 3 of Section 2.1 by executing EUCLID over $\mathbf{Z}_5[x]$ with $a(x) = x^3 + 3$, $b(x) = 2x^2 + 2$.

iteration	q	$a0$	$a1$	$s0$	$s1$	$t0$	$t1$
0	—	$x^3 + 3$	$2x^2 + 2$	1	0	0	1
1	$3x$	$2x^2 + 2$	$4x + 3$	0	1	1	$2x$
2	$3x + 4$	$4x + 3$	0	1	$2x + 1$	$2x$	$4x^2 + 2x + 1$

EUCLID returns $g = 4x + 3 \sim x + 2 = \gcd(x^3 + 3, 2x^2 + 2)$, $s = 1$, $t = 2x$. Check: $4x + 3 = 1(x^3 + 3) + (2x)(2x^2 + 2)$. □

As for the complexity of EUCLID, since the manipulations of $a0, a1$ in Euclid's extended algorithm are the same as those in Euclid's (ordinary) algorithm, it follows that Theorem 2 of Section 2.2 applies with equal force to the extended case:

Theorem 1. For $a, b \in \mathbf{Z}$, let $N \geq a \geq b \geq 0$. Then Euclid's extended algorithm applied to a, b requires $< 2 \log N = O(\log N)$ iterations.

Moreover, it can be shown that when EUCLID is applied to integers a and b with $N \geq a \geq b \geq 0$, the values of $a0, a1, s0, s1, t0, t1$ are all bounded by N (Exercise 2). Hence all values generated by EUCLID are single-precision provided a and b are. Hence further, each of the $O(\log N)$ iterations requires constant $[O(1)]$ time, and we conclude

Theorem 2. For $a, b \in \mathbf{Z}$, let $N \geq a \geq b \geq 0$. If N is a single-precision integer, then Euclid's extended algorithm applied to a, b requires $O(\log N)$ time.

In the polynomial case, it can readily be shown that the manipulations of $(s0, s1)$ and $(t0, t1)$ do not affect the time bound of Theorem 3 of Section 2.2 (Exercise 3), letting us conclude

Theorem 3. For $a(x), b(x) \in F[x]$, let $N \geq \deg a(x) \geq \deg b(x)$. Then Euclid's extended algorithm requires $O(N^2)$ time, assuming that coefficient operations take constant $[O(1)]$ time.

3. COMPUTATION OF MOD m INVERSES

In this section we investigate the solutions of the congruence $au \equiv 1 \pmod{m}$ over a Euclidean domain D, for given $a, m \in D$.

3.1 Theory of mod m Inverses

Over $\mathbf{Z}$ the congruence $5u \equiv 1 \pmod 8$ has a solution $u = 5$, besides others such as $-11, -19, 13, 21$. The congruence $6u \equiv 1 \pmod 8$, on the other hand, has no integer solutions. We want a theorem to explain what is going on.

Theorem 1. Let $a, m \in D$. Then the congruence

$$au \equiv 1 \ (\text{mod } m)$$

has a solution $u \in D$ if and only if $(m, a) = 1$.

Proof. If $(m, a) = 1$, then there exists $s, t \in D$ such that $sm + ta = 1$. Hence $u = t$ satisfies $au \equiv 1 \ (\text{mod } m)$.

Conversely, suppose that $au \equiv 1 \ (\text{mod } m)$ has a solution $u \in D$. Then $au + mv = 1$ for some $v \in D$, whence $(a, m) = 1$ by Exercise 6 of Section IV.4. $\square$

In $D/(m)$, we have $[a + (m)][u + (m)] = au + (m)$. Hence

$$[a + (m)][u + (m)] = 1 + (m) \quad \Leftrightarrow \quad au \equiv 1 \ (\text{mod } m),$$

and we conclude the

Corollary. In $D/(m)$,

$$a + (m) \text{ is invertible} \Leftrightarrow (m, a) = 1.$$

The next theorem tells us how to get all solutions of $au \equiv 1 \ (\text{mod } m)$ from a particular one.

Theorem 2. If $v \in D$ is a particular solution of $au \equiv 1 \ (\text{mod } m)$, then u is a solution if and only if $u \equiv v \ (\text{mod } m)$.

Proof. Let $av \equiv 1 \ (\text{mod } m)$ for some $v \in D$. If $u \equiv v \ (\text{mod } m)$, then $au \equiv 1 \ (\text{mod } m)$, trivially. Conversely, if $au \equiv 1 \ (\text{mod } m)$, then $au \equiv av \ (\text{mod } m)$, so that $m \mid a(u - v)$. But $(m, a) = 1$, so that $m \mid (u - v)$ (Lemma 5 of Section IV.4.2). Hence $u \equiv v \ (\text{mod } m)$, as required. $\square$

We recall the condition that we have imposed (at the beginning of Section 2) on the mod m function for any Euclidean domain:

$$(*) \qquad\qquad a \equiv a' \ (\text{mod } m) \quad \Rightarrow \quad a \bmod m = a' \bmod m.$$

This means that for the congruence $u \equiv a \ (\text{mod } m)$, $a \bmod m$ is the *unique* solution in the range of the mod m function, and we conclude

Corollary 1. Let $m, a \in D$, with $(m, a) = 1$. Then $au \equiv 1 \ (\text{mod } m)$ has a unique solution in the range of the mod m function for D.

We denote this unique solution to $au \equiv 1 \ (\text{mod } m)$ that lies in the range of the mod m function by $a^{-1} \bmod m$.

Our next corollary shows that the argument a of the inverse mod m function, $a^{-1} \bmod m$, can be reduced mod m.

Corollary 2. $a^{-1} \bmod m = (a \bmod m)^{-1} \bmod m$.

Proof. Trivially we have

$$au \equiv 1 \ (\text{mod } m) \quad \Leftrightarrow \quad (a \bmod m)u \equiv 1 \ (\text{mod } m)$$

(substitution property of $\equiv (\text{mod } m)$). Hence the unique solutions to $au \equiv 1$

$(\bmod m)$ and $(a \bmod m)u \equiv 1 \ (\bmod m)$ in the range of the mod m function must be the same. $\square$

The case m a prime is especially important. When m is a prime, then $(a, m) = 1$ if and only if $m \nmid a$, so that $a^{-1} \bmod m$ exists whenever $m \nmid a$. This reflects the fact that if m is prime, then $D/(m)$ is a field: the existence of $a^{-1} \bmod m \ (m \nmid a)$ is tantamount to the existence of a multiplicative inverse for the nonzero coset $a + (m)$ (Corollary to Theorem 1).

3.2 Computation of mod m Inverses

The proof of Theorem 1 of Section 3.1 gives us the key to computing mod m inverses in a Euclidean domain D. For $m, a \in D$, we apply our procedure EUCLID of Section 2.3,

$$\text{EUCLID}(m, a, g, s, t),$$

which returns a g.c.d. g of m and a along with s and t satisfying

$$(m, a) \sim g = sm + ta.$$

By Theorem 1 of Section 3.1, if $g \nsim 1$ (i.e., if g is not a unit in D), then $a^{-1} \bmod m$ does not exist. If $g \sim 1$, then

$$1 = g^{-1}sm + g^{-1}ta,$$

so that $g^{-1}t$ is a solution to the congruence $au \equiv 1 \ (\bmod m)$. Thus $a^{-1} \bmod m$ is given by $g^{-1}t \bmod m$, and we have derived the following

Algorithm 1 (*Computation of $a^{-1} \bmod m$*).

Input. $a, m \in D$, where D is a Euclidean domain.
Output.   If $(m, a) = 1$, then $a^{-1} \bmod m$; otherwise an error message.

> **procedure** INVERSE(a, m);
> **begin**
>   EUCLID(m, a, g, s, t);
>   **if** g is a unit **then**
>   **return** $(g^{-1} \times t) \bmod m$
>   **else**
>   announce "$a^{-1} \bmod m$ does not exist"
> **end**.

Note.   Since only g and t, not s, are required from EUCLID, the computations for $s0, s1$ in EUCLID can be suppressed for the sake of efficiency. $\square$

Since the time to compute INVERSE(a, m) is essentially the time to compute EUCLID(m, a, g, s, t), we conclude from Theorems 2 and 3 of

Section 2.3:

Theorem 1. If a and m are single-precision integers $\leq N$, then INVERSE(a, m) requires $O(\log N)$ time.

Theorem 2. If $a(x), m(x) \in F[x]$ have $\deg a(x), \deg m(x) \leq N$, then INVERSE$(a(x), m(x))$ requires $O(N^2)$ time, assuming that coefficient operations take constant $[O(1)]$ time.

Now for some sample executions and applications of INVERSE.

Example 1. (a) With $m = 21$, $a = 16$, we execute INVERSE(a, m) in order to compute 16^{-1} in $\mathbf{Z}_{21}$. EUCLID$(m, a, g, -, t)$ computes the following tableau (where all computations involving $s0, s1$ have been suppressed).

iteration	q	$a0$	$a1$	$t0$	$t1$
0	—	21	16	0	1
1	1	16	5	1	-1
2	3	5	1	-1	4
3	5	1	0	4	-21

EUCLID thus returns $g = 1$ and $t = 4$. INVERSE then returns $4 \bmod 21 = 4$. Check: $16 \times 4 = 64 \equiv 1 \pmod{21}$, so that $4 = 16^{-1}$ in $\mathbf{Z}_{21}$.

(b) With $m = 21$, $a = 18$, EUCLID$(m, a, g, -, t)$ computes the following tableau:

iteration	q	$a0$	$a1$	$t0$	$t1$
0	—	21	18	0	1
1	1	18	3	1	-1
2	6	3	0	-1	7

EUCLID thus returns $g = 3$. Since 3 is not a unit in $\mathbf{Z}$, INVERSE emits the message "$a^{-1} \bmod m$ does not exist." □

An important application of INVERSE is to compute inverses in the field $\mathbf{Z}_p$ (p prime). When we invoke INVERSE(p, a) with $0 < a < p$, we are assured from the outset that EUCLID returns $g = (p, a) = 1$.

Example 2. With $p = 59$ (a prime), $a = 41$, we execute INVERSE(a, p) in order to compute 24^{-1} in the field $\mathbf{Z}_{59}$. EUCLID$(p, a, g, -, t)$ computes the following tableau:

iteration	q	$a0$	$a1$	$t0$	$t1$
0	—	59	24	0	1
1	2	24	11	1	-2
2	2	11	2	-2	5
3	5	2	1	5	-27
4	2	1	0	-27	59

EUCLID thus returns $g = 1$ and $t = -27$. INVERSE then returns $-27 \bmod 59$ = 32. Check: $32 \times 24 = 768 \equiv 1 \pmod{59}$.

Over $F[x]$, if $m(x)$ is irreducible over F and $\deg a(x) < \deg m(x)$, then INVERSE$(a(x), m(x))$ returns $a(x)^{-1}$ in the field $F[x]_{m(x)} \cong F[x]/(m(x))$.

Example 3. $m(x) = x^5 + x^2 + 1$ is irreducible over $\mathbf{Z}_2$. With $a(x) = x^2 + 1$, we execute INVERSE$(a(x), m(x))$ to compute $(x^2 + 1)^{-1}$ in $\mathrm{GF}(2^5) = \mathbf{Z}_2[x]_{m(x)}$. The following tableau is computed by EUCLID$(m(x), a(x), g, -, t(x))$:

iteration	q	$a0$	$a1$	$t0$	$t1$
0	—	$x^5 + x^2 + 1$	$x^2 + 1$	0	1
1	$x^3 + x + 1$	$x^2 + 1$	x	1	$x^3 + x + 1$
2	x	x	1	$x^3 + x + 1$	$x^4 + x^2 + x + 1$
3	x	1	0	$x^4 + x^2 + x + 1$	$x^5 + x^2 + 1$

EUCLID thus returns $g = 1$, $t(x) = x^4 + x^2 + x + 1$. INVERSE subsequently returns $g^{-1}t(x) \bmod m(x) = t(x) = x^4 + x^2 + x + 1$. Check: $(x^4 + x^2 + x + 1)(x^2 + 1) = x^6 + x^3 + x + 1 \equiv 1 \pmod{x^5 + x^2 + 1}$, so that $x^4 + x^2 + x + 1 = (x^2 + 1)^{-1}$ in $\mathbf{Z}_2[x]_{m(x)}$. □

One of our major results about extension fields is Theorem 3 of Section VI.1.2: If E is some extension field of a given ground field F and $\alpha \in E$ is algebraic over F with minimum polynomial $m_\alpha(x)$ over F, then $F(\alpha)$, the smallest extension of F containing α, is given up to isomorphism by $F[x]/(m_\alpha(x))$. The isomorphism is the natural one: $a(\alpha) \mapsto a(x) + (m(x))$, where $\deg a(x)$ may be taken to be $< \deg m(x)$. Thus $a(\alpha)^{-1}$ in $F(\alpha)$ is essentially given by $a(x)^{-1}$ in $F[x]_{m_\alpha(x)}$, which we can compute by INVERSE as in Example 3.

Example 4. Let us find, in exact terms, the inverse of $2\sqrt[3]{4} - 3$. Letting $\alpha = \sqrt[3]{2}$, then $2\sqrt[3]{4} - 3 = 2\alpha^2 - 3$, which lies in $\mathbf{Q}(\alpha) \cong \mathbf{Q}[x]/(x^3 - 2)$ (Example 2 of Section VI.1.2). Hence $1/(2\sqrt[3]{4} - 3) = a\alpha^2 + b\alpha + c$ for some $a, b, c \in \mathbf{Q}$. To determine a, b, c we execute INVERSE$(a(x), m(x))$ with $a(x) = 2x^2 - 3$, $m(x) = x^3 - 2$.

EUCLID$(m(x), a(x), g, -, t(x))$ computes the following tableau (cf. Example 2 of Section 2.1):

iteration	q	$a0$	$a1$	$t0$	$t1$
0	—	$x^3 - 2$	$2x^2 - 3$	0	1
1	$\frac{1}{2}x$	$2x^2 - 3$	$\frac{3}{2}x - 2$	1	$-\frac{1}{2}x$
2	$\frac{4}{3}x + \frac{16}{9}$	$\frac{3}{2}x - 2$	$\frac{5}{9}$	$-\frac{1}{2}x$	$\frac{2}{3}x^2 + \frac{8}{9}x + 1$
3	$\frac{27}{10}x - \frac{18}{5}$	$\frac{5}{9}$	0	$\frac{2}{3}x^2 + \frac{8}{9}x + 1$	$-\frac{9}{5}$

EUCLID thus returns $g = \frac{5}{9}$, $t(x) = \frac{2}{3} x^2 + \frac{8}{9} x + 1$. INVERSE subsequently returns $g^{-1}t(x) \bmod m(x) = \frac{6}{5} x^2 + \frac{8}{5} x + \frac{9}{5} = (2x^2 - 3)^{-1}$ in $\mathbf{Q}[x]_{m(x)}$, which yields the solution to our original problem: with $\alpha = \sqrt[3]{2}$

$$(2\alpha^2 - 3)^{-1} = \frac{6}{5}\alpha^2 + \frac{8}{5}\alpha + \frac{9}{5}. \quad \Box$$

APPENDIX

THE INVARIANT RELATION THEOREM

The Invariant Relation (IR) Theorem of this appendix gives us a powerful mental aid to reason about computer programs—program loops in particular.

Assertions about program variables. Let σ be a program statement (command) and let p and q be assertions about the values of variables manipulated by the program in which σ appears. We write

$$\{p\}\,\sigma\,\{q\}$$

to mean: if p holds before execution of σ, then q holds after execution of σ. In this context we refer to p as the *pre-condition* of σ and to q as the *post-condition* of σ.

Examples.

1. "$\{i < n\}\ i := i + 1\ \{i \leq n\}$" is true;
2. "$\{i < n\}\ i := i + 1\ \{i < n\}$" is false;
3. "$\{i < n\}$ **begin** $i := i + 1;\ n := n + 1$ **end** $\{i < n\}$" is true.

The* while *statement. The **while** statement, or loop,

$$\textbf{while } \gamma \textbf{ do } \sigma$$

John D. Lipson, Elements of Algebra and Algebraic Computing. ISBN 0-201-04115-4.

causes the statement σ to be repeatedly executed while the predicate γ (the "loop condition") is true, in accordance with the following flow diagram:

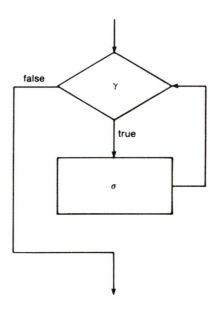

We refer to the **while** statement as an *indefinite* iteration statement, in that the number of times the controlled statement σ is executed is not known a priori on entry to the loop. This is in contrast to *definite* iteration as provided by the **for** statement, an example of which is

for $i := 1$ **until** n **do** σ.

We note that indefinite iteration includes definite iteration as a special case. For example, the above **for** statement is equivalent to

$$i := 1;$$
while $i \leq n$ **do**
begin
$$\sigma;$$
$$i := i + 1$$
end.

The Invariant Relation Theorem.   In a **while** loop

while γ **do** σ

we assume that σ cannot cause a premature transfer from the loop; the loop terminates only by having γ become false. Under this stipulation, we can establish the following theorem, the main result of this appendix, which describes the semantics of a **while** loop in terms of assertions (about program variables):

> **Theorem**   (*Invariant Relation Theorem*).   Let ℓ be a **while** loop
>
> $$\ell: \textbf{while } \gamma \textbf{ do } \sigma$$
>
> and let p be a program assertion. If
>
> 1. (a) p holds on entry to ℓ,
>   (b) $\{p \text{ and } \gamma\}\,\sigma\,\{p\}$,
> 2. the loop ℓ terminates,
>
> then p and $\neg\gamma$ hold on exit from ℓ ($\neg\gamma$ means "not γ").

Proof of the IR Theorem.   By hypothesis 2 the loop terminates, say after n iterations for some $n \geq 0$. Since termination can only occur by having γ false, we have the "$\neg\gamma$" part of the desired conclusion p and $\neg\gamma$. We now establish the "p" part by showing the *invariance* of p: that p holds after k iterations for $k = 0, 1, \ldots, n$. The proof is by induction on k.

Basis $(k = 0)$. p holds by hypothesis 1(a).

Induction.   Assume that p holds after k iterations for $0 \leq k < n$. Since $k < n$, γ must be true at the beginning of the $(k + 1)$st iteration. By hypothesis 1(b), p holds at the end of the $(k + 1)$st iteration, which completes the proof of the invariance of p and completes the proof of the IR theorem. $\square$

Remarks.   1. The assertion p in the IR theorem is called an *invariant relation*, or *invariant*, of the given loop, as characterized by hypotheses 1(a) and 1(b). Hence the name *Invariant Relation* Theorem.

2. Some readers might prefer to regard the IR Theorem as an *axiom*, one which *defines* the semantics of the **while** statement. Fine.

Sample applications of the IR Theorem.

Example 1.   Let us verify that the program of Fig. 3, for integers $a \geq 0$ and $b > 0$, correctly computes q and r satisfying the Division Property:

$$a = bq + r, \qquad 0 \leq r < b.$$

---

	begin $\{a \geq 0,\ b > 0\}$
1.	$q := 0;$
2.	$r := a;$
	$\{p: a = bq + r \text{ and } r \geq 0\}$
3.	$\ell: \textbf{while } r \geq b \textbf{ do}$
	begin
4.	$r := r - b;$
5.	$q := q + 1$
	end
	end

Fig. 3   Division program for **N**.

The comments enclosed by braces { } are assertions about program variables. The assertion preceding line 1 is our assumption about the input integers a and b. The assertion preceding the **while** loop ℓ of line 3 is the crucial one. The claim is that the assertion there,

$$p: \quad a = bq + r \text{ and } r \geq 0,$$

is a loop invariant of ℓ.

But before proving the invariance of p, let us jump to the conclusion p and $\neg\gamma$ of the IR Theorem. Here the loop condition γ is "$r \geq b$," so that

$$p \text{ and } \neg\gamma \Rightarrow a = bq + r \text{ and } r \geq 0 \text{ and } r < b$$

$(*)$
$$\Rightarrow a = bq + r \text{ and } 0 \leq r < b.$$

Thus the conclusion of the IR Theorem is precisely the program correctness verification that we are after: after the loop terminates, q and r satisfy the Division Property ($*$).

To complete the verification, we must of course check out the two hypotheses of the IR Theorem.

1. (*Invariance of p.*)
    (a) On entry to ℓ, $q = 0$ and $r = a \geq 0$ by lines 1 and 2, so that p holds trivially.

    (b) ({p and γ} σ {p}.) Assume p and γ before execution of the loop statement σ (lines 4 and 5). We denote the redefined values of r and q after execution of σ by r' and q'. (We must show that p holds for these primed values.) According to σ we have

$$r' = r - b, \qquad q' = q + 1,$$

so that

$$bq' + r' = b(q + 1) + (r - b)$$
$$= bq + r$$
$$= a,$$

the last equality following because by assumption p holds before execution of σ. Also, $r \geq b$ by γ, so that $r' = r - b \geq 0$. Hence p holds after execution of σ. p is thus a loop invariant as claimed.

2. (*Termination.*) Successive values of r, one value per iteration of the loop ℓ, constitute a strictly decreasing sequence of nonnegative integers $\leq a$ (*strictly* decreasing by line 4 and $b > 0$). Any such sequence must be finite, proving termination.

We have just shown that the two hypotheses of the IR Theorem hold, hence the conclusion (p and $\neg\gamma$) holds, which, as we have already seen, lets us conclude the correctness of our division program. (Incidentally this program correctness proof is a perfectly good substitute for the proof of the Division Property that we presented in Example 5 of Section II.2, insofar as the existence of quotients and remainders is concerned.) $\square$

We now change the viewpoint from program verification to program construction (more specifically, loop derivation). It is in this realm that the IR Theorem finds its most rewarding applications.

In broad outline, the overall strategy is to find a relation p and a condition γ such that p and $\neg \gamma$ together yield a desired computational state of affairs. The IR Theorem tells us that p and $\neg \gamma$ will be achieved by executing the following high level program segment:

> initialize variables so that p holds;
> **while** γ **do**
> σ: maintain p as an invariant
>   and do measurable work.

The initialization statement assures that hypothesis 1(a) of the IR Theorem is satisfied; the specification of the loop statement σ assures that hypotheses 1(b) and 2 are satisfied. In particular, what is meant by the "measurable work" part of σ is this: execution of σ brings the loop closer to termination—one step closer!—and termination is assured after finitely many iterations of σ. Since the hypotheses of the IR Theorem are satisfied, we can conclude p and $\neg \gamma$ on exit from the **while** loop, as already claimed.

Let us see how all this works out in practice.

Example 1.   Devise a program to compute

$$y = x^n \qquad (n \geq 0)$$

in a multiplicative monoid. (A simple problem to be sure, one for which we do not need the help of any high-powered mental aids. However, let us derive a program via the IR Theorem—we promise a payoff in the next example.)

The sequence

$$x^n = x^{n-1}x = x^{n-2}x^2 = \cdots = x^0 x^n$$

suggests that we manipulate program valuables u and v so that

$$p: x^n = x^u v \text{ and } u \geq 0$$

is maintained as an invariant of a loop "**while** γ **do** σ." Our task now is to derive the loop, including its initialization.

Loop condition γ.   On termination we want $u = 0$, since the above loop invariant p then gives $x^n = v$; i.e., v yields the desired computational result. Therefore we take the loop condition γ to be "$u \neq 0$."

Initialization.   If $u = n$ and $v = 1$ initially, then p holds trivially on entry to the loop.

Our partially refined program is then as given in Fig. 4.

begin $\{n \geq 0\}$
    $u := n;$
    $v := 1;$
    $\{p:\ x^n = x^u v$ and $u \geq 0\}$
    while $u \neq 0$ **do**
        σ: redefine u, v by u', v' so that p is maintained and measurable
        work is done
end
$\{x^n = v\}$

Fig. 4 Partially refined program for computing x^n.

Now for the refinement of the loop statement σ. Before execution of σ we can assume p and γ, i.e., $x^n = x^u v$ and $u > 0$. Therefore we have

$$x^n = x^u v = x^{\overbrace{u-1}^{u'}} \underbrace{(xv)}_{v'}.$$

As indicated, p will be maintained by σ if σ redefines u and v by $u' = u - 1$ and $v' = x \times v$. (Note in particular that $u > 0$ implies $u' \geq 0$.) Also, σ performs measurable work: $u' = u - 1$ means that the loop terminates after n iterations.

This completes the derivation of the program of Fig. 5, *correct by construction*, for computing x^n.

begin $\{n \geq 0\}$
    $u := n;$
    $v := 1;$
    $\{p:\ x^n = x^u v$ and $u \geq 0\}$
    while $u \neq 0$ **do**
    begin
        $u := u - 1;$
        $v := x \times v$
    end
end
$\{x^n = v\}$

Fig. 5 Simple program for computing x^n.

We have called the program of Fig. 5 "simple," and indeed it is. However, this simple program requires n iterations to compute x^n, which can be prohibitively costly for *very* large values of n, say in the range $10^5 - 10^8$. (We shall encounter this computational problem—enormous values of n—in Chapter IX.) Our goal, therefore, is to speed up our simple program.

Example 2 (*Fast computation of x^n*). We return to the partially refined program of Fig. 4. Our task is clear: to increase the "measurable work" carried out by each iteration of σ.

In our simple program of Fig. 5 we forced the so-called induction variable u to zero by exploiting

$$u = (u - 1) + 1$$

(which led to $x^u v = x^{(u-1)+1} v = x^{u'} v'$ where $u' = u - 1$, $v' = xv$). Rather than subtract one from u, we divide u by two which gives by the Division Property

$$u = 2(u \operatorname{div} 2) + (u \bmod 2).$$

We now generalize the invariant relation p of Example 1 by the introduction of a variable t:

$$p: x^n = t^u v, \qquad u \geq 0.$$

If $u > 0$, we have

$$x^n = t^u v$$
$$= \left\{ t^{[2(u \operatorname{div} 2) + (u \bmod 2)]} \right\} v$$
$$= \underbrace{(t^2)}_{t'}{}^{\,u \operatorname{div} 2} \underbrace{(t^{u \bmod 2} v)}_{v'}.$$

We see that the loop statement σ of Fig. 4 will preserve our new invariant p if σ redefines t, u, v by

$$t' = t^2,$$
$$u' = u \operatorname{div} 2,$$
$$v' = v \quad \text{if } u \text{ is even}$$
$$= tv \quad \text{otherwise.}$$

These considerations lead to the program of Fig. 6, where we have appropriately initialized t. (Note that the new values t', u', v' of t, u, v are defined in terms of the old values. For this reason, in the **while** loop of Fig. 6 it is crucial that the redefinition of v precede the redefinitions of t and u.)

We have called the program of Fig. 6 "fast." Let us see how fast. The program computes a sequence

$$u_0 = n, \quad u_1 = u_0 \operatorname{div} 2, \quad u_2 = u_1 \operatorname{div} 2, \ldots, \quad u_r = u_{r-1} \operatorname{div} 2 = 0,$$

where u_i is the value of u after the ith iteration. Since $u_0 = n \geq 0$ and $u_i < u_{i-1}$ for $i = 1, 2, \ldots, r$, we have correctly indicated that the sequence of u_i's is finite, which is to say that the program loop terminates after r iterations for r some nonnegative integer.

```
begin {n ≥ 0}
      t := x;
      u := n;
      v := 1;
      {p: xⁿ = tᵘv and u ≥ 0}
      while u ≠ 0 do
      begin
            if u is odd then v := t × v;
            t := t × t;
            u := u div 2
      end
      {xⁿ = v}
end
```

Fig. 6   Fast program for computing x^n.

How big can r be? Since $u_r = 0 = u_{r-1}$ div 2 and $u_{r-1} > 0$, it follows that u_{r-1} must equal unity. Since $u_i \geq 2u_{i-1}$, it must be the case that $n = u_0 \geq 2^{r-1}u_{r-1} = 2^{r-1}$. Hence r, the number of iterations of the loop, satisfies the bound

$$r \leq (\log_2 n) + 1.$$

(One can also show that $r > (\log_2 n) - 1$—exercise for the reader.)

Thus our fast program for x^n requires a number of iterations $\approx \log_2 n$, a vast improvement over the n iterations required by the simple program of Example 2. For example, for n in the billion range, the fast algorithm requires only about thirty iterations.

In Fig. 7 we have presented the results of a sample execution of our fast powering algorithm. □

To recap, at the heart of our development of both the simple (slow) program and the sophisticated (fast) program for computing x^n was an invariant

iteration	u	t	v
0	13	2	1
1	6	4	2
2	3	16	2
3	1	256	32
4	0	65536	$8192 = 2^{13}$

Fig. 7   Fast computation of 2^{13}.

relation, the same one in both cases:

$$p: x^n = t^u v \text{ and } u \geq 0$$

($t = x$ in the simple case). The difference between the two programs lies solely in the way and the speed with which the variable u is forced to zero, the desired terminal value.

If there is a larger message in this appendix, especially in the last two examples, it is to stress the importance of focussing on the relations between values of variables rather than the values themselves when verifying or deriving programs. As human beings we have a much better chance of grasping something static (relations among variables) than something ever changing (values of variables). In particular, the idea of a loop invariant is the key to effective mathematical reasoning about program loops, with the framework for this reasoning being provided by the Invariant Relation Theorem. (Note 1.)

EXERCISES

Section 1

1. (*Exact division of multivariate polynomials.*) Let the polynomials $a(x_1,\ldots,x_r)$, $b(x_1,\ldots,x_r)$ over a field F have degrees m and n, respectively, in each of their r variables. Further suppose that $a(x_1,\ldots,x_r)$ divides $b(x_1,\ldots,x_r)$ in $F[x_1,\ldots,x_r]$. Show that $q(x_1,\ldots,x_r) = b(x_1,\ldots,x_r)/a(x_1,\ldots,x_r)$ can be computed in time $O(n^r(m-n)^r)$.

2. Devise and analyze an algorithm for the base B subtraction of a length n integer b from a length m integer a ($a \geq b \geq 0$).

Section 2

1. (*Recursive g.c.d. procedure.*) Prove the correctness of the given recursive procedure GCD; i.e., show that GCD(a,b) returns a g.c.d. of a and b for a and b in an arbitrary Euclidean domain.

> **procedure** GCD(a,b);
> **return if** $b = 0$ **then** a
> **else** GCD($b, a \text{ div } b$).

[*Hint:* Let $p(n)$ be the assertion "GCD(a,b) returns a g.c.d. of a and b for all b with $b = 0$ or $d(b) \leq n$."] Show how GCD computes GCD(228, 612) (cf. Example 1 of Section 2.1).

2. (*G. E. Collins.*) The objective of this exercise is to bound the values that are computed by Euclid's extended algorithm EUCLID (Algorithm 1 of Section 2.3).

EUCLID, when applied to integers a and b (say, for convenience, $a \geq b \geq 0$), computes sequences a_i, s_i, t_i ($i = 1,\ldots,n$) according to

$$a_{i+1} = a_{i-1} - a_i q_i, \qquad a_{i+1} < a_i,$$
$$s_{i+1} = s_{i-1} - s_i q_i,$$
$$t_{i+1} = t_{i-1} - t_i q_i,$$

where $a_0 = a$, $a_1 = b$, $s_0 = 1$, $s_1 = 0$, $t_0 = 0$, $t_1 = 1$. Also, $a_{n+1} = 0$, $a_n = \gcd(a, b)$. Thus a_i, s_i, t_i are the values of $a0, s0, t0$ after the ith iteration; $a_{i+1}, s_{i+1}, t_{i+1}$ are the values of $a1, s1, t1$ after the ith iteration.

(a) (*Preliminary result.*) In any Euclidean domain, show that if $ax = by$ where $(a, b) = 1$, $(x, y) = 1$, then $x = \pm b$, $y = \pm a$.

(b) Let $\alpha_i = t_{i+1} s_i - s_{i+1} t_i$. Show that $\alpha_{i+1} = -\alpha_i$ ($i = 1, \ldots, n$). Conclude that each $\alpha_i = \pm 1$.

(c) Conclude further that $\gcd(s_i, t_i) = 1$ ($i = 1, \ldots, n + 1$).

(d) Let $\gcd(a, b) = d$. Show that $(a/d)s_{n+1} = -(b/d)t_{n+1}$. Conclude that $s_{n+1} = \pm b/d$, $t_{n+1} = \pm a/d$. [*Hint:* Part (a).]

(e) Finally conclude that each a_i, s_i, t_i is $\leq a$ in absolute value. (*Hence all intermediate values generated by* EUCLID *are single-precision provided a is.*)

3. Let $a(x)$, $b(x) \in F[x]$, with $N \geq \deg a(x) \geq \deg b(x)$. In EUCLID show that the computation time (as measured by coefficient operations in F) for the manipulations of $s0, s1, t0, t1$ is $O(N^2)$.

4. Determine all solutions $x, y \in \mathbf{Z}$ to each of
    (a) $35x - 44y = 18$;
    (b) $84x + 54y = -24$.
    [*Hint:* Exercises 7 and 8 of Section IV.4.]

Section 3

1. (a) Compute in $\mathbf{Z}_{78}$: (i) 35^{-1}; (ii) 36^{-1}.
    (b) Compute in $\mathbf{Z}_5[x]_{m(x)}$ where $m(x) = x^3 + 2$:
    (i) $(2x^3 + 3)^{-1}$; (ii) $(2x^2 + 2)^{-1}$.

2. (a) Prove *Fermat's Theorem:* for integers a and p where p is a prime and $p \nmid a$,

$$a^{p-1} \equiv 1 \pmod{p}.$$

    (b) Now devise and analyze an algorithm based on (a) for computing $a^{-1} \bmod p$. (Note 1.)

Appendix

1. (*Division Algorithm for* $R[x]$.) Let R be a commutative ring, and let $a(x) = a_m x^m + \cdots + a_0$ and $b(x) = b_n x^n + \cdots + b_0$ be polynomials in $R[x]$ where b_n is a unit in R. Show that the following program computes $q(x)$ and $r(x)$ satisfying the Division Property,

$$a(x) = b(x)q(x) + r(x) \quad \text{where } \deg r(x) < \deg b(x).$$

```
begin
    q(x) := 0;
    r(x) := a(x);
    k := deg r(x);
    n := deg b(x);
    c := b_n^{-1};
    while k ≥ n do
    begin
        t(x) := cr_k x^{k-n};
```

$$q(x) := q(x) + t(x);$$
$$r(x) := r(x) - b(x)t(x);$$
$$k := \deg r(x)$$

end

end.

[*Hint:* Show that p: "$a(x) = b(x)q(x) + r(x)$" is an invariant relation of the above loop.]

2. (*Integer long division via a loop invariant.*) Refine the following program for computing q and r satisfying the Division Property: $a = bq + r$ with $0 \leq r < b$. (It is assumed that base B representation is used, so that the operations mod B, div B are efficient.)

begin $\{a \geq 0, b > 0\}$
  $r := a;$
  $q := 0;$
  $w := B^n \times b$ where $B^n \times b > a \geq B^{n-1} \times b;$
  $\{p: w \times q + r = a, 0 \leq r < w\}$
  while $w \neq b$ do
  begin
    $w := w$ div $B;$
    manitain invariance of p
  end
end.

Execute your program (i) for $a = 5423$, $b = 17$ ($B = 10$); (ii) for $a = 11101$, $b = 110$ ($B = 2$). Can your program be simplified if $B = 2$? Is there any complication if B is very large (10^5 say)?

3. (*Horner's scheme for polynomial evaluation.*) Show that the following program computes $y = \sum_{i=0}^{n} a_i x^i$ ($n \geq 0$).

begin
  $y := a_n;$
  $i := n;$
  while $i \neq 0$ do
  begin
    $i := i - 1;$
    $y := x \times y + a_i$
  end
end.

[*Hint:* Show that p: "$y = \sum_{j=i}^{n} a_j x^{j-i}$ and $0 \leq i \leq n$" is an invariant relation of the above loop.]

NOTES

Section 1

1. (*Computer representation of polynomials.*) The obvious method of representing a polynomial $a(x) = \sum_{i=0}^{n} a_i x^i$ is by an array of coefficients $a = (a_0 \ a_1 \ \cdots \ a_n)$ (we

use blanks rather than commas as separators). Often this representation is just fine: every general purpose programming language offers arrays as a built-in data structure, and polynomial manipulation algorithms based on an array representation are especially easy to implement—essentially they are transliterations of mathematical formulas $a_i + b_i$, $\Sigma_{i+j=k} a_i b_j$, etc.

In many applications, however, polynomials are frequently *sparse* in the sense of having many missing (zero coefficient) terms, like $a(x) = x^{15} - 4x^3 + 7$. It then behooves us to adopt a representation that exploits this sparseness, so that storage (and time) requirements grow with the number of terms actually occurring rather than with the degree. The solution is obvious: explicitly represent only those terms that occur (much as we write $x^{15} - 4x^3 + 7$ rather than $x^{15} + 0x^{14} + 0x^{13}$ etc.). Thus we write $a(x) = \sum_{i=1}^{n} a_i x^{e_i}$, where $a_i \neq 0$ and (say) $e_1 > e_2 > \cdots > e_n$, and represent $a(x)$ by the array

$$a = (a_1 \, e_1 \, a_2 \, e_2 \, \cdots \, a_n \, e_n).$$

For example, $a(x) = x^{15} - 4x^3 + 7$ is represented by $a = (1 \; 15 \; -4 \; 3 \; 7 \; 0)$.

For multivariable polynomials, this representation can be extended recursively by the useful notion of a *list*. In computing, a *list* over a set S is an array $(a_1 \, a_2 \, \cdots \, a_n)$ where each a_i either is in S or (recursively) is a list over S. The multivariate polynomial $a(x_1, \ldots, x_r) = \sum_{i=0}^{n} a_i(x_1, \ldots, x_{r-1}) x_r^{e_i}$ is represented by the list

$$a = (a_1 \, e_1 \, a_2 \, e_2 \, \cdots \, a_n \, e_n)$$

where each a_i is the representation of $a_i(x_1, \ldots, x_{r-1})$. Thus for $r > 1$, each a_i is itself a list.

For example,

$$a(x,y) = (2x^3 + 1)y^7 + (4x^5 - 5x^2 + 9)y^4 + 1$$

is represented by the list

$$((2 \; 3 \; 1 \; 0) \quad 7 \quad (4 \; 5 \; -5 \; 2 \; 9 \; 0) \quad 4 \quad (1 \; 0) \quad 0).$$

A major algebraic manipulation programing system, SAC-1, adopts such a representation for multivariate polynomials, calling it the *rational canonical form*. See G. E. Collins, *SYMSAM* 2, 144–152.

Other representations of multivariate polynomials are possible. A polynomial $a(x_1, \ldots, x_r)$ can be expressed in the expanded form

$$a(x_1, \ldots, x_r) = \sum_{e \in \mathbb{N}^r} a_e x_1^{e_1} x_2^{e_2} \cdots x_r^{e_r}$$

(cf. form 3 of Example 5, Section IV.1.1). $a(x_1, \ldots, x_r)$ can then be represented by an array $\{(a_e, e) \mid a_e \neq 0\}$ of nonzero coefficients and exponent vectors, in some preferred order. To illustrate, for the above bivariate polynomial $a(x,y)$ we have

$$a(x,y) = 2x^3 y^7 + 1x^0 y^7 + 4x^5 y^4 - 5x^2 y^4 + 9x^0 y^4 + 1x^0 y^0,$$

which can be represented by the array

a_e	e_1	e_2
2	3	7
1	0	7
4	5	4
-5	2	4
9	0	4
1	0	0

The above representation scheme is essentially the one adopted by ALTRAN, another major algebraic manipulation system (see A. D. Hall Jr., *SYMSAM 2*, 153–157; also *Comm. ACM* **14** (1971), 517–521).

Interestingly enough, even after years of experience, there is still no unanimity as to which representation scheme for multivariate polynomials is superior.

2. Algol is an elegant and popular programming language, dating from 1960 [P. Naur, *Comm. ACM* **6** (1963), 1–17]. The major features of our simple Algol-like programming language are illustrated by the program of Fig. 1, Section 1.2:

1.   A sequence of statements

$$\textbf{begin } \sigma1;\sigma2;\ldots;\sigma n \textbf{ end}$$

is executed by executing $\sigma1, \sigma2, \ldots, \sigma n$ in turn. Syntactically, such a sequence behaves as a single statement—a *compound* statement. A program is a statement (usually a compound statement as in Fig. 1 of Section 1.2).

2.   An *assignment statement*

$$\textsf{var := expr}$$

is a command to assign the value of the right-hand side expression to the left-hand side variable.

3.   An *iterative statement*

$$\textbf{for } i := i_1 \textbf{ until } i_2 \textbf{ do } \sigma$$

is a command to execute the statement σ repeatedly, with the *controlled variable* i taking on the values $i_1, i_1 + 1, \ldots, i_2$ in turn. If $i_2 < i_1$, then σ is not executed at all; the iteration statement behaves as a null (no effect) statement. In Fig. 1 of Section 1.2, σ is the compound statement of lines 3–5.

We mention another important kind of statement.

4.   A *conditional statement*

$$\textbf{if } \gamma \textbf{ then } \sigma1 \textbf{ else } \sigma2$$

is a command to execute the statement $\sigma1$ if the predicate γ is true, the statement $\sigma2$ otherwise. A second form is

$$\textbf{if } \gamma \textbf{ then } \sigma$$

which is a command to execute σ if γ is true. The flow of control for these statements is graphically displayed by the following "flow diagrams":

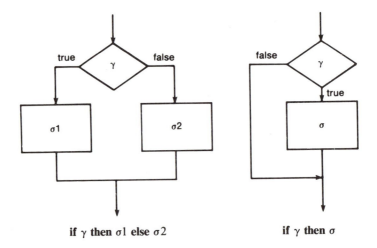

if γ then $\sigma1$ else $\sigma2$ if γ then σ

As well as conditional statements we also have *conditional expressions:*
$$\text{var} := \textbf{if } \gamma \textbf{ then } \textsf{expr1} \textbf{ else } \textsf{expr2}$$
is a command to assign var the value of expr1 if γ is true, the value of expr2 otherwise. For example,
$$x := \textbf{if } a < b \textbf{ then } a \textbf{ else } b$$
assigns x the minimum of a and b.

We shall explain other features of our Algol-like programming language as they are needed.

3. See Knuth (1969), pp. 235 ff., for a method of determining a trial quotient digit based on one leading digit of the divisor and two leading digits of the (current) dividend. This trial quotient is never more than 2 in error, independent of the base B. [Further note: Section 4.3 of Knuth (1969) contains detailed specifications for multiple-precision integer arithmetic algorithms.]

Section 2

1. (*Procedures.*)   A *procedure definition* in our Algol-like programming language has the form
$$\textbf{procedure } \textsf{name}(\textsf{arg1}, \dots, \textsf{argn});$$
$$\sigma.$$
arg1, ..., argn are the *formal arguments* of the procedure name; the *procedure statement* σ specifies the computations to be carried out by name (σ of course may be a compound statement).

The execution of a procedure involves (1) specifying *actual* arguments for the formal arguments and (2) executing the procedure statement using these actual arguments in place of the formal ones. This execution results in the procedure returning values in one of two ways, corresponding to two different kinds of procedures:

1. *Function procedures:* a single value is returned by a **return** statement.

Example.   Let the (very trivial) procedure diff be defined by
$$\textbf{procedure } \text{diff}(a, b);$$
$$\textbf{return } a - b.$$
If $a = 4$ and $b = 3$, then execution of
$$s := \text{diff}(a, b); \quad t := \text{diff}(b, a)$$
results in $s = 1$ and $t = -1$. $\square$

2. *Subroutine procedures:* any number of values are returned by the arguments of the procedure.

Example.   We can organize the above function procedure diff as a subroutine procedure by introducing a "result" or "output" argument, say d:
$$\textbf{procedure } \text{diff}(a, b, d);$$
$$d := a - b.$$
If $a = 4$ and $b = 3$, the execution of
$$\text{diff}(a, b, s); \quad \text{diff}(b, a, t)$$
again results in $s = 1$ and $t = -1$. (Note the method of invocation.) $\square$

As is the case with most modern programming languages, our procedures (either function or subroutine) are permitted to be *recursive* (self-invoking). For example the integer factorial function

$$n! = 1 \qquad \text{if } n = 0$$
$$= n(n-1)! \quad \text{otherwise}$$

is implemented by the following recursive (function) procedure:

> **procedure** factorial(n);
> **return if** $n = 0$ **then** 1
>   **else** $n \times$ factorial($n - 1$).

2. The derivation of this $2 \log_2 N$ bound is due to G. E. Collins [Section 2 of "Computing Time Analyses for some Arithmetic and Algebraic Algorithms," in R. G. Tobey (ed.), *Proceedings of the 1968 Summer Institute on Symbolic Mathematical Computation*, IBM Programming Laboratory Report FSC69-0312, June 1969.] This compares with Lamé's bound (tightest possible) of $1.44 \log_2 N$, a bound intimately wrapped up with Fibonacci numbers. [See, for example, J. V. Uspensky and M. A. Heaslet, *Elementary Number Theory* (New York: McGraw-Hill, 1939), 43–45.]

Section 3

1. A comparison between "Fermat's algorithm" and Euclid's extended algorithm for computing mod p inverses is to be found in G. E. Collins, *Math. Comp.* **23** (1969), 197–200.

Appendix

1. The important idea of treating computer programs on an axiomatic basis—of describing the semantics of programming commands in terms of pre- and post-conditions—was developed by C. A. R. Hoare [*Comm. ACM* **12** (1969), 576–583].

COMPUTATION BY HOMOMORPHIC IMAGES

Overview. Let $\phi\colon A \to A'$ be a morphism of Ω-algebras. As noted in Section III.4.1 we can exploit the definitive property (M) of a morphism,

(M) $$\phi(\omega(a_1, \ldots, a_n)) = \omega(\phi(a_1), \ldots, \phi(a_n)),$$

to obtain information about a desired result $h = \omega(a_1, \ldots, a_n)$ in A: We perform the corresponding computation over A', namely the evaluation of $r = \omega(\phi(a_1), \ldots, \phi(a_n))$; from (M) we know that h (whatever it is) must satisfy $\phi(h) = r$. Again, the impetus for computing r rather than h itself is that computation over A' will in general be easier than computation over A, reflecting the fact that A' is an abstract, less detailed version of A. However, we must heed the fact that r need not, and in general will not, uniquely determine h.

Let us consider now the useful possibility of performing not just a single operation but a sequence of operations. Thus what we are interested in computing is the value of an *expression* over an Ω-algebra A. For example, if $\Omega = \{ +, \cdot \}$, then

$$e(x, y, z) = x + yz$$

is an expression involving two operations, to be interpreted in the obvious way.

John D. Lipson, Elements of Algebra and Algebraic Computing. ISBN 0-201-04115-4.

Let $\phi: A \to A'$ be a morphism of $\{+, \cdot\}$-algebras. Then with the above expression e and for any $a, b, c \in A$ we have

$$\phi(e(a, b, c)) = \phi(a + bc)$$

$$= \phi(a) + \phi(bc) \qquad\qquad \phi \text{ preserves } +$$

$$= \phi(a) + \phi(b)\phi(c) \qquad\qquad \phi \text{ preserves } \cdot$$

$$= e(\phi(a), \phi(b), \phi(c)).$$

So we see, at least for this example, that a morphism ϕ respects not only single operations but expressions as well (sequences of operations):

$$\phi(e(a, b, c)) = e(\phi(a), \phi(b), \phi(c)).$$

We shall soon establish the general case: If $e(x_1, \ldots, x_s)$ is an Ω-expression involving s (formal) arguments $x_1, \ldots, x_s$ combined by finitely many operations from Ω and if $\phi: A \to A'$ is a morphism of Ω-algebras, then for any (actual) arguments $a_1, \ldots, a_s \in A$,

(M′) $$\phi(e(a_1, \ldots, a_s)) = e(\phi(a_1), \ldots, \phi(a_s)).$$

We can now exploit this generalized morphism relation (M′) in exactly the same way that we exploited the ordinary morphism relation (M). If instead of computing $h = e(a_1, \ldots, a_s)$ over A we compute $r = e(\phi(a_1), \ldots, \phi(a_s))$ in the homomorphic image A', we obtain information about h: $\phi(h) = r$. The impetus for computing over A' rather than over A is now even more compelling, since the simpler computational properties of A' will be enjoyed by all of the Ω-operations involved in the evaluation of the expression e.

But again we must recognize that in general only partial information about h is obtained from r. In favorable cases, however, the information rendered about h by r may in fact be sufficient to determine h, as we now illustrate.

Example. Over $\mathbf{Z}$ we wish to compute the value h of $e(x, y, z) = x + yz$

(a) for $x = 45$, $y = -13$, $z = 3$;
(b) for $x = 45$, $y = 13$, $z = 3$.

We assume that we have a priori knowledge, by means of an oracle say, that $0 \le h < 9$ for (a), and $0 \le h < 99$ for (b). Our goal is to exploit this information to elicit the correct answers [(a) $h = 6$ and (b) $h = 84$] via homomorphic image computations.

(a) We map our $\mathbf{Z}$-computation into the corresponding $\mathbf{Z}_9$-computation by applying our generalized morphism relation (M′) with ϕ the remainder mod 9 map:

$$r_9(h) = r_9(e(45, -13, 3))$$

$$= e(r_9(45), r_9(-13), r_9(3)) \qquad \text{(M′)}$$

$$= 0 + 5 \cdot 3 \qquad \text{(over } \mathbf{Z}_9)$$

$$= 6.$$

By this modular computation we have determined that $h \equiv 6 \pmod 9$. Combined with the information $0 \le h < 9$, we have that $h = 6$. Thus we have arrived at the correct integer answer by a single modular computation.

(b) For this case we have a much larger range over which to determine the answer. Of course we could carry out the computation over $\mathbf{Z}_{99}$ and capture the answer just as in (a). But this would largely defeat the purpose of the homomorphic image method, where the key idea is to replace complex operations (in our setting, operations on large integers) by easy operations (operations on small integers). (For the purposes of this example we are asking the reader to regard 99 as a large integer.)

So let us proceed as in (a) and compute $h = e(45, 13, 3)$ over $\mathbf{Z}_9$. We obtain $r_9(h) = 3$, so that $h \equiv 3 \pmod 9$. Since $0 \le h < 99$ we have that

$$h \in \{3, 12, 21, 30, 39, 48, 57, 66, 75, 84, 93\}.$$

Thus by a single modular computation we have obtained partial information about the actual value of h (a cynic might say *very* partial).

How to resolve the uncertainty short of carrying out the computation over $\mathbf{Z}$ or $\mathbf{Z}_{99}$? The solution is simple and ingenious. We choose another "small" modulus, 13 say, and repeat the computation only this time over $\mathbf{Z}_{13}$. The result is $r_{13}(h) = 6$, so that $h \equiv 6 \pmod{13}$. This means that

$$h \in \{6, 19, 32, 45, 58, 71, 84, 97\}.$$

But clearly h must lie in the intersection of the above two solution sets. *This determines h uniquely: $h = 84$.* $\square$

* * *

The above example, simple and contrived though it is, illustrates the essence of the method of homomorphic images: to achieve the result of a complex computation by piecing together the results of perhaps several simpler (homomorphic image) computations. In this chapter we develop the technical apparatus that underlies this important computer algebra technique and explore its applications to integer and polynomial manipulation.

1. COMPUTATION BY A SINGLE HOMOMORPHIC IMAGE

Although our preoccupation in this section is with integer computations, much of the development (especially that of Section 1.1) has more general application (as we shall see in Section 3).

1.1 Ω-Expressions and their Evaluation

Let A be an Ω-algebra. By a (finite) computation over A we mean the evaluation of an expression that combines elements of A under finitely many operations of Ω. Thus a sample computation (i) over $\mathbf{Z}$ is the evaluation of $3 + 4 \cdot 5$ to give 23, (ii) over $\mathbf{Z}_7$ is the evaluation of $3 + 4 \cdot 5$ to give 2 (regarding $\mathbf{Z}$ and $\mathbf{Z}_7$ as $\{+, \cdot\}$-algebras).

Since expressions and their evaluation lie at the core of computing, it behooves us to answer precisely: What is an expression? How is an expression evaluated? This we now do in a universal algebraic setting.

The first question (what is an expression?) has to do with the syntax of expressions—their existence as strings of symbols quite apart from how they are to be interpreted.

Definition. Let Ω be a set of *operation symbols*, meaning that each symbol $\omega \in \Omega$ has an associated arity $n = n(\omega) \in \mathbf{N}$. Let $X_s = \{x_1, \ldots, x_s\}$ be a set of s distinct symbols, or "indeterminates." Then the set $\mathcal{E}(X_s) [= \mathcal{E}_\Omega(X_s)]$ of all (*formal* Ω-) *expressions* $e(x_1, \ldots, x_s) = e(x)$ *in the s indeterminates of* X_s is defined recursively by:

1. $x_1, \ldots, x_s \in \mathcal{E}(X_s)$;
2. For any $\omega \in \Omega$,

$$e_1(x), \ldots, e_n(x) \in \mathcal{E}(X_s) \Rightarrow \omega(e_1(x), \ldots, e_n(x)) \in \mathcal{E}(X_s);$$

3. That's all: there is nothing in $\mathcal{E}(X_s)$ not implied by clauses 1 and 2.

Equality in $\mathcal{E}(X_s)$ is formal equality: two expressions are equal when they are the same as strings of symbols.

For the case of a binary infix operation, say $+$, clause 2 becomes

$$e_1(x), e_2(x) \in \mathcal{E}(X_s) \Rightarrow (e_1(x) + e_2(x)) \in \mathcal{E}(X_s).$$

Example 1. Let $\Omega = \{+, \cdot\}$. Let us check that $(x + (yz)) \in \mathcal{E} = \mathcal{E}(\{x, y, z\})$. First, $x, y, z \in \mathcal{E}$ by clause 1, $(yz) \in \mathcal{E}$ by clause 2, and finally $(x + (yz)) \in \mathcal{E}$ by clause 2. ☐

We note that $(x + (y + z))$ and $((x + y) + z)$ are distinct expressions, since they are distinct as strings of symbols. And strictly speaking, $x + y + z$ is not an expression at all (missing parentheses). However, we usually admit such potentially ambiguous expressions under some convention for association and precedence of operation symbols. For example, $x + y + z$ might be regarded as a shorthand for $((x + y) + z)$ (left-to-right association); $x + yz$ as a shorthand for $(x + (yz))$ (precedence of $\cdot$ over $+$).

Remark. $\mathcal{E}(X_s)$ is clearly an Ω-algebra by its very definition. It is called a (*free*) *word algebra*, free in the sense that its elements are subject to no algebraic laws whatsoever. In particular observe that the word algebra of Example 1 is nonassociative. (Note 1.)

Example 2. We note that any polynomial $a(x) = a_n x^n + \cdots + a_0$ with integer coefficients can be naturally regarded as an Ω-expression, for $\Omega = \{+, -, 0, \cdot, 1\}$. Simply take each a_i as a shorthand for $a_i \cdot 1 = 1 + 1 + \cdots + 1$ (a_i 1's). ☐

The second question that we posed (how is an expression evaluated?) has to do with the semantics of expressions. The semantics are the obvious ones: to evaluate an Ω-expression $e(x)$ over an Ω-algebra A, one plugs in values $a_i \in A$ for each of the indeterminates x_i and then interprets the (formally meaningless) operation symbols as operations over A. All this is made precise by the following

Definition (*Evaluation of Ω-expressions*). Let $e(x_1, \ldots, x_s) = e(x)$ be an Ω-expression, let A be an Ω-algebra, and let $a_1, \ldots, a_s \in A$. Then the *value* $e(a_1, \ldots, a_s) = e(a)$ *of* $e(x)$ *for* $x_i = a_i$ is defined recursively by

1. $e(x) = x_i \Rightarrow e(a) = a_i$;
2. $e(x) = \omega(e_1(x), \ldots, e_n(x)) \Rightarrow e(a) = \omega(e_1(a), \ldots, e_n(a))$.

Example 3. Let $e(x, y, z) = x + yz$. Then the above definition evaluates $e(2, 3, 4)$ over $\mathbf{Z}$ by (i) evaluating $e_1(x, y, z) = x$ to give 2 (one application of clause 1), (ii) evaluating $e_2(x, y, z) = yz$ to give 12 (two applications of clause 1 followed by one application of clause 2), and finally (iii) evaluating $e = e_1 + e_2$ to yield 14 (one application of clause 2). □

Enter a morphism $\phi: A \to A'$ of Ω-algebras. By definition such morphisms preserve the values of operations:

(M) $\phi(\omega(a_1, \ldots, a_n)) = \omega(\phi(a_1), \ldots, \phi(a_n))$.

The main result of this section, indeed the result which lies at the heart of the homomorphic image computational technique, is that morphisms, besides preserving the values of Ω-*operations*, more generally preserve the values of Ω-*expressions*.

Notation. Throughout the chapter we shall abbreviate whenever convenient

$$\text{“}a_1, \ldots, a_s\text{”} \qquad \text{to } a,$$
$$\text{“}\phi(a_1), \ldots, \phi(a_s)\text{”} \quad \text{to } \phi(a),$$
$$\text{“}a'_1, \ldots, a'_s\text{”} \qquad \text{to } a',$$

etc.

Theorem 1 (*Fundamental Morphism Theorem*). Let $\phi: A \to A'$ be a morphism of Ω-algebras, and let $e(x_1, \ldots, x_s) = e(x)$ be an Ω-expression. Then

(M') $\phi(e(a)) = e(\phi(a))$;

or in terms of a mapping diagram:

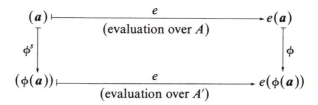

Proof. The proof is by induction on the number k of operation symbols appearing in $e(x)$.

Basis ($k = 0$). Then $e(x) = x_i$, so that

$$\phi(e(a)) = \phi(a_i) = e(\phi(a)),$$

both equalities following from clause 1 of the evaluation definition.

Induction. For any $k \geq 0$, assume the statement of Theorem 1 for all Ω-expressions having $\leq k$ operation symbols, and let $e(x)$ have $k + 1$ operation symbols. Then $e(x)$ has the form

$$e(x) = \omega(e_1(x), \ldots, e_n(x))$$

for some n-ary operation symbol $\omega \in \Omega$, where $e_1(x), \ldots, e_n(x)$ have $\leq k$ operation symbols.

CASE 1. $n = 0$. Then ω is a nullary operation, call it 0. But $\phi(0) = 0$ because ϕ preserves all operations, including nullary ones. Thus Theorem 1 holds for the (trivial) special case of $e(x) = 0$, a nullary operation symbol.

CASE 2. $n > 0$ (the more interesting case). Here we have (writing $\phi[\cdot]$ for $\phi(\cdot)$)

$\phi[e(a)]$

$$\begin{aligned} &= \phi[\omega(e_1(a), \ldots, e_n(a))] &&\text{defn. of evaluation} \\ &= \omega(\phi[e_1(a)], \ldots, \phi[e_n(a)]) &&\text{ϕ is a morphism} \\ &= \omega(e_1(\phi[a]), \ldots, e_n(\phi[a])) &&\text{induction hypothesis} \\ &= e(\phi[a_1], \ldots, \phi[a_s]) &&\text{defn. of evaluation,} \end{aligned}$$

which completes the proof of the induction step. $\square$

Example 4   (*Casting out nines*).   The familiar method of "casting out nines" for checking the results of decimal arithmetic has its logical basis in the Fundamental Homomorphism Theorem, as we now show.

Let $a = (a_n a_{n-1} \cdots a_0)_{10}$ be a decimal integer (each a_i is a decimal digit). Then a denotes the value of

$$a(x) = a_n x^n + a_{n-1} x^{n-1} + \cdots + a_0$$

for $x = 10$; i.e., $a = a(10)$. This polynomial $a(x)$ can naturally be regarded as an Ω-expression, $\Omega = \{+, -, 0, \cdot, 1\}$ (Example 2).

We now apply the remainder $\mod 9$ morphism $\mathbf{Z} \to \mathbf{Z}_9$ to $a = a(10)$. By Theorem 1 we have

$$r_9(a(10)) = a(r_9(10)) = a(1)$$

where $a(1)$ is evaluated over $\mathbf{Z}_9$. Thus we have derived the rule

$$(a_n \cdots a_1 a_0)_{10} \equiv a_n + \cdots + a_1 + a_0 \pmod{9}.$$

In words: a decimal integer is congruent mod 9 to the sum of its digits. Since $a_n + \cdots + a_0 < a = (a_n \cdots a_0)_{10}$ clearly, the rule may be applied repeatedly to obtain $r_9(a)$. Example: $38476 \equiv_9 28 \equiv_9 10 \equiv_9 1$, giving $r_9(38476) = 1$. Thus we have derived a rule for computing mod 9 remainders that involves no explicit division, only the addition of single digit integers. For hand calculation, this *casting-out-nines* method, as it is called, has obvious attractions.

We can now combine casting out nines with yet another application of Theorem 1 to obtain a rapid test for hand checking the correctness of integer arithmetic. To illustrate, let us check that

$$(*) \qquad\qquad 43 \cdot 768 + 9571 = 42595.$$

Let $e(a, b, c) = ab + c$. By Theorem 1 we have $r_9(e(a, b, c)) = e(r_9(a), r_9(b), r_9(c))$ for any integers a, b, c. Thus *if* the calculation $(*)$ is correct we must have $r_9(42595) = e(r_9(43), r_9(768), r_9(9571))$, where the right-hand side is of course evaluated over $\mathbf{Z}_9$. Checking this out:

$$r_9(42595) = r_9(25) = 7;$$

$$e(r_9(43), r_9(768), r_9(9571))$$

$$= e(7, 3, 4)$$

$$= 7 \cdot 3 + 4 \quad \text{(over } \mathbf{Z}_9)$$

$$= 7,$$

which completes the check. (See Exercise 1.) □

Partial algebra considerations. We now wish to extend from total to partial algebras the above concepts and ideas related to evaluation and morphisms. Since we very much have in mind the operation of division in rings, we first introduce the concept of a "ring-with-division."

Let R be a commutative ring. The relation of divisibility which we defined for integral domains in Section IV.4.1 applies also to commutative rings: $b|a$ over R means $a = bc$ for some $c \in R$. *For b neither zero nor a zerodivisor, we say that a/b is defined over R whenever $b|a$.* Then $a/b = c$ where $a = bc$. We call a ring in which division is admitted as a partial operation a *ring-with-division*. Hence a ring-with-division is a partial Ω-algebra with $\Omega = \{+, -, 0, \cdot, 1, /\}$.

Some examples of division: Over $\mathbf{Z}$, $6/3$ is defined (with value 2), $7/3$ is undefined. Over $\mathbf{Z}_m$, $a/b = ab^{-1}$ is defined whenever $(b, m) = 1$ (Theorem 1 of Section VII.3.1); when $(b, m) > 1$, then b is a zerodivisor (Exercise 1 of Section IV.2), hence cannot be used as a divisor. Over a field F, $a/b = ab^{-1}$ is defined whenever $b \neq 0$.

We note that for the case of rings with zerodivisors (like $\mathbf{Z}_m$), the restriction that b not be a zerodivisor is necessary for well-definedness. For example in $\mathbf{Z}_{12}$ we have $3|6$ since $6 = 3 \times 2$. But $6 = 3 \times 6$ as well, so do we define $6/3$ as 2 or 6?

The answer is that we (sensibly) do neither. Division by zerodivisors, like division by zero, is forbidden.

Returning to a universal algebra setting, we require a criterion for saying that the value of an Ω-expression is *defined* over a partial Ω-algebra A.

Definition (*Definedness over partial algebras*). Let A be a (partial) Ω-algebra and $e(x) = \omega(e_1(x), \ldots, e_n(x))$ an Ω-expression. For $x_i = a_i$ we say that $e(a)$ is *defined* (over A) if (i) $e_1(a), \ldots, e_n(a)$ are all defined (recursion) and (ii) for $t_i = e_i(a)$, $\omega(t_1, \ldots, t_n)$ is defined. Of course $e(a) = a_i$ is always defined for any a_i.

Example 5. Consider the $\{+, /\}$-expressions $e(x, y, z) = (x + y)/z$ and $f(x, y, z) = x/z + y/z$. Regarding $\mathbf{Z}$ as a ring-with-division, then over $\mathbf{Z}$, $e(1, 2, 3)$ is defined with value 1, whereas $f(1, 2, 3)$ is undefined. $\square$

We can now extend the all-important definition of morphism from total to partial algebras.

Definition (*Morphism of partial algebras*). Let A and A' be (partial) Ω-algebras. Then $\phi: A \to A'$ is a *morphism of partial Ω-algebras* provided the morphism relation

(M) $$\phi(\omega(a_1, \ldots, a_n)) = \omega(\phi(a_1), \ldots, \phi(a_n))$$

holds *whenever both sides are defined*, i.e., whenever $\omega(a_1, \ldots, a_n)$ is defined over A and $\omega(\phi(a_1), \ldots, \phi(a_n))$ is defined over A'. (We note that if A and A' are total Ω-algebras, then the above definition of morphism reduces to the ordinary one.)

For us the most important class of morphisms of partial algebras is the one of

Theorem 2. Let $\phi: R \to R'$ be a morphism of commutative rings. Then ϕ is a morphism of rings-with-division, i.e.,

$$\phi(a/b) = \phi(a)/\phi(b)$$

whenever both sides are defined.

Proof. Let a/b, $\phi(a)/\phi(b)$ be defined over R, R'. Then $a = bc$ for some $c \in R$, so we have

$(*)$ $$\phi(a) = \phi(b)\phi(c),$$

using only that ϕ is a ring morphism. Since $\phi(b) | \phi(a)$, we can write $\phi(a) = \phi(b)[\phi(a)/\phi(b)]$, so from $(*)$ we have

$(**)$ $$\phi(b)[\phi(a)/\phi(b)] = \phi(b)\phi(c).$$

Since $\phi(a)/\phi(b)$ is defined over R', $\phi(b)$ is neither zero nor a zerodivisor in R'.

Hence we can cancel $\phi(b)$ from both sides of $(**)$ to obtain

$$\phi(a)/\phi(b) = \phi(c) = \phi(a/b). \quad \Box$$

Example 6. From Theorem 2 we can conclude that $r_m: \mathbf{Z} \rightarrow \mathbf{Z}_m$ is a morphism regarding $\mathbf{Z}$ and $\mathbf{Z}_m$ as rings-with-division, meaning in particular that

$$r_m(a/b) = r_m(a)/r_m(b)$$

whenever $b \mid a$ and $(b, m) = 1$, the latter being the condition that $r_m(a)/r_m(b) = r_m(a)r_m(b)^{-1}$ be defined over $\mathbf{Z}_m$.

As an illustration of this result, over $\mathbf{Z}_6$ we have $r_6(154)/r_6(11) = 4/5 = 4 \cdot 5^{-1} = 4 \cdot 5 = 2$ in agreement with $r_6(154/11) = r_6(14) = 2.$ $\Box$

We now extend our Fundamental Morphism Theorem, Theorem 1, from total to partial algebras.

> **Theorem 3** (*Fundamental Morphism Theorem for Partial Algebras*). Let $\phi: A \rightarrow A'$ be a morphism of partial Ω-algebras. Let $e(x_1, \ldots, x_s) = e(x)$ be an Ω-expression. Then
>
> (M′) $\phi(e(a)) = e(\phi(a))$
>
> whenever both sides are defined (i.e., whenever $e(a) = e(a_1, \ldots, a_s)$ is defined over A and $e(\phi(a)) = e(\phi(a_1), \ldots, \phi(a_s))$ is defined over A').

Remark. The proof which follows is typical of many proofs involving partial algebras: it is somewhat complicated by considerations of definedness. If the reader appreciates that the total algebra proof (Theorem 1) is the kernel of the partial algebra proof, that is good enough (especially on a first reading).

Proof of Theorem 3. The proof proceeds as with the proof of Theorem 1, by induction on the number k of operation symbols appearing in $e(x)$.

Basis ($k = 0$). As in the proof of Theorem 1.

Induction. For any $k \geq 0$, assume the statement of Theorem 3 for all Ω-expressions having $\leq k$ operation symbols, and let $e(x)$ have $k + 1$ operation symbols. Then $e(x)$ has the form

$$(*) \qquad\qquad e(x) = \omega(e_1(x), \ldots, e_n(x))$$

for some n-ary operation symbol $\omega \in \Omega$, where $e_1(x), \ldots, e_n(x)$ have $\leq k$ operation symbols.

CASE 1. $n = 0$. As in the proof of Theorem 1.

CASE 2. $n > 0$ (here is where the partial property of the operations of Ω must be dealt with). We may assume that $e(a)$ and $e(\phi(a))$ are defined (over A and A' respectively). Our task is to show that $\phi(e(a)) = e(\phi(a))$. For convenience we denote $\phi(\cdot)$ by $\phi[\cdot]$.

First let us take stock of what is defined. Since $e(a)$ is defined, it follows from $(*)$ and our definition of definedness that

(1) $e_1(a), \ldots, e_n(a)$ are defined;

(2) $\omega(t_1, \ldots, t_n)$ is defined,   where $t_i = e_i(a)$.

And since $e(\phi[a])$ is defined, it likewise follows that

(3) $e_1(\phi[a]), \ldots, e_n(\phi[a])$ are defined;

(4) $\omega(t_1, \ldots, t_n)$ is defined,   where $t_i = e_i(\phi[a])$.

Appealing to (2) and our definition of evaluation, we can write

(5) $\phi[e(a)] = \phi[\omega(e_1(a), \ldots, e_n(a))]$.

Since ϕ is a morphism, we have

(6) $\phi[\omega(e_1(a), \ldots, e_n(a))] = \omega(\phi[e_1(a)], \ldots, \phi[e_n(a)])$

provided the righthand side is defined, which we proceed to show.

Now each $\phi[e_i(a)]$ is defined [(1)] and so too is each $e_i(\phi[a])$ [(3)]. Hence we can apply the induction hypothesis to obtain

(7) $\phi[e_i(a)] = e_i(\phi[a])$.

Since ω applied to the $e_i(\phi[a])$'s is defined [(4)], so is ω applied to the $\phi[e_i(a)]$'s [by (7)]. Thus the right-hand side of (6) is defined, and we have also shown that

(8) $\omega(\phi[e_1(a)], \ldots, \phi[e_n(a)]) = \omega(e_1(\phi[a]), \ldots, e_n(\phi[a]))$.

Finally, by our definition of evaluation, we have

(9) $\omega(e_1(\phi[a]), \ldots, e_n(\phi[a])) = e(\phi[a])$.

The sequence of equalities (5), (6), (8), (9) have yielded the desired result: $\phi[e(a)] = e(\phi[a])$. $\square$

Example 7.   We can apply Theorem 3 to extend the casting-out-nines check to integer calculations involving division. Let us check whether

$$(43 \cdot 768 + 9571)/35 = 1227.$$

Proceeding formally, let $e(a, b, c, d) = (ab + c)/d$. By Theorem 3,

$$r_9(e(a, b, c, d)) = e(r_9(a), r_9(b), r_9(c), r_9(d))$$

whenever both sides are defined (over $\mathbf{Z}$ and $\mathbf{Z}_9$ respectively). Checking, we have

$$r_9(1227) = r_9(12) = 3;$$

$$e(r_9(35), r_9(43), r_9(768), r_9(9571))$$
$$= e(8, 7, 3, 4)$$
$$= (7 \cdot 3 + 4)/8 \quad \text{(over } \mathbf{Z}_9)$$
$$= 2,$$

so we know that $(43 \cdot 768 + 9571)/35$ cannot possibly be equal to 1227. (See Exercise 2.) $\square$

1.2 Solutions to an Integer Congruence

At the heart of the homomorphic image technique for computing over $\mathbf{Z}$ (to be presented in Section 1.3) is the recovery of an integer u from its $\bmod m$ remainder, or "residue," $r_m(u)$. We therefore devote this short and simple section to examining particular solutions U for u in the congruence

$$u \equiv r \;(\mathrm{mod}\, m), \qquad 0 \le r < m.$$

Now one particular solution is evidently $U = r$; just as evidently, if U is a particular solution, then so is $U + km$, k an arbitrary integer. For our purposes there are two preferred solutions of interest:

Theorem 1. The congruence

$$u \equiv r \;(\mathrm{mod}\, m), \qquad 0 \le r < m$$

(a) has a unique *least positive solution* U satisfying $0 \le U < m$, namely $U = r$;
(b) has a unique *least absolute value solution* V satisfying $-m/2 \le V < m/2$, namely $V = \beta_m(r)$ defined by

$$\beta_m(r) = r \qquad \text{if } r < m/2$$

$$= r - m \quad \text{otherwise.}$$

Proof. Clearly $U = r$ is a solution that satisfies $0 \le U < m$, and $V = \beta_m(r)$ is a solution that satisfies $-m/2 \le V < m/2$. The remaining issue is uniqueness, which follows as an easy consequence of the following general result:

Lemma 1. Let U, U' be solutions to the congruence $u \equiv r$ $(\mathrm{mod}\, m)$ in the range $\alpha \le u < \alpha + m$, where α is a fixed but arbitrary real number. Then $U = U'$.

Proof. Say $U \le U'$. Since $\alpha \le U \le U' < \alpha + m$, it follows that $0 \le U' - U < m$. Since $U \equiv r$ $(\mathrm{mod}\, m)$ and $U' \equiv r$ $(\mathrm{mod}\, m)$, it follows that $U \equiv U' (\mathrm{mod}\, m)$, i.e., that $m | (U' - U)$. Thus if $U \ne U'$ we have m dividing a positive integer $< m$, which is impossible. $\square$

Returning to the proof of Theorem 1, the uniqueness of the least positive solution U follows from Lemma 1 with $\alpha = 0$, of the least absolute value solution follows from Lemma 1 with $\alpha = -m/2$. $\square$

Example 1. (a) The congruence $u \equiv 9$ $(\mathrm{mod}\, 15)$ has the unique least positive solution $U = 9$ and the unique least absolute value solution $V = -6$.
   (b) The congruence $u \equiv 5$ $(\mathrm{mod}\, 15)$ has the unique least positive solution $U = 5$ and the unique least absolute value solution $V = 5$. $\square$

1.3   The Homomorphic Image Scheme for Z

Let an integer quantity h be defined by the value of an expression $e(x_1, \ldots, x_s)$ for integer arguments $a_1, \ldots, a_s$, where by "expression" in the context of computing over $\mathbf{Z}$ (or over any other ring for that matter) we shall mean a ring expression or ring-with-division expression. Our goal is to evaluate $h = e(a_1, \ldots, a_s)$.

The obvious, direct method is to simply evaluate $e(a_1, \ldots, a_s)$ over $\mathbf{Z}$. The homomorphic image scheme, on the other hand, proposes to evaluate $e(a_1, \ldots, a_s)$ by a less obvious, indeed roundabout route, but one with compensating advantages which we shall elucidate. In a nutshell, the homomorphic image scheme computes not h itself, but $r = h \bmod m$ for some appropriate modulus m, and then recovers h from its $\bmod m$ residue r. The advantage of this circuitous method (again in a nutshell) is that r can be computed over $\mathbf{Z}_m$ rather than over $\mathbf{Z}$, so that the simplicity of $\bmod m$ arithmetic (bounded operands and results) and structure of $\bmod m$ arithmetic (e.g., $\mathbf{Z}_m$ is a field if m is prime) can be exploited.

Now r by itself does not uniquely determine h. Thus we need some a priori knowledge about h. The knowledge we assume is a bound B on h in one of two forms:

(1) $0 \le h < B$: here we know h to be positive;
(2) $|h| < B$: here we do not know the sign of h.

In practice such bounds can often be readily determined by a rough calculation or order of magnitude analysis of the problem at hand, as we shall shortly illustrate.

Algorithm 1   (*Homomorphic image scheme for* $\mathbf{Z}$).

Input.   Expression $e(x_1, \ldots, x_s)$, defined over $\mathbf{Z}$ for arguments $a_1, \ldots, a_s$.
Output. $h = e(a_1, \ldots, a_s) \in \mathbf{Z}$.

VERSION 1. (Bound B on $h \ge 0$ is available.) Let $0 \le h < B$, and choose a modulus m such that (i) $m \ge B$ and (ii) $e(r_m(a_1), \ldots, r_m(a_s))$ is defined over $\mathbf{Z}_m$. Then proceed according to

Step 1   {Modular evaluation}.

(a) Compute $a_i' = r_m(a_i)$ $(i = 1, \ldots, s)$;
(b) Compute $r = e(a_1', \ldots, a_s')$ (over $\mathbf{Z}_m$).

Step 2   {Congruence solution}.

Solve $u \equiv r \pmod{m}$ for the least positive solution U (i.e., $U := r$—Theorem 1 of Section 1.2).

Step 3   {Return value for h}.

Return $h = U$.

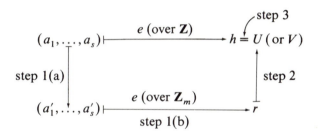

Fig. 1   Integer homomorphic image scheme.

VERSION 2.   (Bound B on $|h|$ is available.) Let $|h| < B$, and choose a modulus m as in Version 1, but with $m \geq 2B$. Then proceed as in Version 1, except in step 2 solve $u \equiv r \pmod{m}$ for the least absolute value solution V (i.e., $V := \beta_m(r)$—Theorem 1 of Section 1.2), and in step 3 return $h = V$. $\square$

The setup of the above homomorphic image scheme for $\mathbf{Z}$ is described by the mapping diagram of Fig. 1.

Some comments on Algorithm 1 before proving its correctness.

1.   Step 1(b), the evaluation of the $\mathbf{Z}_m$ expression $e(a_1', \ldots, a_s')$, is the heart of the homomorphic image scheme. For this computation we can use any algorithm we wish, and of course in any given instance we shall want that algorithm to be an efficient one.

2.   The particular algorithm (operating over $\mathbf{Z}_m$) that we use for the evaluation of $e(\boldsymbol{a}')$ may impose further restrictions on the choice of modulus; e.g., if the algorithm requires a field structure, we would choose the modulus m to be prime. (The first of our examples to follow will illustrate this point.)

3.   If the expression $e(\boldsymbol{x})$ is a ring expression, i.e., does not involve division, then there is no question as to the definedness of $e(\boldsymbol{a})$ over $\mathbf{Z}$ and $e(r_m(\boldsymbol{a}'))$ over $\mathbf{Z}_m$. It is only when division is involved that definedness becomes an issue. (The second of our two examples will illustrate this point.)

4.   Our rather elaborate statement of step 2 in terms of solving an integer congruence anticipates future developments (Section 3).

Now for the proof that Algorithm 1 (either version) computes what we are after, namely $h = e(\boldsymbol{a})$.

Theorem 1   (*The integer homomorphic image scheme works*).   In Algorithm 1, for $h = e(\boldsymbol{a})$,

(a) Version 1 computes $U = h$;
(b) Version 2 computes $V = h$.

Proof.   As noted in Example 6 of Section 1.1, $r_m: \mathbf{Z} \to \mathbf{Z}_m$ is a ring-with-division morphism. By assumption, $e(\boldsymbol{a})$ and $e(r_m(\boldsymbol{a}))$ are both defined, over $\mathbf{Z}$ and $\mathbf{Z}_m$ respectively. Hence by our Fundamental Morphism Theorem (Theorem 3 of

Section 1.1), we have

$$r_m(e(a)) = e(r_m(a)).$$

Thus r computed in step 1 satisfies $r = r_m(e(a))$, which means that $h = e(a)$ is a solution to the congruence $u \equiv r \pmod{m}$.

(a) In Version 1 we have $0 \leq h < B \leq m$. Hence h, like U returned by step 2, is a least positive solution to the congruence $u \equiv r \pmod{m}$. By the uniqueness of this solution (Theorem 1 of Section 1.2), $U = h$.

(b) In Version 2 we have $|h| < B \leq m/2$. Hence h, like V returned by step 2, is a least absolute value solution to the congruence $u \equiv r \pmod{m}$. Again by uniqueness, $V = h$. $\square$

The two examples which follow are intended to illustrate some of the nuances and the utility of the homomorphic image method.

Example 1 (*"Exact" solution of linear equations over* **Z**). First let us briefly review the familiar "Gaussian elimination" algorithm for computing the solution to a system of n linear equations in n unknowns over a field F:

$$a_{11}x_1 + \cdots + a_{1n}x_n = b_1,$$

$$\vdots \qquad\qquad \vdots \qquad\qquad (a_{ij}, b_i \in F)$$

$$a_{n1}x_1 + \cdots + a_{nn}x_n = b_n,$$

or, in matrix terms,

$$Ax = b$$

where $A = (a_{ij})$ is the $n \times n$ coefficient matrix and $\mathbf{b} = (b_i)$ is the right-hand side n-vector. (Note 2.)

Starting from the original $n \times (n+1)$ "augmented matrix" $\tilde{A}^{(0)} = (A : \mathbf{b})$, Gaussian elimination redefines $\tilde{A}^{(0)}$ by a sequence of matrices $\tilde{A}^{(k)} = (A^{(k)} : \mathbf{b}^{(k)})$ $(k = 1, \ldots, n-1)$ by performing a series of *row operations* (on these matrices) of the following two types:

Type 1.   Interchanging two rows.
Type 2.   Adding to a row a scalar multiple of some other row.

For convenience we denote the elements $a_{ij}^{(k)}, b_i^{(k)}$ and rows $R_i^{(k)}$ of $\tilde{A}^{(k)}$ simply by a_{ij}, b_i, R_i (keeping in mind that they depend on k). To compute $\tilde{A}^{(k)}$ from $\tilde{A}^{(k-1)}$, Gaussian elimination proceeds as follows:

Step 1 (Pivotting).   If $a_{kk} = 0$ then R_k is interchanged with some R_i $(i > k)$ having $a_{ik} \neq 0$. (Now the "pivot element" a_{kk} is $\neq 0$.)

Step 2 (Elimination).   Each R_i for $i > k$ is redefined according to

$$(1) \qquad\qquad R_i := R_i - \mu R_k \qquad \text{where } \mu = a_{ik}/a_{kk}.$$

The multiplier μ has been chosen to "eliminate" variable x_k from equations (rows) $k+1$ through n, i.e., to have it appear with zero coefficient in those equations. Thus (induction) we finally obtain $\tilde{A}^{(n-1)}$ in *triangular form:*

(2)
$$\tilde{A}^{(n-1)} = \begin{bmatrix} a_{11} & a_{12} & \cdots & a_{1n} & b_1 \\ & a_{22} & \cdots & a_{2n} & b_2 \\ & & \ddots & \vdots & \vdots \\ O & & & a_{nn} & b_n \end{bmatrix}.$$

(Of course the a_{ij}'s and b_i's of $\tilde{A}^{(n-1)}$ are not the same as those of the original matrix $\tilde{A}$.)

The row operations of Gaussian elimination have the vital property of leaving the solution unchanged, so that the matrices $\tilde{A}^{(k)}$ all represent systems that are equivalent solution-wise to the given system $\tilde{A}$. The solution to $\tilde{A}^{(n-1)}$, hence to the original system $\tilde{A}$, is readily obtained by back-substitution:

(3)
$$x_k = \left(b_k - \sum_{j=k+1}^{n} x_j a_{kj} \right) \Big/ a_{kk} \qquad (k = n, n-1, \ldots, 1).$$

As for the complexity of Gaussian elimination, it is easily seen that triangularization (elimination proper) requires $O(n^3)$ field operations and back-substitution requires $O(n^2)$ field operations. Hence the overall solution requires $O(n^3)$ field operations.

So much for the general problem and its computational solution. We now look at the special case where the a_{ij}'s and b_i's are all integers. To fix ideas we shall consider the integer system

(4)
$$3x_1 - 2x_2 + 6x_3 = 4,$$
$$4x_1 + x_2 + x_3 = 0,$$
$$2x_1 + x_2 + 2x_3 = -1.$$

Now any such integer system can be regarded as a real system, and a computer solution can readily be obtained using real (floating point) arithmetic. But the inevitable accumulation of rounding errors leads us to search for the *exact* solution, i.e., the solution over **Q**.

The obvious approach is to regard the integer system as a rational system and to perform Gaussian elimination and back-substitution over **Q**. This is precisely what we have done in Fig. 2(a). However, the complexity of rational arithmetic (which might not be too apparent in our very small example) leads us to consider an alternative approach. For us this approach is the homomorphic image method.

But immediately we are confronted with a problem. The components x_k of the solution vector are rational numbers, whereas our homomorphic image method is geared to integer results. Somehow we must "integerize" our solution. *Cramer's Rule*, which expresses the solution to a linear system in terms of determinants, provides just the tool we need.

$$\text{Solve } \begin{bmatrix} 3 & -2 & 6 \\ 4 & 1 & 1 \\ 2 & 1 & 2 \end{bmatrix}\begin{bmatrix} x_1 \\ x_2 \\ x_3 \end{bmatrix} = \begin{bmatrix} 4 \\ 0 \\ -1 \end{bmatrix}$$

(a) Over **Q**

$$\tilde{A}^{(0)}: \quad \begin{matrix} 3 & -2 & 6 & 4 \\ 4 & 1 & 1 & 0 \\ 2 & 1 & -2 & -1 \end{matrix}$$

$$\tilde{A}^{(1)}: \quad \begin{matrix} 3 & -2 & 6 & 11 \\ 0 & 11/3 & -7 & -26/3 \\ 0 & 7/3 & -6 & -22/3 \end{matrix}$$

$$\tilde{A}^{(2)}: \quad \begin{matrix} 3 & -2 & 6 & 4 \\ 0 & 11/3 & -7 & -16/11 \\ 0 & 0 & -17/11 & -3/11 \end{matrix}$$

Back-substitution:

$x_3 = -11/17[-3/11] = 3/17,$

$x_2 = 3/11[-16/3 - (-21/3)(3/17)]$

$\quad = -19/17,$

$x_1 = 1/3[4 + 2(-19/17) - 6(3/17)]$

$\quad = 4/17.$

Thus,

$$\mathbf{x} = \begin{bmatrix} 4/17 \\ -19/17 \\ 3/17 \end{bmatrix}$$

(b) Over **Z**$_{97}$

$$\tilde{A}^{(0)}: \quad \begin{matrix} 3 & 95 & 6 & 4 \\ 4 & 1 & 1 & 0 \\ 2 & 1 & 95 & 96 \end{matrix}$$

$$\tilde{A}^{(1)}: \quad \begin{matrix} 3 & 95 & 6 & 4 \\ 0 & 36 & 90 & 27 \\ 0 & 67 & 91 & 61 \end{matrix}$$

$$\tilde{A}^{(2)}: \quad \begin{matrix} 3 & 95 & 6 & 4 \\ 0 & 36 & 90 & 27 \\ 0 & 0 & 69 & 35 \end{matrix}$$

Back-substitution:

$x_3 = 69^{-1}(35) = 45(35) = 23,$

$x_2 = 36^{-1}[27 - (90)(23)]$

$\quad = 62(91) = 16,$

$x_1 = 3^{-1}[4 - (95)(16) - 6(23)]$

$\quad = 65(92) = 63.$

Thus,

$$\mathbf{x} = \begin{bmatrix} 63 \\ 16 \\ 23 \end{bmatrix}$$

Also,

$$\det A = 3(36)(69) = 80$$

Fig. 2 Solution of a linear system (a) over **Q**, (b) over **Z**$_{97}$.

Recall the definition of the *determinant* of an $n \times n$ matrix $A = (a_{ij})$:

(5)
$$\det A = \sum_{(n!)} \pm a_{1i_1} a_{2i_2} \cdots a_{ni_n}$$

where the sum is over all $n!$ permutations $(i_1, \ldots, i_n)$ of $\mathbf{n}$, and each term is appropriately signed (this latter detail need not concern us here). Cramer's Rule expresses the solution $\mathbf{x}$ to a system $A\mathbf{x} = \mathbf{b}$ as follows: Let $d = \det A$, and let

$y_j = \det A_j$ where A_j is the matrix obtained from A by replacing the j-th column by the right-hand side **b**. Then

$$(6) \qquad\qquad x_j = y_j/d \quad (\textit{Cramer's Rule}).$$

Now for a simple, but for us a crucial, observation. According to (5) the determinant $\det A$ of an $n \times n$ matrix $A = (a_{ij})$ is the (integer) value of a ring expression evaluated at integer arguments (the n^2 a_{ij}'s). Hence any determinantal quantity is susceptible to evaluation by our homomorphic image scheme, Algorithm 1, which we propose to use for the computation of d and the y_j's of (6).

So let $Ax = \mathbf{b}$ be a given integer system of n linear equations in n unknowns. In preparation for the application of Version 2 of Algorithm 1, let B be a bound on $\max(|d|, |y_1|, \ldots, |y_n|)$ and let the modulus m be $\geq 2B$. Step 1 of Algorithm 1 then requires the $\bmod m$ reduction of A and **b**:

$$(7) \qquad A' = r_m(A) = (r_m(a_{ij})), \qquad \mathbf{b}' = r_m(\mathbf{b}) = (r_m(b_i))$$

and the computation over $\mathbf{Z}_m$ of the determinants

$$(8) \qquad\qquad d' = \det A', \qquad y'_j = \det A'_j.$$

[By the computation over $\mathbf{Z}_m$ of a determinant (5) we of course mean the evaluation of (5) with operations interpreted in $\mathbf{Z}_m$. We note, trivially, that d' and y'_j are defined over $\mathbf{Z}_m$ for any modulus m, since the definition (5) of a determinant does not involve division—it is a *ring* expression as opposed to a *ring-with-division* expression.]

Steps 2 and 3 of Algorithm 1 then return d and **y** according to

$$(9) \qquad\qquad d = \beta_m(d'), \qquad y_j = \beta_m(y'_j)$$

(here is where the requirement $m \geq 2B$ comes in). The solution over **Q** (the "exact" solution) is then given by Cramer's Rule (6).

The homomorphic image scheme leaves open the method by which the $\mathbf{Z}_m$ determinants d', y'_j of (8) are computed. We of course wish to choose an efficient method, which certainly precludes the explicit evaluation of determinants according to (5)—a process that grows as $n!$. We recall that for the case of matrices over a field, determinants can be evaluated efficiently by reduction to upper triangular form, i.e., essentially by Gaussian elimination. To obtain a field, we need only choose our modulus m to be a prime p (as well as being $\geq 2B$).

We are now working in a field $(\mathbf{Z}_p)$, and we proceed in the standard way. We employ Gaussian elimination (1) over $\mathbf{Z}_p$ to reduce $\tilde{A} = (A' : \mathbf{b}')$ to upper triangular form, the matrix $\tilde{A}^{(n-1)}$ of (2).

Now two fundamental properties of determinants are that (i) interchanging two rows changes the sign of the determinant and (ii) adding to a row a scalar multiple of some other row leaves the determinant unchanged. Thus the determinant d' of the original matrix $A' = r_m(A)$ is related to the determinant of the triangularized matrix $A^{(n-1)}$ by

$$d' = (-1)^\sigma \det A^{(n-1)}$$

where $\sigma = 0$ if the number of row interchanges carried out is even, 1 if odd. Since the determinant of a triangular matrix is given by the product of its diagonal elements, we obtain

(10) $$d' = (-1)^\sigma a_{11} \cdots a_{nn},$$

where the above a_{kk}'s are of course those of $A^{(n-1)}$.

To compute y_j', we need not (explicitly) evaluate a determinant at all. For we can solve for $\mathbf{x}'$ from $A^{(n-1)}$ by back-substitution (3) (thus $\mathbf{x}'$ satisfies $A'\mathbf{x}' = \mathbf{b}'$) and then obtain $\mathbf{y}'$ by Cramer's Rule (6),

(11) $$\mathbf{y}' = d'\mathbf{x}'.$$

Let us examine the homomorphic image scheme *in toto* for the case of our given integer system (4). In Fig. 2(b) we have performed Gaussian elimination over the field $\mathbf{Z}_{97}$ (with hindsight we know that $97 \geq 2B$) to obtain $\mathbf{x}' = (63, 12, 23)$, $d' = 80$. By (11) we then have $\mathbf{y}' = (93, 19, 94)$. This completes step 1 of the homomorphic image scheme.

Steps 2 and 3 then return d and $\mathbf{y}$ in accordance with (9),

$$d = 80 - 97 = -17,$$
$$y_1 = 93 - 97 = -4, \quad y_2 = 19, \quad y_3 = 94 - 97 = -3.$$

Finally, Cramer's Rule yields the desired rational solution $\mathbf{x} = (4/17, -19/17, 3/17)$, in agreement with our solution computed over $\mathbf{Q}$. *Only our homomorphic image solution has been achieved entirely by finite field arithmetic.*

Remarks (about the choice of prime). 1. With the benefit of hindsight (i.e., knowing the answer!) we see that any prime $p \geq 2 \times 20 = 40$ would have yielded the correct answer, according to Theorem 1(b). (See Exercise 3.)

2. In computing practice, rather than using a relatively small prime such as 97 we would use a prime at or near the wordsize of our computer in order to satisfy the $p \geq 2B$ bound for the widest possible class of problems. For example, on a machine with a 32 bit wordsize (31 bits + sign) we would be dealing with a prime in the $2^{31} \approx 2 \times 10^9$ neighborhood.

3. For the integer linear system problem, solving $A\mathbf{x} = \mathbf{b}$ for $d, y_1, \ldots, y_n$, a bound B on the determinantal quantities d, y_j is readily determined. If $|a_{ij}|$, $|b_i| \leq C$, then from the definition of a determinant (5) we have

$$|d|, |y_j| \leq n! C^n.$$

So for our problem ($n = 3$, $C = 6$), we are guaranteed the correct answer if $p \geq 2 \times 1297 = 2594$ (in view of remark 1, a very pessimistic requirement).

4. Insight or bounding may dictate that we use a prime beyond the wordsize of our computer. We shall deal with that problem in Section 3.

In ferretting out the essential idea behind the homomorphic image scheme as illustrated in the integer linear equations example, we find that the algorithmic aspects of **Z** mirror the structural aspects of **Z**. To turn **Z** into a field (mainly to be able to perform division) we can do it in one of two ways: by *extension* of **Z** to its field of quotients **Q** (i.e., by replacing **Z** by a more complicated structure), or by *abstraction* of **Z** to its homomorphic image $\mathbf{Z}_p$ for some prime p (i.e., by replacing **Z** by a simpler structure). The structural simplicity of $\mathbf{Z}_p$ compared with **Q** is very much reflected by their arithmetic: finite field arithmetic is simpler than rational arithmetic because all values are bounded (by the modulus p) and because time-consuming g.c.d. calculations to reduce fractions to lowest terms are avoided. It is this arithmetic simplicity of modular arithmetic that our homomorphic image method exploits when it takes the abstraction route to turn **Z** into a field.

Our next example, though technically much simpler than Example 1, illustrates an application of the homomorphic image scheme to the evaluation of an expression involving division, so that definedness becomes an issue.

Example 2. Here we apply our homomorphic image scheme to compute a binomial coefficient $C = C(n, k)$,

$$(*) \qquad C = \frac{n!}{k!(n-k)!} = \frac{n(n-1)\cdots(n-k+1)}{k!}.$$

First we note that C, though ostensibly rational, is in fact an integer (Exercise 3 of Section II.2), moreover a positive integer. Thus Version 1 of Algorithm 1 is applicable subject to the choice of an appropriate modulus m. Let $B = B(n, k)$ be a bound on C: $C < B$. The requirements for m are (i) that m be $\geq B$ and (ii) that the expression $(*)$ that defines C be defined over $\mathbf{Z}_m$; i.e., that

$$r = e/f, \qquad \text{where } e = r_m(n)r_m(n-1)\cdots r_m(n-k+1),$$
$$f = r_m(k)r_m(k-1)\cdots r_m(1)$$

be defined over $\mathbf{Z}_m$. We are assured that (ii) as well as (i) is satisfied if we choose m to be a prime p such that $p \geq B$ and $p > k$. If $p \geq n$ as well, then r can be simplified to

$$(**) \qquad \left.\begin{array}{l} r = e/f \\ \text{where } e = n(n-1)\cdots(n-k+1), \quad f = k! \end{array}\right\} \text{ evaluated over } \mathbf{Z}_p.$$

With the modulus p so chosen, step 1 of Algorithm 1 computes r according to $(**)$, steps 2 and 3 then return r for the (correct) value of C.

Let us try out this algorithm on a small example: $C(7, 3)$. Suppose that a rough calculation yields the bound $C < 50$. For the sake of motivation let us assume that our computer wordsize is two decimal digits. This would preclude the usual integer calculation for C—$(7 \cdot 6 \cdot 5)/(3 \cdot 2 \cdot 1)$—since the numerator would overflow. (These considerations would obviously scale up for larger problems and computers; we are dealing with small numbers for the sake of simplicity.) Thus for our modulus p we can choose any prime $50 \leq p < 100$. Let us choose $p = 53$.

Step 1 of Algorithm 1 then computes over $\mathbf{Z}_{53}$:

$$e = 7 \cdot 6 \cdot 5 = 51,$$
$$f = 3 \cdot 2 \cdot 1 = 6,$$
$$f^{-1} = 9,$$
$$r = e/f = 51 \cdot 9 = 35.$$

Steps 2 and 3 then return $r = 35$ as the value for $C(7, 5)$. Thus our algorithm has correctly computed $C(7, 3) = 35$ using modular arithmetic, with the intermediate values e, f, f^{-1} all fitting the two decimal digit wordsize of our computer.

Since the validity of our algorithm is insensitive to which prime p is used in the range $50 \leq p < 100$, let us use another such prime, say $p = 97$, and check that we do indeed obtain the same result. Repeating the computation only this time over $\mathbf{Z}_{97}$, we obtain

$$e = 7 \cdot 6 \cdot 5 = 16,$$
$$f = 3 \cdot 2 \cdot 1 = 6,$$
$$f^{-1} = 81,$$
$$r = e/f = 16 \cdot 81 = 35,$$

so again we arrive at the correct answer for $C(7, 5)$, through quite different intermediate results. $\square$

Thus far our homomorphic image scheme has been restricted to the computation of integer-valued expressions involving *integer* arguments. In Exercises 5–8 the homomorphic image scheme is extended to accommodate integer-valued expressions involving *rational* arguments.

$$* \quad * \quad *$$

To recap, our homomorphic image scheme for computing an integer-valued expression $h = e(\cdots)$ is to perform the following two steps for an appropriate (in particular, a sufficiently large) modulus m:

1. Compute $r = e(\cdots)$ over $\mathbf{Z}_m$;
2. Recover h from $r \, (= h \bmod m)$.

The motivation behind this "indirect" method for computing h lies solely in the attractive properties of $\bmod m$ arithmetic: First, all results are bounded by m. Second, by choosing m to be a prime p, we can employ algorithms for the computation of step 1 that require and exploit the structure of a field. In this case the homomorphic image scheme in essence lets us exploit the simplicity of arithmetic in the modular field $\mathbf{Z}_p$, in contrast with the "direct" (pre-image) method which would be encumbered by the complexity of rational arithmetic. (Example 1 in particular illustrated this point.)

But now we must face the following problem: What if, in the homomorphic image scheme, we are forced to use a very large modulus, larger say than the wordsize of our computer? Then the scheme loses much if not all of its attractiveness. After all, the singular appeal of the homomorphic image method is that it allows us to exploit the relative ease of $\bmod\, m$ arithmetic (compared with the relative complexity of unrestricted integer and rational arithmetic). But $\bmod\, m$ arithmetic is only easy when m is not too large.

The solution to this possible quandary is the one we have already sketched in the overview: to compute in step 1 of the homomorphic image scheme not a single residue of h but a *family* of residues

$$r_k = h \bmod m_k \qquad (k = 0, \ldots, n-1)$$

for an appropriately chosen collection of n moduli m_k, where in particular these moduli are chosen to be conveniently small—within the wordsize of our machine. In essence we have replaced a single "big" problem by n "small" problems (a "divide-and-conquer" strategy that pays dividends in many areas of computing).

But now with these r_k's in hand, we must perform a generalization of step 2: recover h from the r_k's. The solution to this problem, completely algebraic in character, goes back to Chinese mathematics of antiquity and is the topic of the next section. From this solution will come the application of the homomorphic image technique to polynomial manipulation.

2. CHINESE REMAINDER AND INTERPOLATION ALGORITHMS

The problem of computing solutions to systems of simultaneous linear congruences over the integers, such as

$$u \equiv 1 \ (\bmod\, 5),$$
$$u \equiv 0 \ (\bmod\, 7),$$
$$u \equiv 4 \ (\bmod\, 8),$$

was studied by Chinese mathematicians as far back as the first century A.D. In deference to this historical background, we call the general problem of solving a system of n linear congruences

$$u \equiv r_k \ (\bmod\, m_k) \qquad (k = 0, \ldots, n-1)$$

over a Euclidean domain the *Chinese remainder problem* (CRP).

It is readily checked that $u = 196$ is a solution to the above integer system. Our goal in this section is to derive and analyze algorithms for computing solutions to CRPs in a variety of abstract and concrete settings.

2.1 A CRA for Euclidean Domains

Throughout, D denotes a Euclidean domain equipped with the usual parapher-nalia—a degree function d, operations mod and div, etc.

The two-congruence case. We consider the CRP over D for the case of two congruences: the problem of finding a solution $U \in D$ to

$$(1) \quad u \equiv r \,(\mathrm{mod}\, m),$$

$$(2) \quad u \equiv s \,(\mathrm{mod}\, n),$$

where $r, s, m, n \in D$. We make the assumption that $(m, n) = 1$. (The CRP makes sense without this relatively-prime assumption for the moduli, but is more complicated and not relevant to our applications.)

The solution to congruence (1) is given by

$$U = r + \sigma m,$$

where $\sigma \in D$ is arbitrary. Our goal is to choose σ so that congruence (2) is satisfied.

We note that $r + \sigma m \equiv s \,(\mathrm{mod}\, n)$ if $\sigma m \equiv s - r \,(\mathrm{mod}\, n)$. Since $(m, n) = 1$, $m^{-1} \bmod n = c$ exists, which satisfies $cm \equiv 1 \,(\mathrm{mod}\, n)$. Thus, if we choose $\sigma = (s - r)c \bmod n$, then

$$U = \quad r + \sigma m$$

$$\equiv_n r + (s - r)cm$$

$$\equiv_n r + (s - r)1$$

$$\equiv_n s,$$

so that $U = r + \sigma m$ satisfies both congruences (1) and (2). We conclude

Theorem 1. Let D be a Euclidean domain and let $m, n \in D$ be relatively prime. Then the two congruence CRP

$$(*) \qquad \begin{aligned} u &\equiv r \,(\mathrm{mod}\, m), \\ u &\equiv s \,(\mathrm{mod}\, n) \end{aligned}$$

has a solution $U \in D$, which can be computed according to the following *Chinese remainder algorithm* (CRA):

Algorithm 1 *(Two-congruence CRA).*

Input.   Two congruence CRP $(*)$ over a Euclidean domain D.
Output.   A solution $U \in D$.

    begin
1.   c := INVERSE(m, n);
2.   σ := [(s − r) × c] mod n;
3.   U := r + σ × m
    end □

Remark. From our derivation of the above algorithm, it is evident that σ in line 2 could be defined by any element congruent $\bmod n$ to $(s - r)c$, including $(s - r)c$ itself. Our choice of $(s - r)c \bmod n$ was predicated on making σ "small" [here $d(\sigma) < d(n)$]. In general, small solutions to congruences in Euclidean domains are to be preferred to reduce the cost (in both space and time) of subsequent computations involving those solutions.

Example 1. We apply Algorithm 1 to compute a solution over $\mathbf{Z}$ to

$$u \equiv 6 \ (\bmod 7),$$

$$u \equiv 3 \ (\bmod 9).$$

Algorithm 1 returns $c = 7^{-1} \bmod 9 = 4$, $\sigma = (3 - 6)4 \bmod 9 = 6$, and finally $U = 6 + 6 \times 7 = 48$, which is readily checked to be a solution. $\square$

The reader can readily check that the system of congruences of Example 1 also has solutions -15 and 111 (besides others). We shall soon see how to get all the solutions to such system of congruences in terms of a particular solution. But first we generalize the setting from two to n congruences.

The n-congruence case. We now consider the n-congruence CRP over a Euclidean domain D:

$$u \equiv r_0 \quad (\bmod m_0),$$

$$u \equiv r_1 \quad (\bmod m_1),$$

$$\vdots$$

$$u \equiv r_{n-1} \ (\bmod m_{n-1}),$$

where the moduli are *pairwise relatively prime*: $(m_i, m_j) = 1$ if $i \neq j$. *Henceforth when referring to the CRP, we shall always make the tacit assumption that the moduli are pairwise relatively prime.* Each r_i is typically (though need not be) in the range of its respective $\bmod m_i$ function (e.g., in the integer case, $0 \leq r_i < m_i$), which accounts for our calling these r_i's "residues" (= remainders).

Let $m_0, \ldots, m_{n-1} \in D$ ($n \geq 1$) be pairwise relatively prime. For $0 \leq k < n$, we define

$$M_k = 1 \quad \text{if } k = 0$$

$$= \prod_{i=0}^{k-1} m_i \quad \text{otherwise}$$

to be used consistently throughout.

Lemma 1. M_k and m_k are relatively prime, $0 \leq k < n$.

Proof. If $k = 0$, then $(M_k, m_k) = (1, m_k) = 1$.

For $k > 0$, suppose to the contrary that M_k and m_k have a nonunit g.c.d. g. Let p be a prime factor of g. Then $p \mid M_k$ and $p \mid m_k$. But $p \mid \prod_{i=0}^{k-1} m_i$ implies that

$p \mid m_i$ for some $0 \le i \le k - 1$ (Corollary 2 to Lemma 5, Section IV.4.2). This gives p as a common divisor of m_i and m_k, which contradicts $(m_i, m_k) = 1$ and forces the desired conclusion $(M_k, m_k) = 1$. $\square$

The main idea that underlies the derivation of our CRA for the solution of

$$u \equiv r_k \,(\mathrm{mod}\, m_k) \qquad (k = 0, \ldots, n - 1)$$

is to devise a program loop for manipulating two program variables U and M so that after k iterations ($k = 0, \ldots, n - 1$), the following property $p(k)$ holds:

$$p(k): \begin{cases} M = M_k, \\ U \equiv r_i \,(\mathrm{mod}\, m_i) \qquad (i = 0, \ldots, k). \end{cases}$$

After the $(n - 1)$st iteration, U will then be a solution to the given CRP.

These considerations lead to the following high-level program:

```
        begin
1.          M := 1;
2.          U := r₀ mod m₀;
3.          for k := 1 until n − 1 do
4.              redefine M, U by M′, U′ so that p(k) holds
        end.
```

After zero iterations of the **for** loop of line 3, the values of M and U are as computed by lines 1 and 2, so that $p(k)$ holds for $k = 0$.

By induction we may assume that after the $(k - 1)$st iteration ($1 \le k < n$), $p(k - 1)$ holds:

$$M = M_{k-1},$$
$$U \equiv r_i \,(\mathrm{mod}\, m_i) \qquad (i = 0, \ldots, k - 1).$$

Our task is to determine M' and U' so that $p(k)$ holds after the kth iteration. M' is easy: $M' = M_k = M_{k-1} m_{k-1}$, so that M' can be computed as $M m_{k-1}$.

U': We note that $(m_k, M_k) = 1$ by Lemma 1. Hence we can use Algorithm 1 to compute a solution U' to the two-congruence CRP

$$(1) \quad u \equiv U \,(\mathrm{mod}\, M_k),$$

$$(2) \quad u \equiv r_k \,(\mathrm{mod}\, m_k).$$

We claim that U' satisfies $p(k)$. For by (1), $U' \equiv U \,(\mathrm{mod}\, M_k)$, so that $U' \equiv U$ $(\mathrm{mod}\, m_i)$ for $i = 0, \ldots, k - 1$. By the induction hypothesis, we then have $U' \equiv r_i$ $(\mathrm{mod}\, m_i)$ for $i = 0, \ldots, k - 1$, which when combined with (2) completes the proof of the claim that U' satisfies $p(k)$.

We have essentially proved

Theorem 2. Over a Euclidean domain D, the n-congruence CRP

$$u \equiv r_k \,(\mathrm{mod}\, m_k) \qquad (k = 0, \ldots, n - 1)$$

has a solution $U \in D$, which can be computed according to the following Chinese remainder algorithm:

Algorithm 2 (*n-congruence CRA*).

Input. CRP over a Euclidean domain D:

$$u \equiv r_k \pmod{m_k} \qquad (k = 0, \dots, n-1).$$

Output. A solution $U \in D$.

```
    begin
1.      M := 1;
2.      U := r₀ mod m₀;
3.      for k := 1 until n − 1 do
        begin
4.          M := M × m_{k−1};
5.          c := INVERSE(M, m_k);
6.          σ := [(r_k − (U mod m_k)) × c] mod m_k;
7.          U := U + σ×M
            {M = M_k; U ≡ r_i (mod m_i) for i = 0, ..., k}
        end
    end {U ≡ r_k (mod m_k) for k = 0, ..., n − 1}  □
```

Note that lines 5-7 redefine U by a solution U' to the two-congruence CRP

$$u \equiv U \pmod{M = M_k},$$

$$u \equiv r_k \pmod{m_k},$$

as required.

A useful property of Algorithm 2 is given by

Theorem 3. After the kth iteration $(k = 0, \dots, n-1)$, Algorithm 2 has computed

$$M = M_k,$$

$$U = U_k = \sigma_0 M_0 + \sigma_1 M_1 + \cdots + \sigma_k M_k$$

where each σ_i is in the range of the corresponding m_i function.

Proof. Induction. In particular, $\sigma_0 = r_0 \bmod m_0$ (line 2) and σ_k is the value of σ computed at the kth iteration (line 6). □

Corollary. The solution U computed by Algorithm 2 has the form

$$U = \sigma_0 M_0 + \sigma_1 M_1 + \cdots + \sigma_{n-1} M_{n-1}$$

(each σ_i is in the range of the corresponding m_i function).

Proof. $U = U_{n-1}$. □

Example 2. We compute a solution U to the system of congruences over $\mathbf{Z}$

$$u \equiv 1 \pmod 3,$$
$$u \equiv 3 \pmod 5,$$
$$u \equiv 0 \pmod 7,$$
$$u \equiv 10 \pmod{11}.$$

Since the moduli are distinct primes, they are pairwise relatively prime as required. Our n-congruence CRA computes the following tableau:

k	r_k	m_k	M	c	σ	U
0	1	3	1	—	—	1
1	3	5	3	2	4	13
2	0	7	15	1	1	28
3	10	11	105	2	8	868

Thus $U = 868$ is a solution to the given CRP. This solution can also be expressed as

$$U = \sigma_0 M_0 + \sigma_1 M_1 + \sigma_2 M_2 + \sigma_3 M_3$$
$$= \sigma_0 + \sigma_1 m_0 + \sigma_2 m_0 m_1 + \sigma_3 m_0 m_1 m_2$$

where $\sigma_0 = 1$, $\sigma_1 = 4$, $\sigma_2 = 1$, $\sigma_3 = 8$. $\square$

Algorithm 2 gives us a particular solution to the n-congruence CRP. Now for the general solution.

Lemma 2. If $U \equiv V \pmod{m_i}$ for $i = 0, \ldots, k - 1$, then $U \equiv V \pmod{M_k}$. (As usual the m_i's are pairwise relatively prime in a Euclidean domain, and $M_k = \prod_{i=0}^{k-1} m_i$.)

Proof. By induction on k.

Basis. The case $k = 1$ holds trivially. We prove the case $k = 2$. If $U \equiv V \pmod{m_0}$ and $U \equiv V \pmod{m_1}$, then $m_0 | (U - V)$ and $m_1 | (U - V)$. But $(m_0, m_1) = 1$. Hence $m_0 m_1 | (U - V)$ by Lemma 5(b) of Section IV.4.2, so that $U \equiv V \pmod{M_2 = m_0 m_1}$ as required.

Induction. Assume Lemma 2 for k moduli ($k \geq 2$): $U \equiv V \pmod{m_i}$ for $0 \leq i \leq k - 1$ implies $U \equiv V \pmod{M_k}$. Now assume $U \equiv V \pmod{m_i}$ for $0 \leq i \leq k$. (Our task is to show that $U \equiv V \pmod{M_{k+1}}$.)

We have (1) $U \equiv V \pmod{M_k}$ by the induction hypothesis, (2) $U \equiv V \pmod{m_k}$ by assumption, and (3) $(M_k, m_k) = 1$ by Lemma 1. Hence $U \equiv V \pmod{M_k m_k = M_{k+1}}$ by the two-moduli basis, which completes the proof by induction of Lemma 2. $\square$

Theorem 4. Over a Euclidean domain D, any system of congruences

$(*)$ $\qquad\qquad u \equiv r_k \pmod{m_k} \qquad (k = 0, \ldots, n - 1)$

is equivalent to a single congruence

$(**)$ $\qquad\qquad\qquad\qquad u \equiv U \,(\mathrm{mod}\,M)$

where $U \in D$ is a particular solution to $(*)$ and $M = \prod_{i=0}^{n-1} m_i$ (m_i's pairwise relatively prime). [By *equivalent* is mean that any solution to $(*)$ is a solution to $(**)$ and conversely.]

Proof. If V is a solution to $(**)$, then $V \equiv U \,(\mathrm{mod}\,M)$, so that $V \equiv U$ $(\mathrm{mod}\,m_k)$, trivially. But then $V \equiv r_k \,(\mathrm{mod}\,m_k)$, since U is a solution to $(*)$, which is to say that V is a solution to $(*)$.

In the other direction, if V is a solution to $(*)$, then $V \equiv r_k \equiv U \,(\mathrm{mod}\,m_k)$. But the m_k's are pairwise relatively prime. So by Lemma 2, we obtain $V \equiv U$ $(\mathrm{mod}\,\prod_{i=0}^{n-1} m_i = M)$, which is to say that V is a solution to $(**)$. $\square$

Since the solutions to the CRP $(*)$ are all given by the solutions to the single congruence $(**)$, we immediately conclude the

Corollary. If $U \in D$ is a solution to the CRP $(*)$, then so is $U + tM$ for arbitrary $t \in D$; any solution U is unique mod M.

2.2 A CRA for Z

We continue to focus on the CRP

$(*)$ $\qquad\qquad\qquad u \equiv r_k \,(\mathrm{mod}\,m_k) \qquad (k = 0, \ldots, n-1)$

for relatively prime moduli m_k, only now we take our Euclidean domain to be the integers. For convenience we assume that the moduli are positive.

Theorem 1 (*Preferred solutions to the integer CRP*).

(a) There is a unique solution U to $(*)$ that satisfies $0 \le U < M = \prod_{i=0}^{n-1} m_i$ ("the least positive solution").

(b) There is a unique solution V to $(*)$ that satisfies $-M/2 \le V < M/2$ ("the least absolute value solution"), namely $V = \beta_M(U)$ where U is the (unique) least positive solution of (a). (Recall:

$$\beta_M(U) = U \qquad \text{if } U < M/2$$

$$= U - M \quad \text{otherwise.)}$$

Proof. This theorem generalizes Theorem 1 of Section 1.2; our goal is to reduce the present theorem to that one.) Let T be any solution to the given CRP $(*)$. Then $U = T \,\mathrm{mod}\,M$ is also a solution (Corollary to Theorem 4 of Section 2.1), moreover a solution that satisfies $0 \le U < M$. By Theorem 4 of Section 2.1, the given CRP $(*)$ is equivalent to the single congruence

$$u \equiv U \,(\mathrm{mod}\,M) \quad \text{where } 0 \le U < M.$$

The present theorem now follows immediately from Theorem 1 of Section 1.2.

$\square$

When M is odd (as will always be the case if the moduli m_i are distinct primes > 2—the usual case for us), then it is often convenient to replace the condition $-M/2 \leq V < M/2$ by the equivalent $|V| \leq (M-1)/2$.

Example 1. Let us return to Example 2 of Section 2.1. There we computed a solution $U = 868$ to a CRP with $M = 3 \times 5 \times 7 \times 11 = 1155$. Thus

(a) $U = 868$ is the least positive solution—the unique solution satisfying $0 \leq U < M$;

(b) $\beta_M(U) = 868 - 1155 = -287$ is the least absolute value solution—the unique solution V satisfying $|V| \leq (M-1)/2 = 577$. $\square$

We note from the above example that our CRA, Algorithm 2 of Section 2.1, has computed the least positive solution. We now establish this useful property in general.

According to Theorem 3 of Section 2.1, the successive values U_k ($k = 0, \ldots, n-1$) taken on by the program variable U are given by

$$U_k = \sigma_0 + \sigma_1 m_0 + \sigma_2 m_0 m_1 + \cdots + \sigma_k m_0 m_1 \cdots m_{k-1}$$

where each σ_i is in the range of the $\bmod\, m_i$ function. Hence $0 \leq \sigma_i < m_i$, and we obtain

$$0 \leq U_k \leq (m_0 - 1) + (m_1 - 1)m_0 + (m_2 - 1)m_0 m_1 + \ldots$$

$$+ (m_k - 1)m_0 m_1 \cdots m_{k-1}$$

$$= m_0 m_1 \ldots m_k - 1$$

which we record as

Lemma 1. In Algorithm 2, the value U_k of U after the kth iteration satisfies

$$0 \leq U_k < M_{k+1} \left(= \prod_{i=0}^{k} m_i \right) \qquad (k = 0, \ldots, n-1).$$

Since the solution U returned by Algorithm 2 of Section 2.1 is U_{n-1}, we conclude

Theorem 2. Let our CRA (Algorithm 2 of Section 2.1) be applied to the integer CRP $u \equiv r_k \pmod{m_k}$ ($k = 0, \ldots, n-1$). Then the computed solution U satisfies $0 \leq U < M$ ($M = \prod_{i=0}^{n-1} m_i$) i.e., U is the (unique) least positive solution.

In computing practice the two preferred solutions to the integer CRP, least positive and least absolute value, are about equally useful. It therefore behooves us in the integer case to organize our CRA, Algorithm 2 of Section 2.1, in the form of two procedures: (1) CRA($\mathbf{r}, \mathbf{m}, n$) and (2) βCRA($\mathbf{r}, \mathbf{m}, n$) that return (1) the least positive and (2) the least absolute value solutions to the integer CRP

$u \equiv r_k \pmod{m_k}$ $(k = 0, \ldots, n - 1)$. The formal arguments $\mathbf{r}$ and $\mathbf{m}$ designate the lists of values $(r_0, \ldots, r_{n-1})$ and $(m_0, \ldots, m_{n-1})$.

Algorithm 1 (*Least positive CRA*).

     procedure CRA($\mathbf{r}, \mathbf{m}, n$);
       execute Algorithm 2 of Section 2.1;
       return U $\{0 \le U < M\}$
    end $\square$

Algorithm 2 (*Least absolute value CRA*).

     procedure βCRA($\mathbf{r}, \mathbf{m}, n$);
       execute Algorithm 2 of Section 2.1;
       $V :=$ **if** $U < M/2$ **then** U **else** $U - M$;
       return V $\{-M/2 \le V < M/2\}$
    end $\square$

For an alternative least absolute value CRA, see Exercise 8.

Timing analysis of the integer CRA algorithm. In our algebraic computing applications of the CRA, we shall deal with CRPs of the form

$$u \equiv r_k \pmod{m_k} \qquad (k = 0, \ldots, n - 1)$$

for which the r_k's and m_k's are $\le W$, our computer's wordsize. Typically the m_k's are distinct wordsize ($\le W$) primes, hence satisfying the pairwise-relatively-prime assumption. Under the assumption that the r_k's and m_k's are single-precision integers, we now wish to bound the time required by our CRA, Algorithm 2 of Section 2.1, for computing a solution U to the above n-congruence CRP.

Referring to Algorithm 2 of Section 2.1, we see that the time to compute the solution U is essentially the time to execute the **for** loop of line 3. Let us examine the kth iteration of this loop:

Line 4: Before execution of line 4, M has value $M_{k-1} = \prod_{i=0}^{k-2} m_i$ (Theorem 3 of Section 2.1), hence is $\le (k-1)$-precision. The computation $M \times m_{k-1}$ therefore requires $O(k)$ time.

Line 5: M has value $M_k = \prod_{i=0}^{k-1} m_i$, hence is $\le k$-precision. INVERSE(M, m_k) is essentially computed as INVERSE($M \bmod m_k, m_k$). The computation $M \bmod m_k$ requires time $O(k)$ (the time to divide a k-precision integer by a single-precision integer). INVERSE then requires constant (i.e., independent of n) time to compute $M^{-1} \bmod m_k$ (actually $O(\log W)$ time by Theorem 1 of Section VII.3.2). Hence line 5 requires $O(k)$ time overall.

Line 6: At line 6, U has the value U_{k-1} computed at the $(k-1)$st iteration. By Lemma 1, $U_{k-1} < \prod_{i=0}^{k-1} m_i$, hence is $\le k$ precision. The computation $U \bmod m_k$ therefore requires $O(k)$ time. Since r_k, $U_{k-1} \bmod m_k$, and c are all

$\leq W$, the $\bmod m$ subtraction and multiplication require constant time. Hence line 6 requires $O(k)$ time overall.

Line 7: The computation $\sigma \times M$ requires $O(k)$ time as does the indicated addition. Hence line 7 requires $O(k)$ time.

Therefore the kth iteration of the **for** loop of line 3 requires time $T_k \leq Ck$ for some constant C. Thus the time T for the execution of the entire algorithm is $\leq \Sigma_{k=1}^n Ck = O(n^2)$ (where we have used $\Sigma_{k=1}^n k = n(n+1)/2 = O(n^2)$).

Noting that the time to compute V from U in the least absolute value CRA (Algorithm 2 above) is $O(n)$, we conclude

Theorem 3. Procedures CRA and βCRA compute the least positive and least absolute value solutions to the single-precision integer CRP

$$u \equiv r_k \ (\bmod m_k) \qquad (k = 0, \ldots, n-1)$$

in time $O(n^2)$.

Systems of polynomial congruences over $\mathbf{Z}[x]$. We are interested in finding a solution $U(x) \in \mathbf{Z}[x]$ to a system of polynomial congruences of the form

$$u(x) \equiv \rho_k(x) \ (\bmod m_k) \qquad (k = 0, \ldots, n-1)$$

where $\rho_k(x) = \Sigma_{j=0}^q \rho_{jk} x^j \in \mathbf{Z}[x]$.

The following lemma serves to reduce this problem to a family of integer CRPs.

Lemma 2. Let $a(x) = \Sigma_i a_i x^i$, $b(x) = \Sigma_i b_i x^i \in \mathbf{Z}[x]$, $m \in \mathbf{P}$. Then

$$a(x) \equiv b(x) \ (\bmod m) \quad \Leftrightarrow \quad a_i \equiv b_i \ (\bmod m).$$

Proof. $m \mid (a(x) - b(x)) \quad \Leftrightarrow \quad m \mid (a_i - b_i)$. $\square$

Theorem 4. Let $\rho_k(x) = \Sigma_{j=0}^q \rho_{jk} x^j \in \mathbf{Z}[x]$ and $m_k > 0$ be given $(k = 0, \ldots, n-1)$, where the m_k's are pairwise relatively prime. Let $M = \Pi_{i=0}^{n-1} m_i$. Then the system of polynomial congruences

$$(*) \qquad\qquad u(x) \equiv \rho_k(x) \ (\bmod m_k) \qquad (k = 0, \ldots, n-1)$$

has unique solutions $U(x) = \Sigma_{j=0}^q U_j x^j$, $V(x) = \Sigma_{j=0}^q V_j x^j$ that satisfy

(a) $0 \leq U_j < M$ ("least-positive coefficient solution"),
(b) $-M/2 \leq V_j < M/2$ ("least-absolute-value coefficient solution").

Either of these solutions can be computed in $O(qn^2)$ time.

Proof. By Lemma 2, the jth coefficient of any solution polynomial to $(*)$ is a solution to the integer CRP

$$u \equiv \rho_{jk} \ (\bmod m_k) \qquad (k = 0, \ldots, n-1).$$

The existence and uniqueness of U_j and V_j satisfying (a) and (b) then follows from Theorem 1. By Theorem 3, each U_j or V_j can be computed (using procedure CRA or βCRA) in time $O(n^2)$; hence $U(x)$ or $V(x)$ can be computed in time $O(qn^2)$. $\square$

Example 2. Solve:

$$u(x) \equiv x + 2 \pmod 3,$$

$$u(x) \equiv x^2 + 4 \pmod 5.$$

Solution: Any solution $ax^2 + bx + c$ satisfies

$$a \equiv 0 \pmod 3, \quad b \equiv 1 \pmod 3, \quad c \equiv 2 \pmod 3,$$

$$a \equiv 1 \pmod 5, \quad b \equiv 0 \pmod 5, \quad c \equiv 4 \pmod 5.$$

The least positive solutions to these systems of integer congruences are readily computed to be $a = 6$, $b = 10$, $c = 14$; the least absolute value solutions are $a = 6$, $b = -5$, $c = -1$. Hence $U(x) = 6x^2 + 10x + 14$ and $V(x) = 6x^2 - 5x - 1$ are the (unique) least-positive and least-absolute-value coefficient solutions (having degree ≤ 2) to the given system of polynomial congruences. $\square$

2.3 A CRA for F[x]: Interpolation

In Section IV.3.3 we posed and solved the n-point *interpolation problem* over a field F: for given $\alpha_k, \beta_k \in F$ $(k = 0, \ldots, n - 1$; α_k's distinct), find $U(x) \in F[x]$ that satisfies

$$U(\alpha_k) = \beta_k \qquad (k = 0, \ldots, n - 1).$$

We presented a solution of degree $< n$, Lagrange's Interpolation Formula, and also showed the uniqueness of that solution.

What has all this got to do with the Chinese remainder problem? From Section IV.3.3 we recall the Remainder Theorem that tells us that the remainder on dividing $u(x) \in F[x]$ by $x - \alpha$ $(\alpha \in F)$ is $u(\alpha)$:

$$u(x) = (x - \alpha)q(x) + u(\alpha).$$

Hence $x - \alpha$ divides $u(x) - u(\alpha)$, and we immediately conclude

Lemma 1. $u(\alpha) = \beta \iff u(x) \equiv \beta \pmod{x - \alpha}$.

From Lemma 1, we see that the above interpolation problem over F can be formulated as a special case of the CRP over (the Euclidean domain) $F[x]$: Find $U(x) \in F[x]$ that is a solution to

$$u(x) \equiv \beta_k \pmod{x - \alpha_k} \qquad (k = 0, \ldots, n - 1).$$

This polynomial CRP is special in that the moduli are linear. Since the α_k's are distinct, the $x - \alpha_k$'s are distinct irreducible polynomials over F. Hence the moduli satisfy our pairwise-relatively-prime restriction.

The payoff in transforming our original interpolation problem to a Chinese remainder problem is this: *we can now compute a solution to the interpolation problem by our general Chinese remainder algorithm* (Algorithm 2 of Section 2.1).

The following two lemmas lead to a conceptual simplification of our general CRA.

Lemma 2. Let $u(x) \in F[x]$, $\alpha \in F$. Then

$$u(x) \bmod x - \alpha = u(\alpha).$$

(In words: Polynomial remaindering by $x - \alpha$ is equivalent to polynomial evaluation at $x = \alpha$.)

Proof. Factor Theorem. $\square$

Lemma 3. Let $u(x) \in F[x]$, $\alpha \in F$, with $u(\alpha) \neq 0$. Then

$$u(x)^{-1} \bmod x - \alpha = u(\alpha)^{-1}.$$

Proof. By definition, $u(x)^{-1} \bmod x - \alpha$ is the (unique) polynomial $v(x)$ of degree $< \deg x - \alpha$ that satisfies $u(x)v(x) \equiv 1 \pmod{x - \alpha}$. Hence $v(x) = v$, a constant. Let us check that $v = u(\alpha)^{-1}$ satisfies $u(x)v \equiv 1 \pmod{x - \alpha}$:

$$u(x)u(\alpha)^{-1} \equiv_{x-\alpha} [u(x) \bmod x - \alpha] u(\alpha)^{-1}$$

$$= u(\alpha)u(\alpha)^{-1} \qquad \text{Lemma 2}$$

$$= 1. \ \square$$

Lemmas 1, 2, and 3 serve to "translate" statements about polynomial congruence and remaindering (with respect to a linear modulus $x - \alpha$) to corresponding statements about polynomial evaluation (at the point $x = \alpha$).

We now translate (specialize) our general CRA (Algorithm 2 of Section 2.1) to our interpolation problem setting, replacing $\bmod x - \alpha$ remaindering by evaluation at $x = \alpha$ (Lemma 2) and $\bmod x - \alpha$ inversion by evaluation at $x = \alpha$ followed by inversion in F (Lemma 3). We call the resulting algorithm *Newton's Interpolation Algorithm* in honor of its discoverer, Sir Isaac Newton. Its correctness is an immediate consequence of the correctness of our general CRA.

Theorem 1. A solution $U(x)$ to the n-point interpolation problem over F, find $U(x) \in F[x]$ such that

$$U(\alpha_k) = \beta_k \qquad (k = 0, 1, \ldots, n - 1),$$

is computed by

Algorithm 1 (*Newton's Interpolation Algorithm*).

> **begin**
> 1. $U(x) := \beta_0$;
> 2. $M(x) := 1$;
> 3.   **for** $k := 1$ **until** $n - 1$ **do**
>   **begin**
> 4. $M(x) := M(x) \times (x - \alpha_{k-1})$;
> 5. $c := M(\alpha_k)^{-1}$;
> 6. $\sigma := (\beta_k - U(\alpha_k)) \times c$;
> 7. $U(x) := U(x) + \sigma \times M(x)$
>   **end**
> **end** $\square$

Example 1. Compute $U(x) \in Z_7[x]$ that interpolates to the samples

$\alpha_k = k$	$\beta_k = U(\alpha_k)$
0	3
1	0
2	6
3	2

Newton's Interpolation Algorithm computes the following tableau:

k	α_k	β_k	$M(x)$	c	σ	$U(x)$
0	0	3	1	—	—	3
1	1	0	x	1	4	$4x + 3$
2	2	6	$x^2 + 6x$	4	1	$x^2 + 3x + 3$
3	3	2	$x^3 + 4x^2 + 2x$	6	5	$5x^3 + 6x + 3$

Thus $U(x) = 5x^3 + 6x + 3$ is a solution to the given interpolation problem. $\square$

Theorem 3 of Section 2.1 becomes

Theorem 2. After the kth iteration $(k = 0, \ldots, n - 1)$ Newton's Interpolation Algorithm computes

$$M(x) = M_k(x) = \prod_{i=0}^{k-1} (x - \alpha_i),$$

$$U(x) = U_k(x) = \sigma_0 M_0 + \sigma_1 M_1 + \cdots + \sigma_k M_k$$
(where $\sigma_0 = \beta_0 \in F$, and $\sigma_k \in F$ is the value of σ computed at the kth iteration).

Corollary. For the n-point interpolation problem $U(\alpha_k) = \beta_k$ $(k = 0, \ldots, n - 1)$, Newton's Interpolation Algorithm computes the unique solution having degree $< n$. Furthermore this solution can be expressed in the form

$$U(x) = \sigma_0 + \sigma_1(x - \alpha_0) + \sigma_2(x - \alpha_0)(x - \alpha_1) + \cdots$$
$$+ \sigma_{n-1}(x - \alpha_0)(x - \alpha_1) \cdots (x - \alpha_{n-2}).$$

Proof. Since the solution $U(x)$ returned by Newton's Interpolation Algorithm is $U_{n-1}(x)$, the corollary apart from uniqueness follows immediately from Theorem 2.

As for uniqueness, since deg $U(x) < n$, $U(x)$ must be the unique interpolating polynomial (having degree $< n$) that solves the given n-point interpolation problem (Theorem 5 of Section IV.3.3). $\square$

Remark. Uniqueness can also be argued as a consequence of our general CRP uniqueness result of the Corollary to Theorem 4, Section 2.1—Exercise 5.

Example 2. From the tableau of Example 1, the (unique) interpolating polynomial $U(x)$ of degree < 4 can be expressed as

$$U(x) = 3 + 4x + 1x(x-1) + 5x(x-1)(x-2). \quad \square$$

We have now derived two different solutions to the n-point interpolation problem over a field F, finding $U(x) \in F[x]$ of degree $< n$ satisfying $U(\alpha_k) = \beta_k$ $(k = 0, \ldots, n-1)$:

(1)   The *Lagrangian solution* of Theorem 5, Section IV.3.3:

$$U(x) = \sum_{k=0}^{n-1} \beta_k L_k(x) \quad \text{where } L_k(x) = \prod_{\substack{i=0 \\ i \neq k}}^{n-1} \frac{x - \alpha_i}{\alpha_k - \alpha_i}.$$

(2)   The *Newtonian solution* as computed by Newton's Interpolation Algorithm:

$$U(x) = \sigma_0 + \sigma_1(x - \alpha_0) + \sigma_2(x - \alpha_0)(x - \alpha_1) + \cdots$$
$$+ \sigma_{n-1}(x - \alpha_0)(x - \alpha_1) \cdots (x - \alpha_{n-2}).$$

These two solutions of degree $< n$, though ostensibly quite different (at least in form), must of course yield the same unique interpolating polynomial. Both solutions are theoretically noteworthy. However, the Newtonian solution has a significant practical advantage over the Lagrangian solution of being *extensible:* an n-point solution $U(x)$ can be extended to accommodate a new $(n + 1)$st point (α_n, β_n) by executing another iteration of Newton's Interpolation Algorithm. The Lagrangian solution, on the other hand, is not extensible; the Lagrangian coefficients $L_k(x)$ would all have to be recomputed with the addition of a new point.

Timing analysis of Newton's Interpolation Algorithm.   Our goal is to establish a bound on the time required by Newton's Interpolation Algorithm to compute a solution to an n-point interpolation problem $U(\alpha_k) = \beta_k$ $(k = 0, \ldots, n-1)$ over a field F. We assume that the field operations $+, -, \times, ^{-1}$ of F can be carried out in time bounded by a constant.

We require a celebrated result about the work required to evaluate a polynomial.

Lemma 4.   A polynomial $a(x) = \sum_{i=0}^{n} a_i x^i$ of degree n over a ring can be evaluated in n additions and n multiplications.

Proof.   Use *Horner's rule* (nested multiplication):

$$a(x) = (\cdots ((a_n x + a_{n-1})x + a_{n-2})x + \cdots + a_1)x + a_0.$$

(See Exercise 3 of the Appendix to Chapter VII for an implementation of this formula.) $\square$

Thus a polynomial of degree n can be evaluated in $O(n)$ time if coefficient addition and multiplication can be carried out in constant $[O(1)]$ time.

Referring to Newton's Interpolation Algorithm (Algorithm 1), we see that the time to compute the solution $U(x)$ is essentially the time to compute the **for** loop of line 3. We examine the kth iteration of that loop.

Line 4: Before execution of line 4, $M(x)$ has value $M_{k-1}(x) = \prod_{i=0}^{k-2}(x - \alpha_i)$ (Theorem 2); hence $\deg M(x) = k - 1$. The computation $M(x) \times (x - \alpha_{k-1})$ therefore requires $O(k)$ time.

Line 5: Here $M(x)$ has value $\prod_{i=0}^{k-1}(x - \alpha_i)$, hence has degree k. By Lemma 4, the evaluation $M(\alpha_k)$ requires $O(k)$ time; by assumption the inversion in F requires $O(1)$ time. Hence the overall time for line 5 is $O(k)$.

Line 6: By Theorem 2, $U(x)$ has degree $\leq k$; hence the evaluation $U(\alpha_k)$ requires $O(k)$ time. The overall time for line 6 is also clearly $O(k)$.

Line 7: The indicated (polynomial) addition and multiplication require $O(k)$ time, clearly.

Thus the time to perform the kth iteration is $O(k)$, or Ck for some constant C, so that the time to perform the entire algorithms is $C\sum_{k=1}^{n}k = O(n^2)$. We have proved

Theorem 3.   Let F be a field with field operations computable in $O(1)$ time. Then for any n-point interpolation problem over F, Newton's Interpolation Algorithm computes the unique (degree $< n$) solution in $O(n^2)$ time.

* * *

In Note 1 we have highlighted the interesting history of our Chinese remainder and interpolation algorithms.

3. COMPUTATION BY MULTIPLE HOMOMORPHIC IMAGES

To compute a quantity h defined by some integer-valued expression, our homomorphic image scheme, Algorithm 1 of Section 1.3, essentially computes $r = h \bmod m$ over $\mathbf{Z}_m$ and recovers h from its residue r. A vital requirement for the recovery step to work is that the modulus m be sufficiently large. But if this "sufficiently large" requirement forces the use of an inconveniently large modulus, greater say than the wordsize of our computer, then as earlier remarked, the homomorphic image scheme loses much if not all of its attractiveness. After all, the main purpose of using the homomorphic image scheme is to avoid complicated arithmetic.

Our solution to this potentially inhibiting problem is the ingenious one that we first mentioned in the overview to this chapter. We extend the *single* homomorphic image scheme to a *multiple* homomorphic image (MHI) scheme in computing a family of residues r_k of h, $r_k = h \bmod m_k$, with respect to an appropriately chosen family of moduli m_k; in particular, each m_k is chosen to be conveniently small. The recovery step now is to obtain h from these r_k's; this we can now do by our Chinese remainder algorithm.

Perhaps the most interesting feature of the multiple homomorphic image scheme is that it provides a powerful computational tool for polynomial manipulation (something which the single image scheme did not do). But first we deal with the integer case.

3.1   The MHI Scheme for Z

In the MHI scheme for **Z** presented below, the product $M = \prod_{i=0}^{n-1} m_i$ of the n moduli m_i plays much the same role as the single modulus m of the single homomorphic image scheme, Algorithm 1 of Section 1.3.

Algorithm 1   (*MHI scheme for* **Z**).

Input.   Expression $e(x_1, \ldots, x_s) = e(x)$, defined over **Z** for arguments $a_1, \ldots, a_s \in \mathbf{Z}$.
Output. $h = e(a) \in \mathbf{Z}$.

VERSION 1.   (Bound B on $h \geq 0$ is available.) Let $0 \leq h < B$ and choose n pairwise relatively prime moduli $m_0, \ldots, m_{n-1}$ such that (i) their product $M = \prod_{i=0}^{n-1} m_i$ is $\geq B$ and (ii) $e(r_{m_k}(a))$ is defined over $\mathbf{Z}_{m_k}$ ($k = 0, \ldots, n-1$). Then proceed according to

Step 1   {Modular evaluation}.

For $k = 0, \ldots, n-1$:
(a) Compute $a_i^{(k)} = r_{m_k}(a_i)$ ($i = 1, \ldots, s$);
(b) Compute $r_k = e(a^{(k)})$ (over $\mathbf{Z}_{m_k}$).

Step 2   {CRA}.
Solve $u \equiv r_k \pmod{m_k}$ ($k = 0, \ldots, n-1$) for the least positive solution U (Algorithm 1 of Section 2.2).

Step 3   {Return value for h}.
Return $h = U$.

VERSION 2.   (Bound B on $|h|$ is available.) Let $|h| < B$, and choose moduli $m_0, \ldots, m_{n-1}$ as in Version 1, but with $M \geq 2B$. Then proceed as in Version 1, except in step 2 solve the system of congruences for the least absolute value solution V (Algorithm 2 of Section 2.2), and in step 3 return $h = V$. $\square$

In Fig. 3 we have described the setup of the integer MHI scheme by a mapping diagram (cf. Fig. 1 of Section 1.3). This diagram suggests that the MHI scheme is essentially a *transform* scheme, wherein the problem of computing an expression e over a "hard" domain **Z** is mapped into the problem of computing e over the product domain $\prod_k \mathbf{Z}_{m_k}$, i.e., computing e componentwise over the "easy" domains $\mathbf{Z}_{m_k}$. From this viewpoint, step 1(a) is the *forward transform* which converts each integer argument α to a *residue form* $\{\alpha^{(k)} = r_{m_k}(\alpha)\}_k$. Step

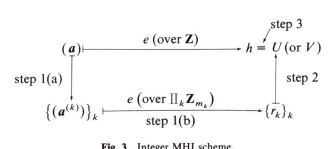

Fig. 3   Integer MHI scheme.

2(a) is then the *reverse transform*, converting a residue representation back to the normal form.

Remarks (about the moduli).   The n moduli are typically chosen to be the n largest primes $\leq$ the wordsize of our computer. (In Appendix 1 we take up the problem of computing such lists of primes.)

   1.   By choosing each modulus m_k to be a prime p_k, each component homomorphic image computation of step 1(b) is carried out over a field, namely $\mathbf{Z}_{p_k}$, with all that that implies for being able to perform division.

   2.   By choosing each prime p_k to be $\leq$ wordsize, we ensure that the $\bmod p_k$ operations are simple: a very few machine operations on wordsize integers.

   3.   By choosing large primes (though $\leq$ wordsize) rather than small ones we minimize the number of homomorphic image computations that must be carried out.

   4.   By choosing the primes to be distinct, we trivially satisfy the pairwise-relatively-prime restriction.

   Now for the proof of validity of Algorithm 1. Much of the work has been done in the validity proof of the single homomorphic image scheme.

   Theorem 1   (*The integer MHI scheme works*).   In Algorithm 1, for $h = e(a)$,

   (a) Version 1 computes $U = h$;
   (b) Version 2 computes $V = h$.

Proof.   As in the proof of Theorem 1, Section 1.3, $h = e(a)$ is a solution to the CRP

$$(*)\qquad\qquad u \equiv r_k \,(\bmod\, m_k)\qquad (k = 0,\ldots, n-1).$$

   (a) In Version 1, $0 \leq h < B \leq M$. Hence h, like U returned by step 2, is a least positive solution to the CRP $(*)$. By the uniqueness of this solution (Theorem 1(a) of Section 2.2), $U = h$.

   (b) In Version 2, $|h| < B \leq M/2$. Hence h, like V returned by step 2, is a least absolute value solution the CRP $(*)$. Again by uniqueness, $V = h$. $\square$

Example 1. The MHI scheme finds immediate application to the problem of solving integer systems of linear equations exactly (i.e., over **Q**), for which we developed a single homomorphic image method of solution in Example 1 of Section 1.3. (We shall appeal here to much of that development.)

Let $A\mathbf{x} = \mathbf{b}$ be a given system of n linear equations in n unknowns x_i with a_{ij}, b_i integers. Again recall Cramer's Rule: the solution to this system is given by $x_j = y_j/d$ where $d = \det A$ and $y_j = \det A_j$; A_j is the matrix obtained from A by replacing its jth column by the righthand side vector **b**.

In preparation for the application of Version 2 of Algorithm 1, let B be a bound on $\max(|d|, |y_1|, \dots, |y_n|)$ (see remark 3 at the conclusion of Example 1, Section 1.3). Now choose K distinct primes $p_0, \dots, p_{K-1}$ so that $M = \prod_{i=0}^{K-1} p_i$ is $\geq 2B$.

Step 1 of Algorithm 1 then requires the computation: for $k = 0, \dots, K-1$

$$(1) \qquad\qquad A' = r_{p_k}(A), \quad \mathbf{b}' = r_{p_k}(\mathbf{b}) \qquad \text{(step 1(a))}$$

and then

$$(2) \qquad\qquad d^{(k)} = \det A', \quad y_j^{(k)} = \det A'_j \ (j = 1, \dots, n) \qquad \text{(step 1(b))}.$$

Steps 2 and 3 then return d as the least absolute value solution to the CRP

$$(3) \qquad\qquad u \equiv d^{(k)} \ (\mathrm{mod}\, p_k) \qquad (k = 0, \dots, K-1)$$

and $y_j \ (j = 1, \dots, n)$ as the least absolute value solution to the CRP

$$(4) \qquad\qquad u \equiv y_j^{(k)} \ (\mathrm{mod}\, p_k) \qquad (k = 0, \dots, K-1).$$

Since $M \geq 2B$, we are assured from the validity of the general MHI scheme (Theorem 1) that steps 2 and 3 do indeed return d and y_j as claimed. The rational solution to the given system of equations is then $x_j = y_j/d$.

Just as in the single homomorphic image method of Example 1, Section 1.3, we compute the determinantal quantities of step 1 by Gaussian elimination over the fields $\mathbf{Z}_{p_k}$.

Computing time analysis. We now derive a bound on the time T required to solve an integer system $A\mathbf{x} = \mathbf{b}$ of n linear equations in n unknowns by the MHI scheme, where the a_{ij}'s and b_i's are $\leq m$ precision.

Note. In the context of the integer MHI scheme involving K moduli m_k, it is convenient to measure the precision of multiple (> 1) precision integers not with respect to the machine wordsize W but with respect to some integer $w \leq m_k$ for all k. Schematically:

$$W \left. \begin{array}{c} \\ \rule{0pt}{20pt} \end{array} \right\} m_k\text{'s lie here}.$$

Thus an integer x is $\leq m$ precision if $x < w^m$. A particularly convenient choice of w is a power of ten, say the largest power of ten that is $< W$. The precision of a decimal integer is then available almost by inspection.

Referring to the general MHI scheme of Algorithm 1 as applied to our linear equations example, let K moduli be used (K to be bounded shortly).

Step 1(a), in accordance with eq. (1), requires for each modulus (prime) p_k the division of the $n(n+1)$ elements ($\leq m$ precision) of $(A : \mathbf{b})$ by p_k, hence requires $O(Kmn^2)$ time overall.

Step 1(b) requires for each prime p_k the computation of the $n+1$ determinantal quantities $d^{(k)}, y_j^{(k)}$ of eq. (2). This is done by Gaussian elimination over $\mathbf{Z}_{p_k}$, an $O(n^3)$ algorithm. (We shall shortly resolve the complication that arises when the coefficient matrix $\bmod p_k$ is singular.) Hence step 1(b) requires $O(Kn^3)$ time overall.

Step 2 requires the solution of $n+1$ CRPs for d and the y_j's (eqs. (3) and (4)), hence requires $O(nK^2)$ time.

Thus

$$(*) \qquad \begin{aligned} T &= T_{1(a)} + T_{1(b)} + T_2 \\ &= O(Kmn^2 + Kn^3 + nK^2). \end{aligned}$$

For a more informative time bound, we wish to find a bound for K in terms of the two size parameters, m and n.

As already noted, we require $\prod_{k=0}^{K-1} p_k \geq 2B$ where $B > \max(|d|, |y_1|, \ldots, |y_n|)$. Since each of d and the y_j's are $n \times n$ determinants with entries $< w^m$, we can take $B = n! w^{mn}$ (appealing only to the $n!$ term definition of a determinant). Thus K primes will be sufficient under the following conditions:

$$\text{if } \prod_{k=0}^{K-1} p_k \geq 2n! w^{mn};$$

hence,

$$\text{if } w^K \geq 2n! w^{mn} \quad (\text{since } w < \text{each } p_k);$$

hence,

$$\text{if } K \geq mn + \log_w n! + \log_w 2;$$

hence,

$$\text{if } K \geq mn + n \log_w n + \log_w 2.$$

We see that $O(mn + n \log_w n + \log_w 2)$ primes are required.

Now in applications n is small (certainly < 50) while w is large (in the billions). Hence $\log_w n$ over the quite restricted range of n behaves essentially like a constant. Thus the number of primes required is essentially $O(mn)$. Substituting this for K in $(*)$ yields

$$T = T(m, n) = O(m^2 n^3 + mn^4).$$

For constant m we see that $T = T(n)$ is $O(n^4)$.

Now for an example to illustrate the workings of the MHI scheme for solving integer linear systems. Solve

$$
\begin{aligned}
9x_1 + x_2 - 5x_3 &= 1, \\
7x_1 + 9x_2 &= -1, \\
-4x_1 - 5x_2 + 2x_3 &= -2.
\end{aligned}
$$

For the convenience of hand computation, we choose three small primes: $p_0 = 5$, $p_1 = 7$, $p_2 = 11$. Thus $M = 5 \times 7 \times 11 = 385$, and our three image scheme will yield the correct answer provided 385 is $\geq 2B$ where B is $> \max(|d|, |y_1|, |y_2|, |y_3|)$. (This is indeed the case; in fact, $d = 143$, $y_1 = -95$, $y_2 = 58$, $y_3 = -188$. The purpose of this example, then, is to see how the MHI scheme computes these values.)

In step 1, the computations using the moduli 5 and 7 proceed without novelty, using the Gaussian elimination scheme (over the fields Z_5 and Z_7) as developed in Example 1 of Section 1.3. We obtain

$$
\begin{aligned}
d^{(0)} &= 3, \quad y_1^{(0)} = 0, \quad y_2^{(0)} = 3, \quad y_3^{(0)} = 2, \\
d^{(1)} &= 3, \quad y_1^{(1)} = 3, \quad y_2^{(1)} = 2, \quad y_3^{(1)} = 1.
\end{aligned}
$$

But the third modulus, 11, gives rise to an interesting complication, to which we alluded in our computing time analysis. We now investigate.

Working in Z_{11}, our augmented matrix $\tilde{A}^{(0)}$ is given by

$$
\tilde{A}^{(0)}: \quad
\begin{matrix}
9 & 1 & 6 & 1 \\
7 & 9 & 0 & 10 \\
7 & 6 & 2 & 9
\end{matrix}
$$

By Gaussian elimination over Z_{11}, we obtain

$$
\tilde{A}^{(1)}: \quad
\begin{matrix}
9 & 1 & 6 & 1 \\
0 & 7 & 10 & 8 \\
0 & 0 & 0 & 4
\end{matrix}
$$

$$
\tilde{A}^{(2)}: \quad
\begin{matrix}
9 & 1 & 6 & 1 \\
0 & 7 & 10 & 8 \\
0 & 0 & 0 & 4
\end{matrix}
$$

From $\tilde{A}^{(2)}$ we see that the original coefficient matrix $A' = r_{11}(A)$ is *singular*, i.e., has zero determinant. Over Z_{11}, the system $A'x' = b'$ has no solution.

But it is not x' that we are after; it is y' ($= y^{(2)}$ of (2)). And y' is a determinant which exists quite independent of x'. Indeed from $\tilde{A}^{(2)}$ we can compute $\mathbf{y}'$ by explicit determinant evaluation:

$$y_1' = \det \begin{bmatrix} 1 & 1 & 6 \\ 8 & 7 & 10 \\ 4 & 0 & 0 \end{bmatrix} = 4,$$

$$y_2' = \det \begin{bmatrix} 9 & 1 & 6 \\ 0 & 8 & 10 \\ 0 & 4 & 0 \end{bmatrix} = 3,$$

$$y_3' = \det \begin{bmatrix} 9 & 1 & 1 \\ 0 & 7 & 8 \\ 0 & 0 & 4 \end{bmatrix} = 10,$$

which completes the computation of step 1 for the modulus 11.

Of course for larger systems we would be very reluctant to use explicit determinant expansion in order to handle the somewhat accidental case of singular systems. *Interestingly enough, this is not necessary.* Singular systems can be solved (for the y_j's) just as efficiently as the usual nonsingular ones by invoking some elementary results from the theory of determinants. This is the topic of Appendix 2.

To complete the example, steps 2 and 3 return d, y_1, y_2, y_3 as the least absolute value solutions to the following CRPs:

d	y_1	y_2	y_3	
$u \equiv 3$	$u \equiv 0$	$u \equiv 3$	$u \equiv 2$	(all mod 5)
$u \equiv 3$	$u \equiv 3$	$u \equiv 2$	$u \equiv 1$	(all mod 7)
$u \equiv 0$	$u \equiv 4$	$u \equiv 3$	$u \equiv 10$	(all mod 11)

The CRA of Algorithm 2, Section 2.2, yields

$$d = 143, \quad y_1 = -95, \quad y_2 = 58, \quad y_3 = -188,$$

and Cramer's Rule gives the desired rational solution:

$$x_1 = -95/143, \quad x_2 = 58/143, \quad x_3 = -188/143.$$

3.2 The MHI Scheme for F[x]

Our goal now is to adapt our integer MHI scheme to the problem of evaluating an expression $h(x) = e(a_1(x), \ldots, a_s(x))$ over a polynomial domain $F[x]$, F a field.

At the heart of the integer MHI scheme is a series of computations over various homomorphic images $\mathbf{Z}_{m_k}$ of $\mathbf{Z}$. We make these computations "easy" by choosing the m_k's to be conveniently small ($\leq$ wordsize). To exploit the MHI idea in a polynomial ($F[x]$) setting, we want to choose homomorphic images $F[x]_{m_k(x)}$ of $F[x]$ that are easy to compute over. An especially propitious choice is for each $m_k(x)$ to be of the form $x - \alpha_k$ with $\alpha_k \in F$. Then $F[x]_{m_k(x)}$ is essentially F itself. In particular, the arithmetic of $F[x]_{m_k(x)}$ will then be the "easy" arithmetic of F rather than the "hard" arithmetic (especially when large degree polynomials are involved) of $F[x]$.

These considerations underlie the following polynomial MHI scheme, which is completely analogous (in the Euclidean domain sense) to the integer scheme. Whereas the integer scheme requires a bound on the (absolute) value of h, the polynomial scheme requires a bound on the degree of $h(x)$.

Notation. We abbreviate "$a_1(x), \ldots, a_s(x)$" to $a(x)$.

Algorithm 1 (*MHI scheme for $F[x]$*).

Input.   Ring or ring-with-division expression $e(t_1, \ldots, t_s) = e(t)$, defined over $F[x]$ for arguments $a_1(x), \ldots, a_s(x) \in F[x]$.
Output. $h(x) = e(a(x)) \in F[x]$.

Let n be a bound on $\deg h(x)$—$\deg h(x) < n$—and choose n distinct points $\alpha_k \in F$ that satisfy: for $a_i^{(k)} = a_i(\alpha_k)$, $e(a^{(k)})$ is defined over F ($k = 0, \ldots, n - 1$). Then proceed according to

Step 1 {Evaluation (over F)}.

For $k = 0, \ldots, n - 1$:
(a) Compute $a_i^{(k)} = a_i(\alpha_k)$ $(i = 1, \ldots, s)$;
(b) Compute $\beta_k = e(a^{(k)})$ (over F).

Step 2 {Interpolation}.
Compute $U(x)$ having $\deg < n$ such that $U(\alpha_k) = \beta_k$ ($k = 0, \ldots, n - 1$).

Step 3 {Return value for h}.
Return $h(x) = U(x)$. $\square$

As with the integer MHI scheme, the setup of the polynomial MHI scheme is aptly described by a mapping diagram (Fig. 4).

We note that the polynomial MHI scheme, like the integer one, is essentially a transform scheme—from the "hard" domain $F[x]$ to the "easy" domain F^n, over which expressions are evaluated componentwise (over F). The multipoint evaluation of step 1 is the forward transform, which converts a polynomial

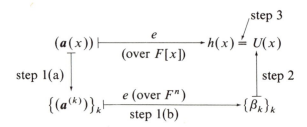

Fig. 4 Polynomial MHI scheme .

$a(x)$ from its usual *coefficient representation* $\Sigma a_i x^i$ to its *sample-value representation* $\{a^{(k)} = a(\alpha_k)\}_k$. The interpolation algorithm of step 2 is the reverse transform.

Terminology.   For obvious reasons we also refer to the polynomial MHI scheme as the method of *evaluation-interpolation*.

Remarks.   1.   Step 1(b) is the computational heart of the polynomial MHI scheme, just as it was in the integer case. We can use any algorithm we wish (hopefully an efficient one) for the evaluation over F of each $e(a^{(k)})$.

2.   Questions of definedness of the expression e over $F[x]$ or over F only arise when e involves division (again, as in the integer case).

3.   In the integer MHI scheme we had two versions corresponding to the two natural solutions to the integer CRP (least positive and least absolute value). In the polynomial MHI scheme, on the other hand, we have only one version, corresponding to the unique natural solution to the n-point interpolation problem (the unique interpolating polynomial of degree $< n$). For this reason (and others), the polynomial MHI scheme tends to be conceptually simpler than the integer one.

4.   The polynomial evaluations of step 1(a) are modular reductions (Lemma 2 of Section 2.3) and the interpolation step, step 2, is the solution to a CRP (Lemma 1 of Section 2.3). Thus our polynomial MHI scheme really is the Euclidean domain analog of our integer MHI scheme.

Theorem 1   (*The polynomial MHI scheme works*).   Algorithm 1 computes $U(x) = h(x)$, where $h(x) = e(a(x))$.

Proof.   The evaluation map $\phi_{\alpha_k}: a(x) \mapsto a(\alpha_k) = a^{(k)}$ is a morphism $F[x] \to F$, regarding $F[x]$ and F as rings or as rings-with-division (for the latter case we appeal to Theorem 2 of Section 1.1). By assumption, $e(a(x))$ and $e(a^{(k)})$ are both defined, over $F[x]$ and F respectively. By the Fundamental Morphism Theorem (Theorem 3 of Section 1.1) we have

$$\phi_{\alpha_k}[e(a(x))] = e\big(\phi_{\alpha_k}[a(x)]\big),$$

which is to say that β_k computed in step 1(b) satisfies

$$h(\alpha_k) = \beta_k.$$

But then $h(x)$ and $U(x)$ computed in step 2 are two polynomials of degree $< n$ that interpolate to the n points $(\alpha_k, \beta_k) \in F \times F$ ($k = 0, 1, \ldots, n-1$). By uniqueness of the interpolating polynomial, $U(x) = h(x)$ as required. $\square$

Example 1   (*Polynomial multiplication by evaluation-interpolation*).   Let $a(x)$, $b(x) \in F[x]$ have degree $\leq n$. To compute their product $c(x) = a(x)b(x)$, we

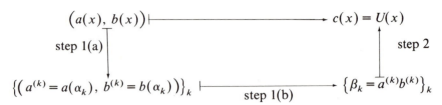

Fig. 5 Polynomial multiplication by evaluation-interpolation.

can execute Algorithm 1 with $e(t_1, t_2) = t_1 t_2$ and arguments $t_1 = a(x)$, $t_2 = b(x)$. Since $c(x) = e(a(x), b(x))$ has degree $\leq 2n$, evaluation-interpolation requires the use of $N = 2n + 1$ distinct points $\alpha_k \in F$. The general evaluation-interpolation scheme of Fig. 4 now becomes that of Fig. 5. $\square$

The classical "school" method for multiplying two polynomials of degree n requires $O(n^2)$ time. How does evaluation-interpolation compare?

Step 1(a) requires the evaluation of two polynomials of degree n at $N = 2n + 1$ points: $O(n^2)$ time by repeated use of Horner's rule (Lemma 4 of Section 2.3).

Step 1(b) requires N multiplications over F: $O(n)$ time.

Step 2 requires the solution of an N-point interpolation problem: $O(n^2)$ time using Newton's Interpolation Algorithm (Theorem 3 of Section 2.3).

Thus evaluation-interpolation, like the classical method, requires $O(n^2)$ time to multiply two degree n polynomials. A finer analysis is therefore necessary to choose between them. We take the required number of multiplications, $M(n)$, as the basis of comparison.

For the classical method, we have $M(n) = (n + 1)^2 = n^2 + O(n)$. For the method of evaluation-interpolation, we require

Lemma 1.   Newton's Interpolation Algorithm for the n-point interpolation problem over a field F requires $2n^2 + O(n)$ multiplications (over F).

Proof.   Refining the analysis of Newton's Interpolation Algorithm leading to Theorem 3 of Section 2.3, we see that at the kth iteration of Algorithm 1, Section 2.3, each of lines 4, 5, 6, and 7 requires k multiplications. (For lines 5 and 6 we use the fact that Horner's Rule requires k multiplications to evaluate a polynomial of degree k.) Thus the number of multiplications required by Newton's Interpolation Algorithm is $4\sum_{k=1}^{n} k = 2n^2 + O(n)$. $\square$

Now for the analysis of the evaluation-interpolation scheme for the multiplication of two degree n polynomials:

Step 1(a) requires the evaluation of two polynomials of degree n at $N = 2n + 1$ points: $nN = 2n^2 + O(n)$ multiplications.

Step 1(b) requires $N = 2n + 1$ multiplications.

Step 2 requires the solution of an N-point interpolations problem: $2N^2 + O(N)$ multiplications by the above lemma, or $8n^2 + O(n)$ multiplications.

Thus for evaluation-interpolation, we have $M(n) = 10n^2 + O(n)$—ten times worse than the familiar school algorithm!

Of course we must conclude that our "sophisticated" evaluation-interpolation algorithm is vastly inferior to the usual school algorithm. But we note, even from our $O(n^2)$ analysis, that *if* we could speedup the transform and reverse transform steps (steps 1(a) and 2), say from $O(n^2)$ to $O(n^{1.99})$, then our evaluation-interpolation method would indeed yield a superior algorithm, at least asymptotically for large n. Our concluding chapter is devoted to a remarkable algorithm that achieves just such a speedup.

3.3 The MHI Scheme for Z[x]

Our problem here is to evaluate an expression $h(x) = e(a_1(x), \ldots, a_s(x))$ over $\mathbf{Z}[x]$, where the arguments $a_i(x)$ and the value $h(x)$ are all in $\mathbf{Z}[x]$. Of course in principle we can regard this as an evaluation problem over $\mathbf{Q}[x]$ and use evaluation-interpolation. But the prospect of performing evaluation-interpolation over $\mathbf{Q}$, with all that that implies regarding the manipulation of large integers or rational numbers, leads us to consider another approach, one that combines the ideas of Sections 3.1 and 3.2.

As a preliminary step, we extend the remainder mod m map from $\mathbf{Z}$ to $\mathbf{Z}[x]$ in the obvious way: for $a(x) = \sum_i a_i x^i \in \mathbf{Z}[x]$, $r_m: \mathbf{Z}[x] \to \mathbf{Z}_m[x]$ is defined by

$$r_m(a(x)) = \sum_i r_m(a_i) x^i.$$

We also denote $r_m(a(x))$ by $a(x) \bmod m$—"the mod m residue of $a(x)$." (Example: $(8x^2 + 10x - 3) \bmod 5 = 3x^2 + 2$.) It is easily verified that $r_m: \mathbf{Z}[x] \to \mathbf{Z}_m[x]$ is a morphism, and that

$$b(x) = r_m(a(x)) \Rightarrow a(x) \equiv b(x) \pmod{m}.$$

The algorithm that follows is essentially a generalization (from $\mathbf{Z}$ to $\mathbf{Z}[x]$) of the integer MHI scheme of Algorithm 1, Section 1.3. It utilizes an a priori bound on the coefficients h_i of the polynomial $h(x)$ to be computed, and there are two versions: one for nonnegative h_i's and one for signed h_i's. For simplicity we consider only the case of $h(x)$ being defined by a ring expression. By now there is no novelty in dealing with rings-with-division.

Algorithm 1 (*MHI scheme for* $\mathbf{Z}[x]$).

Input.   Ring expression $e(t_1, \ldots, t_s) = e(t)$, arguments $a_1(x), \ldots, a_s(x) \in \mathbf{Z}[x]$.

Output. $h(x) = e(a(x)) \in \mathbf{Z}[x]$.

VERSION 1.   (Bound B on coefficients $h_i \geq 0$ is available.) Let $0 \leq h_i < B$ and choose n pairwise relatively prime moduli $m_0, \ldots, m_{n-1}$ such that

$M = \prod_{i=0}^{n-1} m_i$ is $\geq B$. Then proceed according to

Step 1 {Modular evaluation}.

For $k = 0, 1, \ldots, n - 1$:
(a) Compute $a_i^{(k)}(x) = r_{m_k}(a_i(x))$ $(i = 1, \ldots, s)$;
(b) Compute $\rho_k(x) = e(a^{(k)}(x))$ (over $\mathbf{Z}_{m_k}[x]$).

Step 2 {CRA}.
Solve the polynomial CRP

$$u(x) \equiv \rho_k(x) \pmod{m_k} \qquad (k = 0, \ldots, n - 1)$$

for the least-positive coefficient solution $U(x)$ (Theorem 4 of Section 2.2).

Step 3 {Return value for $h(x)$}.
Return $h(x) = U(x)$.

VERSION 2. (Bound b on $|h_i|$ is available.) Let $|h_i| < B$, and choose moduli $m_0, \ldots, m_{n-1}$ as in Version 1, but with $M \geq 2B$. Then proceed as in Version 1, except in step 2 solve the system of congruences for the least-absolute-value coefficient solution $V(x)$ (Theorem 4 of Section 2.2), and in step 3 return $h(x) = V(x)$. $\square$

Theorem 1 (*The $\mathbf{Z}[x]$ MHI scheme works*). In Algorithm 1, for $h(x) = e(a(x))$,

(a) Version 1 computes $U(x) = h(x)$;
(b) Version 2 computes $V(x) = h(x)$.

Proof. The remainder mod m map is a morphism $\mathbf{Z}[x] \to \mathbf{Z}_m[x]$. Hence by our Fundamental Morphism Theorem (Theorem 1 of Section 1.1), we have

$$r_m(e(a(x))) = e(r_m(a(x)))$$

for each modulus $m = m_k$ $(k = 0, \ldots, n - 1)$. Thus each $\rho_k(x)$ computed in step 1 satisfies $\rho_k(x) = r_{m_k}(e(a(x)))$ which means that $h(x) = e(a(x))$ is a solution to the polynomial CRP

$(*)$ $\qquad\qquad u(x) \equiv \rho_k(x) \pmod{m_k} \qquad (k = 0, \ldots, n - 1)$.

(a) In Version 1, $0 \leq h_i < B \leq M$. Hence $h(x)$, like $U(x)$ returned by step 2, is a least-positive coefficient solution to the polynomial CRP $(*)$. By the uniqueness of this solution (Theorem 4 of Section 2.2), $U(x) = h(x)$.

(b) In Version 2, $|h_i| < B \leq M/2$. Hence $h(x)$, like $V(x)$ returned by step 2, is a least-absolute-value coefficient solution to the polynomial CRP $(*)$. Again by uniqueness, $V(x) = h(x)$. $\square$

The crux of Algorithm 1 is step 1(b), the computations over the $\mathbf{Z}_{m_k}[x]$'s of the $\rho_k(x)$'s. If we choose each m_k to be a prime p_k, then each $\mathbf{Z}_{m_k}$ is a field, which means that we can use *evaluation-interpolation to compute the* $\rho_k(x)$*'s*. Thus

Algorithm 1 constitutes an "outer" algorithm in which evaluation-interpolation serves for step 1(b) as an "inner" algorithm.

The overall effect of this MHI scheme for $\mathbf{Z}[x]$ is to replace the single (hard) evaluation of the input ring expression $e(t)$ over $\mathbf{Z}[x]$ by many (easy) evaluations of e over the finite fields $\mathbf{Z}_{p_k}$. The moduli p_k should be chosen to be conveniently small (in particular, less than computer wordsize) so that the $\mathbf{Z}_{p_k}$ evaluations of e are indeed easy.

Example 1. Let us use Algorithm 1 to compute the determinant $d(x) = \det A(x)$ of the $\mathbf{Z}[x]$ matrix

$$A(x) = \begin{pmatrix} 1 - 2x & -8 + 3x \\ 6 + 9x & 7 - x \end{pmatrix}.$$

Of course the computation of a 2×2 determinant, even one with polynomial entries, does not require a special technique. Indeed, by the usual formula, we have

$$d(x) = (1 - 2x)(7 - x) - (6 + 9x)(-8 + 3x)$$
$$= 55 + 39x - 25x^2.$$

However, the problem is still large enough to exhibit the dynamics of the MHI scheme for $\mathbf{Z}[x]$, and that is our sole objective.

Suppose by a rough analysis or calculation that we have determined a bound $|d_i| < 100$ for the coefficients of $d(x) = d_0 + d_1 x + d_2 x^2$. Thus Version 2 of the $\mathbf{Z}[x]$ MHI scheme is applicable, and we require moduli with product $M \geq 200$. We choose two moduli $p_0 = 13$, $p_1 = 17$ as being conveniently small (for hand calculations). Then $M = 13 \times 17 = 221$, and Theorem 1 guarantees that Algorithm 1 will yield the correct answer.

Step 1 requires the computation of

$$\rho_0(x) = \det A_0(x) \quad \text{where } A_0(x) = r_{13}(A(x)) = \begin{pmatrix} 1 + 11x & 5 + 3x \\ 6 + 9x & 7 + 12x \end{pmatrix},$$

$$\rho_1(x) = \det A_1(x) \quad \text{where } A_1(x) = r_{17}(A(x)) = \begin{pmatrix} 1 + 15x & 9 + 3x \\ 6 + 9x & 7 + 16x \end{pmatrix}.$$

Step 2 requires the solution of the system of polynomial congruences

$$u(x) \equiv \rho_0(x) \pmod{13},$$
$$u(x) \equiv \rho_1(x) \pmod{17}$$

for the least-absolute-value coefficient solution $V(x)$. Step 3 then returns $V(x)$ for $d(x)$. These three steps, then, constitute the outer algorithm of Algorithm 1.

To compute $\rho_0(x), \rho_1(x)$ in step 1, we use our inner algorithm of evaluation-interpolation—over the fields $\mathbf{Z}_{13}$ and $\mathbf{Z}_{17}$ using three points (0, 1, and 2) (three points since $\rho_0(x), \rho_1(x)$ have degree < 3).

$\rho_0(x)$ (all computations over $\mathbf{Z}_{13}$):

$$\beta_0 = \det A_0(0) = \det\begin{pmatrix} 1 & 5 \\ 6 & 7 \end{pmatrix} = 3,$$

$$\beta_1 = \det A_0(1) = \det\begin{pmatrix} 12 & 8 \\ 2 & 6 \end{pmatrix} = 4,$$

$$\beta_2 = \det A_0(2) = \det\begin{pmatrix} 10 & 11 \\ 11 & 5 \end{pmatrix} = 7.$$

Interpolation to $\{(0,3),(1,4),(2,7)\}$ yields $\rho_0(x) = 3 + x^2$.

$\rho_1(x)$ (all computations over $\mathbf{Z}_{17}$):

$$\beta_0 = \det A_1(0) = \det\begin{pmatrix} 1 & 9 \\ 6 & 7 \end{pmatrix} = 4,$$

$$\beta_1 = \det A_1(1) = \det\begin{pmatrix} 16 & 12 \\ 15 & 6 \end{pmatrix} = 1,$$

$$\beta_2 = \det A_1(2) = \det\begin{pmatrix} 14 & 15 \\ 7 & 5 \end{pmatrix} = 16.$$

Interpolation to $\{(0,4),(1,1),(2,16)\}$ yields $\rho_1(x) = 4 + 5x + 9x^2$. This completes step 1 and the inner algorithm.

Step 2 thus requires the solution of

$$u(x) \equiv 3 + x^2 \ (\mathrm{mod}\ 13),$$

$$u(x) \equiv 4 + 5x + 9x^2 \ (\mathrm{mod}\ 17)$$

for the least-absolute-value coefficient solution $V(x) = v_0 + v_1 x + v_2 x^2$. In accordance with the proof of Theorem 4, Section 2.2, this is tantamount to solving the following three CRPs for the least absolute value solutions:

(0)	(1)	(2)	
$u \equiv 3$	$u \equiv 0$	$u \equiv 1$	(all mod 13)
$u \equiv 4$	$u \equiv 5$	$u \equiv 9$	(all mod 17)

The solution to (0) yields $v_0 = 55$, to (1) yields $v_1 = 39$, to (2) yields $v_2 = -25$. Hence $V(x) = 55 + 39x - 25x^2$, which is the (correct) value returned for $d(x)$.

□

*   *   *

In Note 1 we have gathered together some historical and bibliographical material on the method of homomorphic images.

Appendix 1: Computing lists of primes. A time-honoured method for computing a list of all primes $\le n$ is the so-called *sieve method* which goes back to the Greek mathematician of antiquity Eratosthenes (276 − 196 B.C.).

From the list of integers $2, 3, \ldots, n$, all multiples of the first prime, $p_1 = 2$, are crossed out (excluding 2 itself). Inductively, suppose that all multiples of the first r primes $p_1, \ldots, p_r$ have been crossed from the list. The claim is that the first integer $x > p_r$ not crossed out, if there is such an integer, is the next prime p_{r+1}. For assume to the contrary that x is composite. Then it must have a prime factor $< x$, hence would have been crossed out as a multiple of some p_k ($1 \leq k \leq r$). All multiples of this prime p_{r+1} are then crossed out. The process terminates when, after crossing out all multiples of some p_r, there are no integers $> p_r$ in the list yet uncrossed. The integers that remain uncrossed (i.e., that survive the sieve) are, by construction, the primes $\leq n$.

A significant efficiency can be incorporated into the sieve method by exploiting the

Simple but important fact: If an integer x is composite, then it has a factor, hence a prime factor, $\leq \sqrt{x}$.

So in the sieve method, when crossing all multiples of p_{r+1} from our list, it is sufficient to consider only those integers $\geq p_{r+1}^2$. Why? Because any integer $x < p_{r+1}^2$ that is composite must, by our "simple but important fact," have a prime factor $\leq \sqrt{x} < p_{r+1}$, i.e., must have a prime factor p_k for $1 \leq k \leq r$. Hence this x would have already been crossed out as a multiple of this p_k. Also, this means that the sieve method can be terminated once the next prime, p_{r+1}, is such that $p_{r+1}^2 > n$.

To get some feeling for the dynamics and efficiency of the sieve method, let us use it to determine all primes ≤ 39. Crossing out multiples of 2 starting with $2^2 = 4$ gives

```
 2   3   4   5   6   7   8   9  10  11  12  13  14  15  16  17  18  19  20
21  22  23  24  25  26  27  28  29  30  31  32  33  34  35  36  37  38  39.
```

Crossing out all multiples of 3 starting with $3^2 = 9$ gives

```
 2   3   4   5   6   7   8   9  10  11  12  13  14  15  16  17  18  19  20
21  22  23  24  25  26  27  28  29  30  31  32  33  34  35  36  37  38  39.
```

(Several integers, e.g. 12 and 18, have been in effect crossed out twice. Of course there is no need to indicate this fact; being crossed out once is sufficient to indicate that an integer is composite.) Crossing out all multiples of 5 starting with $5^2 = 25$ gives

```
 2   3   4   5   6   7   8   9  10  11  12  13  14  15  16  17  18  19  20
21  22  23  24  25  26  27  28  29  30  31  32  33  34  35  36  37  38  39.
```

The next prime to be dealt with is 7, but $7^2 = 49 > 39$, so we are done. The integers that have survived the sieve—2, 3, 5, 7, 11, 13, 17, 19, 23, 29, 31, 37—are the primes ≤ 39.

The sieve method is readily implemented by a computer program (Exercise). An appealing feature of the sieve method is that it does not require any divisions, which are relatively slow operations on most computers.

The sieve method is readily adaptable to the problem of finding the first n (10 or 20, say) *largest* primes that are $\leq W$, the wordsize of our computer. (Such a list of primes is what we need for our multiple homomorphic image scheme over **Z**.) First, an auxiliary table of all primes $p_1 = 2$, $p_2 = 3, \ldots, p_r \leq \sqrt{W}$ is drawn up (using the above sieve method). Then the multiples of each of these primes are crossed from the list of integers $W, W - 1, \ldots, L$, where L is chosen so that $\Delta W = W - L$ is large enough to contain $\geq n$ primes. Again appealing to our above "simple but important fact," the integers that remain uncrossed from this list must be prime.

Is this sieve method practical? It will be, provided (for moderate values of n) (i) the size $\Delta W = W - L$ of the interval over which sieving takes place need not be too large and (ii) the size r of the auxiliary table of primes $\leq \sqrt{W}$ is not too large. A powerful insight into both these provisos is rendered by the celebrated

Prime Number Theorem.   The number of primes $\leq x$ is asymptotically $x/\log x$.

[See Hardy and Wright, *The Theory of Numbers*, 4th ed. (Oxford, 1960), Section 1.8; note that $x/\log x$ is a slight underestimator of the actual number of primes $\leq x$.]

The Prime Number Theorem suggests (but does not guarantee) that about one in every $\log W$ of the ΔW integers $W, W - 1, \ldots, L$ is prime. So for prescribed n, we take ΔW to be "safely" in excess of $n \log W$. The Prime Number Theorem also tells us that the size r of the auxiliary table will not be too much larger than $\sqrt{W} / \log \sqrt{W}$. For example, with $W = 2^{31} - 1$ (corresponding to the wordsize of a certain popular class of machines) and $n = 20$ (say), we would take ΔW to be in excess of $20 \log 2^{31} = 429$, say $\Delta W = 1000$. The size r of the auxiliary table of primes is not much larger than $\sqrt{2^{31}} / \log \sqrt{2^{31}} \approx 4310$. (The actual value of r required turns out to be 4691.)

Such values for ΔW and r are quite manageable insofar as a computer implementation of this modified sieve procedure is concerned. Indeed we see that for fixed n the running time of the modified sieve procedure grows with the square root of W rather than with W itself.

[See Knuth (1969), p. 354, for a table of useful primes related to the wordsizes of a spectrum of current computers.]

Appendix 2: "Adjoint solution" to $A\mathbf{x} = \mathbf{b}$.   Let $A\mathbf{x} = \mathbf{b}$ be a system of n linear equations in n unknowns. The purpose of this note is to show how to compute $y_j = \det A_j$ even in the singular case, $\det A = 0$. (When $\det A = 0$, we cannot compute $\mathbf{y}$ according to $(\det A)\mathbf{x}$, since $\mathbf{x}$ does not exist.) We first recall some elementary theory of determinants.

A *minor* M_{ij} of an $n \times n$ matrix A is the $(n-1) \times (n-1)$ matrix obtained from A by striking out row i and column j. The *cofactor* A_{ij} of a_{ij} is then defined by

$$A_{ij} = (-1)^{i+j} \det M_{ij}.$$

The importance of cofactors is wrapped up with the following familiar formula for evaluating $\det A$ by *cofactor expansion along the jth column* [Birkhoff and Mac Lane (1977), p. 319]:

(1) $$\det A = \sum_{k=1}^{n} A_{kj} a_{kj}.$$

A symmetric formula applies to expansion along the ith row.

The *adjoint* of A is defined by

(2) $$\operatorname{adj} A = (A_{ji}).$$

[*Note:* The (i,j) element of $\operatorname{adj} A$ is A_{ji}, not A_{ij}.]

Lemma 1. $(\operatorname{adj} A)A = A(\operatorname{adj} A) = dI$, where $d = \det A$, $I = n \times n$ identity matrix. [See Birkhoff and Mac Lane (1977), p. 325.]

Example. Over $\mathbf{Z}_7$ consider

$$A = \begin{bmatrix} 4 & 2 & 3 \\ 0 & 1 & 1 \\ 6 & 5 & 1 \end{bmatrix}.$$

Then

$$\operatorname{adj} A = \begin{bmatrix} 3 & 6 & 6 \\ 6 & 0 & 3 \\ 1 & 6 & 4 \end{bmatrix}, \qquad \det A = 6,$$

and

$$(\operatorname{adj} A)A = A(\operatorname{adj} A) = \begin{bmatrix} 6 & 0 & 0 \\ 0 & 6 & 0 \\ 0 & 0 & 6 \end{bmatrix}. \ \square$$

For $A\mathbf{x} = \mathbf{b}$ a given system of n linear equations in n unknowns, the following lemma provides a useful matrix expression for $y_j = \det A_j$.

Lemma 2. $\mathbf{y} = (\operatorname{adj} A)\mathbf{b}$.

Proof. By definition, $y_j = \det A_j$, where A_j is the matrix A with the jth column replaced by $\mathbf{b}$. Expanding $\det A_j$ along the jth column according to (1) gives

(∗) $$y_j = \sum_{k=1}^{n} A_{kj} b_k.$$

The right-hand side of (∗) is seen to be the jth component of the matrix product $(\operatorname{adj} A)\mathbf{b}$ (remember: A_{kj} is element (j,k) of $\operatorname{adj} A$). $\square$

Theorem.   y satisfies $A\mathbf{y} = d\mathbf{b}$, where $d = \det A$.

Proof. $A\mathbf{y} = A(\text{adj } A)\mathbf{b}$ Lemma 2
$\qquad\quad = d\mathbf{b}$ Lemma 1. $\square$

Corollary 1.   (*Cramer's Rule.*)   If $d \neq 0$, the (unique) solution to $A\mathbf{x} = \mathbf{b}$ is given by $\mathbf{x} = \mathbf{y}/d$.

Corollary 2.   If $d = 0$, then y satisfies $A\mathbf{y} = 0$.

Remark.   We refer to the pair $(\mathbf{y}, d)$ as the "adjoint solution" to $A\mathbf{x} = \mathbf{b}$. This adjoint solution exists even when x does not.

Now for the computation of y.

By a simple modification of the Gaussian elimination scheme of Example 1, Section 1.3, the augmented matrix $\tilde{A} = (A : \mathbf{b})$ corresponding to a given linear system $A\mathbf{x} = \mathbf{b}$ can be reduced to "echelon" form [Birkhoff and Mac Lane (1977), Section 7.6]. $\tilde{A}$ is said to be in *echelon form* if in A the number of zeros preceding the first nonzero entry on any nonzero row is larger than the corresponding number of the previous row. Sample 5×6 echelon forms are displayed in Fig. 6.

Reduction to echelon form is achieved by performing the same types of row operations as in ordinary Gaussian elimination (interchanging rows, adding to a row a scalar multiple of some other row), only now allowing as necessary the kth pivot element to be to the right of the kth column, i.e., to be of the form $a_{kk'} \neq 0$ with $k' \geq k$.

As with ordinary Gaussian elimination, these row operations leave solutions and determinants (up to sign) unchanged. The advantage (to us) of echelon form is that it permits efficient computation of the desired determinantal values $d = \det A$, $y_j = \det A'_j$, as we now investigate.

Let $(A : \mathbf{b})$ be an $n \times (n + 1)$ matrix in echelon form.

CASE 1. A has no nonzero rows (as illustrated by Fig. 6(a)). Then $d \neq 0$ and the echelon form is the usual triangular form (with nonzero diagonal elements) produced by ordinary Gaussian elimination. Here we know how to proceed: solve for the solution x to $A\mathbf{x} = \mathbf{b}$ by back-substitution, then compute y according to $d\mathbf{x}$.

```
 *  $  $  $  $  $        *  $  $  $  $  $        *  $  $  $  $  $
 0  *  $  $  $  $        0  *  $  $  $  $        0  0  *  $  $  $
 0  0  *  $  $  $        0  0  0  *  $  $        0  0  0  0  0  $
 0  0  0  *  $  $        0  0  0  0  *  $        0  0  0  0  0  $
 0  0  0  0  *  $        0  0  0  0  0  $        0  0  0  0  0  $

        (a)                    (b)                    (c)
```

Fig. 6   Sample 5×6 echelon forms. ($* = $ nonzero, $-arbitrary)

CASE 2.   The last row only of A is zero (as illustrated by Fig. 6(b)). Here $d = 0$. In $(A : \mathbf{b})$, there is a k such that $a_{ii} = 0$ for all $i \geq k$ (e.g., in Fig. 6(b), $k = 3$). For $y_j = \det A_j$ we consider three cases:

(i) $j > k$: Then $\det A_j = a_{11} \cdots a_{kk}\delta$ where δ is an $(n - k) \times (n - k)$ subdeterminant of A. Hence $\det A_j = 0$ since $a_{kk} = 0$.

(ii) $j = k$. To compute $\det A_k$ we interchange column k (the b_i's) with column $k + 1$, then column $k + 1$ with column $k + 2$, and so on until the b_i's are in column n ($n - k$ interchanges in all). With $k = 3$, $n = 5$ as in Fig. 6(b), A_k would be transformed to

$$
\begin{bmatrix}
a_{11} & a_{12} & a_{14} & a_{15} & b_1 \\
0 & a_{22} & a_{14} & a_{25} & b_2 \\
0 & 0 & a_{34} & a_{35} & b_3 \\
0 & 0 & 0 & a_{45} & b_4 \\
0 & 0 & 0 & 0 & b_5
\end{bmatrix}.
$$

Here we have $\det A_3 = (-1)^{5-3}a_{11}a_{22}a_{34}a_{45}b_5$; in general we have

$$\det A_k = (-1)^{n-k}a_{11} \cdots a_{k-1,k-1}b_n a_{k,k+1} \cdots a_{n-1,n}.$$

(iii) $j < k$: Solve for $y_{k-1}, \ldots, y_1$ from $A\mathbf{y} = 0$ (essentially by backsubstitution).

Example.   Using the configuration of Fig. 6(b), let us solve for $\mathbf{y}$ over $\mathbf{Z}_7$ from

$$
(A : \mathbf{b}) =
\begin{bmatrix}
2 & 4 & 3 & 4 & 2 & 1 \\
0 & 6 & 5 & 1 & 5 & 3 \\
0 & 0 & 0 & 3 & 6 & 0 \\
0 & 0 & 0 & 0 & 1 & 4 \\
0 & 0 & 0 & 0 & 0 & 4
\end{bmatrix}.
$$

Here we have

(i) $y_5 = y_4 = 0$,
(ii) $y_3 = (-1)^2 2 \cdot 6 \cdot 3 \cdot 1 \cdot 4 = 4$,
(iii) $6y_2 = -5y_3 \Rightarrow y_2 = 6$,
    $2y_1 = -4y_2 - 3y_3 \Rightarrow y_1 = 3$.

Hence $\mathbf{y} = (3, 6, 4, 0, 0)$. □

CASE 3.   Two or more of the last rows of A are zero. Here again $d = 0$. Each A_j, when triangularized by the $n - k$ column interchanges as in case 2(ii), is seen to have at least one zero diagonal element. Thus $\det A_j = 0$ for all j, i.e., $\mathbf{y} = 0$. (For the reader familiar with the notion of *rank* of a matrix or determinant, we can provide a more elegant argument. Since A in case 3 has rank $< n - 1$, all minors M_{ij} of order $n - 1$ must vanish. Hence adj $A = 0$ and $\mathbf{y} = (\text{adj } A)\mathbf{b} = 0$.)

Concluding remark.   The reader is forgiven for being chagrined that so much theory is required to cover a case ($\det A = 0$) that hardly ever arises in the use of the homomorphic image scheme. But for completeness we need it.

EXERCISES

Section 1

1. (a) Is the casting-out-nines check of Example 4, Section 1.1, conclusive? Explain.
    (b) Observing that

$$10, 10^3, 10^5, \ldots \equiv -1 \ (\mathrm{mod}\ 11),$$

$$10^0, 10^2, 10^4, \ldots \equiv 1 \quad (\mathrm{mod}\ 11),$$

    develop a "casting-out-elevens" rule for computing remainders mod 11, and use this rule to check the calculation of Example 4, Section 1.1:

$$43 \cdot 768 + 9571 = 42625.$$

2. (a) Check by casting out elevens: $(43 \cdot 768 + 9571)/35 = 1217$.
    (b) Can the computation below be checked by casting out nines? Casting out elevens? Explain.

$$(94 \cdot 87 - 105)/117 = 69.$$

3. Solve the linear system of Example 1, Section 1.3, over $\mathbf{Z}_{41}$, and use this solution to obtain the solution over $\mathbf{Q}$.

4. Solve over $\mathbf{Q}$ by the homomorphic image method, using $\mathbf{Z}_p$ arithmetic for an appropriate prime p:
    (a) $2x - 7y = 4,$ (b) $x - 2y + 3z = -2,$
    $\quad\ 5x - 3y = -1;$ $\ 2x + y - 5z = 4,$
    $\qquad\qquad\qquad\qquad\quad\ 4x - 3y + z = 0.$

Exercises 5–8 extend the homomorphic image method to computing integer-valued expressions involving *rational* arguments.

5. Let $\mathbf{Q}_p$ be the set of rational numbers which when expressed in lowest terms have the form a/b where $p \nmid b$. Prove:
    (a) $\mathbf{Q}_p$ is a subring of $\mathbf{Q}$.
    (b) $\phi \colon \mathbf{Q}_p \to \mathbf{Z}_p$ defined by $a/b \mapsto r_p(a)/r_p(b)$ is a morphism of rings or rings-with-division.
    (c) ϕ is an extension of r_p: if $a = a/1 \in \mathbf{Z}$, then $\phi(a) = r_p(a)$.

6. Generalize (effortlessly) the homomorphic image scheme for $\mathbf{Z}$ and its correctness proof (Algorithm 1 and Theorem 1 of Section 1.3) to accommodate the computation of integer-valued expressions $e(a_1, \ldots, a_s)$ involving *rational* arguments a_i.

7. Show how the binomial coefficient $C(n, k) = n!/[k!(n-k)!]$ can be computed according to the formula

$$C(n, k) = \left(\frac{n}{k}\right)\left(\frac{n-1}{k-1}\right) \cdots \left(\frac{n-k+1}{1}\right)$$

but using modular (rather than rational) arithmetic. As a specific illustration of your algorithm and for purposes of comparison, compute $C(7,3)$ using the same moduli 53 and 97 of Example 2, Section 1.3.

8. The $n \times n$ *Hilbert matrix* H_n, with elements $h_{ij} = 1/(i+j-1)$, is known to have an integer inverse matrix $T_n = (t_{ij})$. Compute T_2 using modular arithmetic. You may use the fact that each $t_{ij} < 50$.

Section 2

1. (*Chinese Remainder Theorem.*)   Let D be a Euclidean domain, and let $m, n \in D$ be relatively prime. Establish the isomorphism

$$D/(mn) \cong D/(m) \times D/(n).$$

[*Hint:* Consider the map ϕ: $a \mapsto (a + (m), a + (n))$; use the Ring Isomorphism Theorem.]

2. Find the least positive and least absolute value solutions of
    (i) $u \equiv 1 \pmod 3$,   (ii) $u \equiv 0 \pmod 3$,
         $u \equiv 3 \pmod 5$, $u \equiv 3 \pmod 5$,
         $u \equiv 2 \pmod 8$; $u \equiv 7 \pmod 8$.

3. Find *all* solutions to
$$u(x) \equiv x \qquad \pmod 3,$$
$$u(x) \equiv 3x + 3 \pmod 5,$$
$$u(x) \equiv 2x + 7 \pmod 8.$$

4. Find the interpolating polynomial $U(x) \in F[x]$ that satisfies $U(1) = 2$, $U(3) = 0$, $U(4) = 3$ where (i) $F = \mathbf{Q}$, (ii) $F = \mathbf{Z}_{11}$.

5. Show that the uniqueness of the (degree $< n$) interpolating polynomial for the n-point interpolation problem follows from the Corollary to Theorem 4 of Section 2.1.

6. (*Generalized Interpolation.*)   Solve over $\mathbf{Z}_7[x]$:
$$u(x) \equiv x + 2 \qquad (\mathrm{mod}\, x^2 + 2),$$
$$u(x) \equiv 4x^2 + 4x + 1 \ (\mathrm{mod}\, x^3 + 2x + 5).$$

7. Modify the n-congruence CRA of Algorithm 2, Section 2.1, to compute the coefficients $\sigma_0, \sigma_1, \ldots, \sigma_{n-1}$ of the "mixed-radix" form of the solution U (rather than U itself):
$$U = \sigma_0 + \sigma_1 m_0 + \sigma_2 m_0 m_1 + \cdots + \sigma_{n-1} m_0 m_1 \cdots m_{n-2}.$$

Illustrate your algorithm on the CRP of Example 2, Section 2.1:
$$u \equiv 1 \pmod 3,$$
$$u \equiv 3 \pmod 5,$$
$$u \equiv 0 \pmod 7,$$
$$u \equiv 10 \pmod{11}.$$

What is the advantage in the integer case of your modified algorithm over Algorithm 2, Section 2.1?

8. (*Another least absolute value CRA.*)   For the case of the integer CRP, add line 6a to the CRA of Algorithm 2, Section 2.1:

   6a.   **if** $\sigma \geq m_k/2$ **then** $\sigma := \sigma - m_k$;

   Show that the resulting algorithm then computes the least absolute value solution to the n-congruence integer CRP. Again illustrate your algorithm on the CRP of Example 2, Section 2.1.

9. (*Lagrangian solution to the CRP.*)   For the general (Euclidean domain) CRP $u \equiv r_k$ (mod m_k) $(k = 0, \ldots, n-1)$, devise a solution U of the form

$$U = \sum_{k=0}^{n-1} r_k L_k$$

   that generalizes Lagrange's Interpolating Polynomial. Illustrate your solution for the systems of congruences of Exercise 2.

10. (*Bivariate Interpolation.*)
    (a) Devise an algorithm for computing $U(x,y) \in F[x,y]$ having degree $< n$ in y such that

    $$U(x, \alpha_k) = \beta_k(x) \qquad (k = 0, \ldots, n-1)$$

    for prescribed $\alpha_k \in F$ (α_k's distinct) and $\beta_k(x) \in F[x]$. Show that $U(x,y)$ is unique.
    (b) If each $\beta_k(x)$ has degree $< m$, derive a bound on the time $T(m,n)$ required to compute $U(x,y)$, assuming that operations of F require $O(1)$ time.
    (c) Compute $U(x,y) \in Z_7[x,y]$ such that

    $$U(x,0) = x^3 + 5,$$
    $$U(x,1) = x^3 + 4x^2 + 6,$$
    $$U(x,2) = x^3 + 2.$$

11. (*Multivariate Interpolation.*)
    (a) Devise an algorithm for computing $U(x_1, \ldots, x_r) \in F[x_1, \ldots, x_r]$ having degree $< n$ in x_r such that

    $$U(x_1, \ldots, x_{r-1}, \alpha_k) = \beta_k(x_1, \ldots, x_{r-1}) \qquad (k = 0, \ldots, n-1)$$

    for prescribed $\alpha_k \in F$ (α_k's distinct) and $\beta_k(x_1, \ldots, x_{r-1}) \in F[x_1, \ldots, x_{r-1}]$. Show that $U(x_1, \ldots, x_r)$ is unique.
    (b) If each $\beta_k(x_1, \ldots, x_{r-1})$ has degree $< n$ in each x_i, derive a bound on the time $T_r(n)$ required to compute the above interpolating polynomial, again assuming that operations of F require $O(1)$ time.

Section 3

1. Solve over Q by the MHI method, using Z_p arithmetic for an appropriate number of primes $p \leq 13$ (cf. Exercise 4 of Section 1).
   (a) $2x - 7y = 4,$ (b) $x - 2y + 3z = -2,$
       $5x - 3y = -1;$ $2x + y - 5z = 4,$
                                  $4x - 3y + z = 0.$

2. Let $A(x)$ be an $n \times n$ matrix with polynomial entries $a_{ij}(x)$ having degree $\leq m$. Derive a bound on the time $T(m,n)$ required to compute $\det A(x)$.

3. Use evaluation-interpolation to compute $\det A(x)$ (i) over $\mathbf{Z}_{11}[x]$, (ii) over $\mathbf{Z}[x]$.

$$A(x) = \begin{pmatrix} 3 + 8x & 1 + 4x^2 \\ 7 & 2x \end{pmatrix}.$$

4. The *characteristic polynomial* of an $n \times n$ matrix $A = (a_{ij})$ is defined by

$$c_A(\lambda) = \det(A - \lambda I)$$
$$= c_n \lambda^n + c_{n-1} \lambda^{n-1} + \cdots + c_0.$$

   Assume that $C(A, F)$ is an algorithm that computes $c_A(\lambda)$ over a field F. Now let A be an integer matrix, and assume that a bound B on the absolute value of the coefficients is known: $|c_i| < B$, $0 \le i \le n$. Argue that $c_A(\lambda)$ can be computed by executing algorithm C over a number of finite fields $\mathbf{Z}_p$ (rather than performing C over $\mathbf{Q}$ or $\mathbf{R}$). (Note 2.)

5. (*Multiplication over* $\mathbf{Z}[x]$.)
   (a) Develop an evaluation-interpolation based algorithm for computing the product of two polynomials $a(x) = \sum_{i=0}^n a_i x^i$, $b(x) = \sum_{i=0}^n b_i x^i \in \mathbf{Z}[x]$ where the coefficients a_i, b_i have precision $\le m$.
   (b) Derive a bound $T(m, n)$ on the time required by your algorithm (assuming, as usual, unit cost for single precision operations).

6. (*Trial division over* $F[x]$ *by divisions over* F.)
   (a) Let $a(x) = \sum_{i=0}^m a_i x^i$, $b(x) = \sum_{i=0}^n b_i x^i$ $(m \ge n)$ be polynomials over a field F, and let $\alpha_k \in F$ $(k = 0, \ldots, m)$ be distinct and such that $b(\alpha_k) \ne 0$. Establish the validity of the following algorithm for the trial division of $a(x)$ by $b(x)$ (cf. Algorithm 1 of Section 3.2):

   Step 1 {Evaluation (divisions over F)}.
     For $k = 0, \ldots, m$:
     (a) Compute $a^{(k)} = a(\alpha_k)$, $b^{(k)} = b(\alpha_k)$;
     (b) Compute $\beta_k = a^{(k)}/b^{(k)}$.

   Step 2 {Interpolation}.
     Compute $U(x)$ having degree $\le m$ such that $U(\alpha_k) = \beta_k$ $(k = 0, \ldots, m)$.

   Step 3 {Return result}.
     If $\deg U(x) = m - n$ then return $U(x)$ for $a(x)/b(x)$; otherwise announce "$b(x) \nmid a(x)$".

   (b) What is a bound on the computing time $T(m, n)$ required by the above algorithm? Conclusion?
   (c) Show how the above algorithm computes the trial divisions $a(x)/b(x)$ over $\mathbf{Z}_7[x]$ for
     (i) $a(x) = x^2 + 2x + 6$, $b(x) = 3x + 5$;
     (ii) $a(x) = x^2 + 2x + 6$, $b(x) = 3x + 4$.

7. Develop an algorithm analogous to that of the previous exercise for carrying out a trial (exact) division a/b over $\mathbf{Z}$ by divisions over an appropriate number of (conveniently small) finite fields $\mathbf{Z}_p$. Illustrate your algorithm by performing the trial divisions $1961/37$, $1962/37$ using only divisions over $\mathbf{Z}_p$ for $p < 20$.

8. Develop an abstract multiple homomorphic image algorithm that generalizes the integer MHI scheme of Algorithm 1, Section 3.1 (both Versions 1 and 2) and the polynomial MHI scheme of Algorithm 1, Section 3.2.

NOTES

Section 1

1. Word algebras play an important role in structural as well as computational aspects of algebra. For a discussion of the former in a universal algebra setting, see G. Birkhoff, *Lattice Theory*, 3rd ed. (Providence, R.I.: Amer. Math. Soc., 1967), Chapter VI.

2. A convenient reference for the theory of simultaneous linear equations (involving matrices and determinants), see Birkhoff and Mac Lane (1977), especially Sections 2.3, 7.5, 7.6, 10.1 and 10.2. Note well on p. 48: "*It is a remarkable fact that the entire theory of simultaneous linear equations applies to fields in general.*"

3. A list of all primes $< 10^5$ is to be found in M. Abromowitz and I. Stegun (eds.), *Handbook of Mathematical Functions* (New York: Dover, 1965), 870–873.

Section 2

1. The origins of the integer CRP are not at all certain, with dates ranging from 200 B.C. to 400 A.D. L. E. Dickson, in his *History of the Theory of Numbers* (New York: Chelsea, 1952), p. 57, attributes the solution (in the form of a verse) of a very special case of the integer CRP to the Chinese mathematician Sun-Tsu around the first century A.D., and goes on (pp. 57–64) to mention the work of some fifty mathematicians (including Euler) in connection with the integer CRP. Without doubt the most comprehensive and algorithmic of these early treatments of the integer CRP was due to Gauss in his *Disquisitiones Arithmeticae* (English translation by A. A. Clarke, New Haven: Yale University Press, 1966), Articles 32 and 36.

The theory and practice of polynomial interpolation (which, as we have seen, is a special case of the CRP over a polynomial domain) owes its definitive beginning to Isaac Newton. In Lemma 5, Book 3, of the *Principia* we find what is essentially Newton's Interpolation Formula along with a method of computation that is essentially Newton's Interpolation Algorithm (thus the names). Newton's own comments on the interpolation problem and his solution are noteworthy: "To describe a geometrical curve which passes through any given points...although the problem may seem intractable at first sight it is nevertheless the contrary. Perhaps it is indeed one of the prettiest problems that I can hope to solve." [W. J. Greenstreet (ed.), *Isaac Newton 1642–1727* (London: G. Bell and Sons, 1927), p. 45].

Lagrange's Interpolation Formula was given by Lagrange in his *Oeuvres* (Vol. 7, p. 286). This formula, however, had been discovered earlier by E. Waring [*Phil. Trans. Roy. Soc. London* **59** (1779), 59–67].

Gauss also investigated polynomial interpolation formulas, both Newtonian and Lagrangian, in *Nachlass Werke*, III, "Theoria Interpolationis Methodo Nova Tractata" (Gottingen, 1876), p. 265. It is interesting that Gauss considered in depth both the integer and the polynomial cases of the CRP yet failed to perceive any relationship between these two cases.

Section 3

1. (*Historical and bibliographical remarks on the method of homomorphic images.*)

A. *The integer case.* The first significant computational application of the integer homomorphic image scheme seems to have occurred as late as the mid 1950's when two Czechoslovakian computer designers, A. Svoboda and M. Valach, revealed the advantages of machines whose arithmetic was carried out modularly (for several small "hard-wired" moduli), the foremost of these advantages being the carry-free nature of addition and multiplication. The integer CRA in this application was used to convert numbers from internal (modular) form to external (decimal) form. Independently and slightly later these same ideas occurred to H. Aiken and H. Garner in America.

An extensive account of the research conducted in the area of modular arithmetic computers (carried out mainly from the late fifties to the mid sixties) is to be found in N. S. Szabo and R. I. Tanaka, *Residue Arithmetic and its Applications to Computer Technology*, (New York: McGraw-Hill, 1967). The prevailing opinion concerning the modular arithmetic approach to general-purpose hardware design seems to be that the experiment was largely a failure. Although modular arithmetic computers afforded a few advantages over traditional design, it turned out that certain basic operations, trivially carried out using a standard representation for integers, became intrinsically difficult when a modular representation was used— sign detection (is $x > 0$?) is a notable example.

Much more successful have been the software (as opposed to hardware) applications of the integer homomorphic image method. The seminal paper in this area is H. Takahasi and Y. Ishibashi, "A 'new method' for exact calculation by a digital computer," *Information Processing in Japan* **1** (1961), 28–42—a splendid paper both for its simplicity and the power of the technique that it advocates (essentially the integer MHI scheme). The integer MHI scheme was taken up by other authors as well, mainly in the context of solving exactly integer systems of linear equations [I. Borosh and A. S. Fraenkel, *Math. Comp.* **20** (1966), 107–112; M. Newman, *J. Res. Nat. Bur. Standards* **71B** (1967), 171–179; J. Howell, *Comm. ACM* **14** (1971), 180–184]. The MHI solution to this problem that we presented in Example 1 of Section 3.1, in particular the use of the adjoint method of solution, is due to S. Cabay [*SIGSAM 2*, 392–398; see also S. Cabay and T. P. L. Lam, *ACM Trans. Math. Software* **3** (1977), 386–398, 404–410].

B. *The polynomial case.* The polynomial MHI scheme—i.e., evaluation-interpolation—was also part of Takahasi and Ishibashi's "new method" (Section 4 of their paper, cited above), although the idea had long been present in the numerical computing literature, specifically the method for computing the characteristic polynomial $c(\lambda) = \det(\lambda I - A)$ of an $n \times n$ matrix A by interpolating to the values $c(\lambda_i)$ for n distinct values λ_i of the eigenvalue parameter λ [see V. N. Fadeeva, *Computational Methods of Linear Algebra* (New York: Dover, 1959), Section 27].

In a series of papers, M. T. McClellan has extensively generalized and analyzed the evaluation-interpolation scheme for the purpose of solving systems of linear equations with multivariate polynomial and rational function coefficients [*J. ACM* **20** (1973), 563–588; *ACM Trans. Math. Software* **3** (1977), 1–25, 147–158]. Application of evaluation-interpolation to the computation of flowgraph generating functions is to be found in J. D. Lipson, *Proc. IFIP Congress 74* (Amsterdam: North Holland, 1974), 489–493.

Evaluation-interpolation, its elegance and utility notwithstanding, does suffer from one notable disadvantage: it is largely unable to exploit sparseness (= many zero-coefficient terms) in polynomial data. For example consider the multiplication of $3x^{100} + 2$ by $4x^{97} + 5x^{20}$. Only four coefficient multiplications are required using the obvious direct method, whereas evaluation-interpolation would require many more (in the thousands). It is precisely this phenomenon that makes "direct" methods—methods that operate on polynomial data using classical algorithms—an attractive alternative to evaluation-interpolation. The use of specially tailored Gaussian elimination schemes for integer and polynomial systems of linear equations is the topic of E. H. Bareiss [*Math. Comp.* **22** (1968), 565–578; *J. Inst. Math. Applic.* 10 (1972), 68–104] and J. D. Lipson [in R. G. Tobey (ed.), *Proc. 1968 Summer Institute on Symbolic Mathematical Computation*, IBM Programming Laboratory Report FSC69-0312, June 1969, 235–303]. The issue of sparseness in the development and analysis of polynomial matrix algorithms (determinant calculation, linear equations solution) is the central topic of E. Horowitz and S. Sahni, *J. ACM* **22** (1975), 38–50; W. M. Gentleman and S. C. Johnson, *ACM Trans. Math. Software* **2** (1976), 232–241; M. L. Griss, *ACM Trans. Math. Software* **2** (1976), 31–49.

2. For one efficient such algorithm $C(A, F)$, see *Danilevsky's Method* in V. N. Fadeeva, *Computational Methods of Linear Algebra* (New York: Dover, 1959), Section 24. This method computes the characteristic polynomial of an $n \times n$ matrix A in $O(n^3)$ field operations.

THE FAST FOURIER TRANSFORM: ITS ROLE IN COMPUTER ALGEBRA

An algorithm may be appreciated on a number of grounds: on technological grounds because it efficiently solves an important practical problem, on aesthetic grounds because it is elegant, or even on dramatic grounds because it opens up new and unexpected areas of application. The *fast Fourier transform* (popularly referred to as the "FFT"), perhaps because it is strong in *all* of these departments, has emerged as one of the "super" algorithms of Computer Science since its discovery in the mid sixties. This concluding chapter is devoted to this remarkable algorithm and some of its major applications to algebraic computing.

1. WHAT IS THE *FAST* FOURIER TRANSFORM?

Recall our application of evaluation-interpolation to polynomial multiplication: to compute the product $c(x)$ over $F[x]$ of $a(x) = \sum_{i=0}^{n} a_i x^i$ and $b(x) = \sum_{i=0}^{n} b_i x^i$, evaluation-interpolation requires that we choose (at least) $N = 2n + 1$ distinct points $\alpha_k \in F$ and then proceeds according to the mapping diagram below.

John D. Lipson, Elements of Algebra and Algebraic Computing. ISBN 0-201-04115-4.

$$(a(x), b(x)) \xrightarrow{\text{(polynomial multiplication)}} c(x) = U(x)$$

(multipoint evaluation) $\uparrow$ (interpolation: $\deg U(x) < N;\ U(\alpha_k) = C_k$) $\uparrow$

$$\{(A_k = a(\alpha_k),\quad B_k = b(\alpha_k))\}_k \xrightarrow[\text{(pointwise multiplication)}]{} \{C_k = A_k B_k\}_k$$

As we discovered, this application was far from successful. Although evaluation-interpolation yields an $O(n^2)$ algorithm, the constant is considerably larger than that of the school method. But we did lay open the tantalizing prospect of performing the evaluation and interpolation steps proper in time $< O(n^2)$, which would then yield a superior algorithm (at least asymptotically— for large n). This is the departure point for this chapter.

How can such a speedup be achieved? The key idea, the one that lies at the heart of the FFT, is simply this: the evaluation-interpolation points (the α_k's), though they must be distinct, are otherwise completely arbitrary. *So let us choose them wisely.*

1.1   The Forward Transform: Fast Multipoint Evaluation

The forward transform of evaluation-interpolation is multipoint polynomial evaluation over a field F. We shall focus our attention on the following

Problem P_N of "size" N: Evaluate a polynomial $a(x) = \sum_{i=0}^{N-1} a_i x^i$ of "length" N (length = degree + 1) at each of a set $E_N = \{\alpha_k\}_{k=0}^{N-1}$ of N distinct points $\alpha_k \in F$ (the "evaluation points").

The solution to P_N is the collection of polynomial values $A_k = a(\alpha_k)$ ($k = 0, \dots, N-1$).

To analyze and compare algorithms for solving P_N, we shall count $M(N)$, the required number of multiplications over F. ($M(N)$ is a valid figure of merit, since multiplications dominate the arithmetic work of the algorithms to be discussed.)

To show off our more inspired solutions to P_N, we record the pedestrian

Proposition 1. For arbitrary evaluation points, P_N can be solved in $M(N) = N^2 + O(N)$.

Proof.   Compute each $a(\alpha_k)$ ($k = 0, \dots, N-1$) by Horner's rule. $\square$

The idea now is to impose some structure on the evaluation points that can be exploited to speedup the solution to P_N.

Definition. Let $N = 2n$ be even. A collection of N distinct points $E_N = \{\alpha_k\}_{k=0}^{N-1}$ is said to have *Property S* if E_N can be written as

$$E_N = \{\pm\alpha_k\}_{k=0}^{n-1}.$$

('S' thus stands for symmetry of sign; if β is in E_N, then so is $-\beta$.)

Let E_N have Property S. Then, since $(-\beta)^2 = (+\beta)^2$, we see that only $N/2$ of the *squares* of points in E_N are distinct. This little observation is the key to speeding up multipoint evaluation.

Proposition 2. Let N be even, $N = 2n$, and let $E_N = \{\alpha_k\}_k$ have property S. Then P_N can be solved in $M(N) = N^2/2 + O(N)$.

Proof. We can decompose $a(x) = \sum_{i=0}^{N-1} a_i x^i$ according to

$(*)$ $a(x) = b(y) + xc(y)$

where $y = x^2, \quad b(y) = \sum_{i=0}^{n-1} a_{2i} y^i, \quad c(y) = \sum_{i=0}^{n-1} a_{2i+1} y^i.$

[Example ($N = 4$):

$$a(x) = a_0 + a_1 x + a_2 x^2 + a_3 x^3$$

$$= \underbrace{(a_0 + a_2 y)}_{b(y)} + x\underbrace{(a_1 + a_3 y)}_{c(y)} \quad \text{where } y = x^2.]$$

Thus $A_k = a(\alpha_k)$, $A_{-k} = a(-\alpha_k)$ can be evaluated according to

Algorithm 1 (*Binary splitting scheme*).

Step 1. Compute $\beta_k = \alpha_k^2$ $(k = 0, \ldots, n-1)$.

Step 2. With $b(y), c(y)$ given by $(*)$, use Horner's rule to compute

$$B_k = b(\beta_k),$$
$$C_k = c(\beta_k) \quad (k = 0, \ldots, n-1).$$

Step 3. Return $A_k = B_k + \alpha_k C_k$,

$$A_{-k} = B_k - \alpha_k C_k \quad (k = 0, \ldots, n-1).$$

Steps 1 and 3 require $O(n)$ multiplications while step 2 requires the evaluation of two polynomials at n points, which using Horner's rule requires $2n^2 + O(n)$ multiplications (Proposition 1). Thus the overall number of multiplications is $M(2n) = 2n^2 + O(n)$, or $M(N) = N^2/2 + O(N)$. $\square$

Thus, by exploiting the small amount of structure of points E_N enjoying Property S, we have achieved a speedup of a factor of two over the solution of Proposition 1—a solid improvement even though the asymptotic character of the solution to our size N multipoint evaluation problem still remains $O(N^2)$.

We note that the above algorithm has essentially solved the original problem P_N of size N in terms of two problems $P_{N/2}$ of half the size. Since the evaluation points for these two subproblems are not special (in particular they will not in general have Property S), Algorithm 1 must be content to use the Horner's rule solution of Proposition 1 to solve these two subproblems. Now wouldn't it be nice (we muse) if we could speed up the solution to these two subproblems as well? To this end we make the leap to a very special class of evaluation points.

In field theory, we refer to a field element ω that has multiplicative order N as a *primitive Nth root of unity:* an Nth root of unity in that ω satisfies $x^N - 1 = 0$; a *primitive N*th root of unity in that ω satisfies $x^k - 1 = 0$ for no positive $k < N$. (Thus a primitive element in a field of order r (Section VI.2) is a primitive $(r-1)$st root of unity.)

Example 1. In **C**, $e^{2\pi i/N}$ is a primitive Nth root of unity (Example 3 of Section III.3.4). In $\mathbf{Z}_{13}$, 8 is a primitive 4-th root of unity. (Check: $8^2 = 12$, $8^3 = 5$, $8^4 = 1$). $\square$

We refer to the N distinct integral powers of a primitive Nth root of unity ω,

$$[\omega] = \{1, \omega, \ldots, \omega^{N-1}\},$$

as a set of N *Fourier points.* These are the points that we shall choose for our multipoint evaluation problem. (We shall shortly elucidate their connection with Fourier transforms.) Let us call a multipoint evaluation problem P_N at N Fourier points a *Fourier evaluation problem F_N (of size N).*

Lemma 1. Let $N = 2n$, ω a primitive Nth root of unity. Then the N Fourier points $[\omega]$ have Property S, with $\omega^{k+n} = -\omega^k$ $(k = 0, \ldots, n-1)$.

Proof. First we note that

$$(\omega^{k+n})^2 = (\omega^k)^2 \omega^N = (\omega^k)^2,$$

using only that ω is an Nth root of unity. But $x^2 = y^2$ in a field implies that $x = \pm y$. Since $\omega^{k+n} \neq \omega^k$ (otherwise we would have $\omega^n = 1$ contradicting the fact that ω is a *primitive N*th root of unity), it must be the case that $\omega^{k+n} = -\omega^k$, as required. $\square$

Consequence. We can apply the binary splitting scheme of Algorithm 1 to solve a Fourier evaluation problem F_N.

Now for the crucial additional property of Fourier points that will let us speed up the solution to the subproblems arising in step 2 of Algorithm 1.

Lemma 2. Let $N = 2n$, ω a primitive Nth root of unity. Then ω^2 is a primitive $(N/2)$th root of unity.

Proof. Since $(\omega^2)^n = \omega^N = 1$, ω^2 is an nth root of unity. But $(\omega^2)^j \neq 1$ for $0 < j < n$, otherwise we would have $\omega^k = 1$ with $0 < k < 2n = N$ in contradiction to ω being a *primitive* Nth root of unity. Thus ω^2 is a primitive nth root of unity, as required. $\square$

Consequence. Provided $n = N/2$ is even, we can apply the binary splitting scheme of Algorithm 1 not only to solve a given Fourier evaluation problem F_N at N Fourier points $[\omega]$ *but also to solve the two subproblems P_n arising in step 2 of Algorithm 1.* For these two subproblems each involve the multipoint evaluation of a length n polynomial at the n points $\{1, \omega^2, \ldots, (\omega^2)^{n-1}\}$. By Lemma 2, these points are a set $[\omega^2]$ of n Fourier points. Hence the two subproblems P_n are in fact *Fourier* evaluation problems F_n, so that Lemma 1 and its consequence applies to these subproblems.

Moreover, the same argument applies to the sub-subproblems that arise, and so on, inductively, for as long as the number of evaluation points remains even. Thus, if we choose $N = 2^m$, then the binary splitting scheme can be carried out (through m levels of recursion) until a trivial problem is reached: the evaluation of a polynomial of length one at one point.

These considerations lead to the abstract recursive FFT procedure of Algorithm 2 for evaluating a polynomial of length N at N Fourier points. By "abstract" we mean that the procedure is operational over any field having the requisite $N = 2^m$ th root of unity.

Algorithm 2   (FFT—*fast Fourier transform*).

Input arguments.
    integer $N = 2^m$,
    polynomial $a(x) = \sum_{i=0}^{N-1} a_i x^i$,
    primitive Nth root of unity ω.

Output argument.
    array $\mathbf{A} = (A_0, \ldots, A_{N-1})$ where $A_k = a(\omega^k)$.

Auxiliary data.
    integer $n = N/2$,
    polynomials $b(x) = \sum_{i=0}^{n-1} b_i x^i$, $c(x) = \sum_{i=0}^{n-1} c_i x^i$,
    arrays $\mathbf{B} = (B_0, \ldots, B_{n-1})$, $\mathbf{C} = (C_0, \ldots, C_{n-1})$.

Procedure FFT is displayed in Fig. 1. (Of course the procedure assumes a data type corresponding to the field over which it is to operate.)

The following correctness proof for procedure FFT makes explicit the intuitive inductive argument that we used in its derivation.

---

procedure FFT($N, a(x), \omega, \mathbf{A}$);
  if $N = 1$
   then
    {Basis.} $A_0 := a_0$
   else
   begin
    {Binary split.}
     $n := N/2$;
     $b(x) := \sum_{i=0}^{n-1} a_{2i} x^i$;
     $c(x) := \sum_{i=0}^{n-1} a_{2i+1} x^i$;
    {Recursive calls.}
     FFT($n, b(x), \omega^2, \mathbf{B}$);
     FFT($n, c(x), \omega^2, \mathbf{C}$);
    {Combine.}
     for $k := 0$ **until** $n - 1$ **do**
     begin
      $A_k \quad := B_k + \omega^k \times C_k$;
      $A_{k+n} := B_k - \omega^k \times C_k$
     end
   end

Fig. 1 FFT procedure.

Theorem 3 (*Procedure FFT works*).   If

$N = 2^m$,
$a(x) = \sum_{i=0}^{n-1} a_i x^i \in F[x]$ is a polynomial of length N,
ω is a primitive Nth root of unity over F,

then FFT($N, a(x), \omega, \mathbf{A}$) returns $A_k = a(\omega^k)$ for $k = 0, \ldots, N - 1$.

Proof.   Let $p(m)$ be the assertion of Theorem 3. Then the correctness proof of FFT is tantamount to the proof of $p(m)$ for all $m \in \mathbf{N}$.

Basis ($m = 0$).   For this case, $N = 1$, FFT returns $A_0 = a_0 = a(\omega^0)$, the latter equality holding because $a(x)$ is the constant polynomial a_0. Thus $p(0)$ holds.

Induction.   Assume $p(m)$ where m is an arbitrary natural number. We now establish $p(m + 1)$. If $N = 2^{m+1}$, then $N > 1$, and the "binary split" step of FFT gives $n, b(x), c(x)$ such that $n = N/2$ and

$$(*) \qquad\qquad a(x) = b(x^2) + xc(x^2).$$

By Lemma 2, ω^2 is a primitive nth root of unity. Moreover $b(x)$ and $c(x)$ are polynomials of length $n = 2^m$. Hence by the induction hypothesis, the "recursive calls" step of FFT returns

$$(**) \qquad B_j = b(\omega^{2k}), \quad C_j = c(\omega^{2k}) \qquad (k = 0, \ldots, n - 1).$$

The "combine" step of FFT then yields, for $k = 0, \ldots, n - 1$,

$$A_k = B_k + \omega^k C_k$$
$$= b((\omega^k)^2) + \omega^k c((\omega^k)^2) \qquad \text{by } (**)$$
$$= a(\omega^k) \qquad \text{by } (*),$$

$$A_{k+n} = B_k - \omega^k C_k$$
$$= b(\omega^{2k}) - \omega^k c(\omega^{2k}) \qquad \text{by } (**)$$
$$= b((\omega^{k+n})^2) + \omega^{k+n} c((\omega^{k+n})^2) \qquad \text{by Lemma 1}$$
$$= a(\omega^{k+n}) \qquad \text{by } (*).$$

Thus $\text{FFT}(N = 2^{m+1}, a(x), \omega, \mathbf{A})$ returns $A_k = a(\omega^k)$ $(k = 0, \ldots, N - 1)$, which establishes $p(m + 1)$ and completes the proof by induction of the correctness of procedure FFT. $\square$

We now show that our FFT is indeed a *fast* Fourier transform.

Theorem 4. Procedure FFT requires

$$M(N) = (N/2) \log_2 N$$

field multiplications to solve a Fourier evaluation problem F_N.

Proof. According to the "else" clause of procedure FFT, $M(N)$ satisfies

$$M(N) = 2M(N/2) + N/2$$

(the "$2M(N/2)$" term due to the two recursive calls, the "$N/2$" term due to the "combine" step), which for $N = 2^m$ becomes

$$M(2^m) = 2M(2^{m-1}) + 2^{m-1}.$$

Iterating this relationship m times gives

$$M(2^m) = m2^{m-1} + M(1)2^m.$$

But $M(1) = 0$, in accordance with the "then" clause (basis) of FFT. Thus we have $M(2^m) = m2^{m-1}$, or $M(N) = (N/2) \log_2 N$ as required. $\square$

Take, for example, $N = 1000$. Then classical multipoint evaluation requires 10^6 multiplications, whereas the FFT requires only about 10^4 multiplications. In more global terms, the FFT has the pleasing quasilinear property that doubling the problem size roughly doubles the computation time (more precisely, $M(2N)/M(N) \to 2$ for large N). This is in contrast with the classical $O(N^2)$ algorithm, which has the distressing property that doubling the problem size quadruples the required computation time.

So our FFT procedure is indeed fast. The factor of two speedup of the binary splitting scheme (Proposition 2 vs. Proposition 1) has been enjoyed at every level of the FFT's recursion, resulting in an asymptotic speedup from

$O(N^2)$ to $O(N \log N)$. This, then, is what the *fast* Fourier transform is all about (Note 1).

We now turn to the companion problem of *interpolation* with respect to Fourier points.

1.2   The Inverse Transform: Fast Interpolation

Let $\alpha_0, \ldots, \alpha_{N-1}$ be N points in a field F, to be used for evaluation and interpolation. Our size N multipoint evaluation problem (with respect to these points) is the problem of computing, for a given polynomial $a(x) = \sum_{i=0}^{N-1} a_i x^i \in F[x]$, the values $b_k = a(\alpha_k)$ $(k = 0, \ldots, N-1)$. Our size N interpolation problem, on the other hand, is the inverse problem of computing, for given $b_k \in F$ $(k = 0, \ldots, N-1)$, the coefficients of the (unique) interpolating polynomial $a(x) = \sum_{i=0}^{N-1} a_i x^i$ which satisfies $b_k = a(\alpha_k)$. In brief: with respect to the relationship $b_k = a(\alpha_k)$ $(k = 0, \ldots, N-1)$, evaluation determines the b_k's from the a_i's, interpolation determines the a_i's from the b_k's.

This inverse relationship between (multipoint) evaluation and interpolation becomes especially transparent when examined in matrix terms. To this end we introduce the $N \times N$ *Vandermonde matrix* $V(\alpha_0, \ldots, \alpha_{N-1})$ *associated with* $\alpha_0, \ldots, \alpha_{N-1}$:

$$V(\alpha_0, \ldots, \alpha_{N-1}) = \begin{bmatrix} 1 & \alpha_0 & \alpha_0^2 & \cdots & \alpha_0^{N-1} \\ 1 & \alpha_1 & \alpha_1^2 & \cdots & \alpha_1^{N-1} \\ \vdots & & & & \\ 1 & \alpha_{N-1} & \alpha_{N-1}^2 & \cdots & \alpha_{N-1}^{N-1} \end{bmatrix}.$$

Let $\mathbf{a} = (a_0, \ldots, a_{N-1})$, $\mathbf{b} = (b_0, \ldots, b_{N-1})$. The definition of matrix-vector multiplication immediately gives

Proposition 1.   For $a(x) = \sum_{i=0}^{N-1} a_i x^i$ and $V = V(\alpha_0, \ldots, \alpha_{N-1})$,

$$V\mathbf{a} = \mathbf{b} \quad \Leftrightarrow \quad b_k = a(\alpha_k).$$

In Proposition 1, interpolation theory guarantees that if the α_k's are distinct then the coefficients a_i of $a(x)$ can be uniquely determined from the b_k's, which by Proposition 1 is to say that $V\mathbf{a} = \mathbf{b}$ can be solved uniquely for $\mathbf{a}$. But by elementary matrix theory this means that V is nonsingular (invertible). Thus we have shown that the Vandermonde matrix $V(\alpha_0, \ldots, \alpha_{N-1})$ for distinct α_k's is nonsingular, which allows us to embellish Proposition 1 to

Proposition 1'.   For $a(x) = \sum_{i=0}^{N-1} a_i x^i$ and $V = V(\alpha_0, \ldots, \alpha_{N-1})$ (α_k's distinct),

$$V\mathbf{a} = \mathbf{b} \quad \Leftrightarrow \quad \mathbf{a} = V^{-1}\mathbf{b} \quad \Leftrightarrow \quad b_k = a(\alpha_k).$$

Example 1.   In $\mathbf{Z}_7$, let $\alpha_0 = 5$, $\alpha_1 = 2$, $\alpha_2 = 3$, and let $a(x) = 2 + 6x + x^2$. Then with $V = V(5, 2, 3)$ we have

$$Va = \begin{bmatrix} 1 & 5 & 5^2 = 4 \\ 1 & 2 & 2^2 = 4 \\ 1 & 3 & 3^2 = 2 \end{bmatrix} \begin{bmatrix} 2 \\ 6 \\ 1 \end{bmatrix} = \begin{bmatrix} 1 \\ 4 \\ 1 \end{bmatrix} = \begin{bmatrix} a(5) \\ a(2) \\ a(3) \end{bmatrix}.$$

And if $b_0 = 1$, $b_1 = 4$, $b_2 = 1$, then the solution to $Va = b$ is $a_0 = 2$, $a_1 = 6$, $a_2 = 1$ —the coefficients of $a(x)$ satisfying $a(\alpha_k) = b_k$ ($k = 0, 1, 2$). $\square$

So, in the light of Proposition 1' we conclude:

(1) Multipoint evaluation (the forward transform) corresponds to a matrix product of the form Va where V is a Vandermonde matrix. Thus the FFT can be regarded as a fast algorithm for computing Va when V is a Vandermonde matrix $V(1, \omega, \ldots, \omega^{N-1})$ associated with a set of N Fourier points—$O(N \log N)$ for $V(1, \omega, \ldots, \omega^{N-1})$ vs. $O(N^2)$ for an arbitrary Vandermonde matrix.

(2) Interpolation (the inverse transform) corresponds to a matrix product of the form $V^{-1}b$ where V is a Vandermonde matrix. Thus Newton's Interpolation Algorithm can be regarded as a fast algorithm for solving a system of linear equations $Va = b$ for $a = V^{-1}b$ when V is a Vandermonde matrix (associated with any set of distinct points, not necessarily Fourier points)—$O(N^2)$ in the Vandermonde case versus $O(N^3)$ for an arbitrary linear system.

But (2) is not what we are after (noteworthy though it might be). We want a *faster* than $O(N^2)$ interpolation algorithm, which exploits the case where V is a Vandermonde matrix associated with Fourier points.

Notation. If ω is a primitive Nth root of unity, then we write $V([\omega])$ for $V(1, \omega, \ldots, \omega^{N-1})$.

Theorem 2. Let ω be a primitive Nth root of unity in a field F in which $N^{-1} [= (N \cdot 1)^{-1}]$ exists. Then

$$V([\omega])^{-1} = N^{-1} V([\omega^{-1}]).$$

Proof. First, it is trivially shown (do it) that ω^{-1}, like ω, is a primitive Nth root of unity. Thus $V([\omega^{-1}])$ denotes the $(N \times N)$ Vandermonde matrix $V(1, \omega^{-1}, \ldots, (\omega^{-1})^{N-1})$. If we show that

$$V([\omega])V([\omega^{-1}]) = NI = \begin{bmatrix} N & & 0 \\ & \ddots & \\ 0 & & N \end{bmatrix},$$

then we are done.

So let $W = V([\omega])V([\omega^{-1}])$. Then

$$w_{ij} = \sum_{k=0}^{N-1} \omega^{ik}\omega^{-kj} \qquad (0 \le i, j < N).$$

CASE 1. $i = j$. Then $w_{ii} = \sum_{k=0}^{N-1} 1 = N$.

CASE 2. $i \neq j$. Then $w_{ij} = \sum_{k=0}^{N-1} (\omega^{i-j})^k$. Since $0 < |i-j| < N$, it follows that $\omega^{i-j} \neq 1$, otherwise we would have a contradiction to $o(\omega) = N$. Hence we can apply the identity $\sum_{k=0}^{N-1} x^k = (x^N - 1)/(x - 1)$ $(x \neq 1)$ to obtain

$$w_{ij} = \frac{(\omega^{i-j})^N - 1}{\omega^{i-j} - 1} = \frac{(\omega^N)^{i-j} - 1}{\omega^{i-j} - 1} = 0. \quad \square$$

Example 2.   In Z_{13}, 8 is a primitive 4th root of unity, with $8^{-1} = 5$. Here we have

$$V([8]) = \begin{pmatrix} 1 & 1 & 1 & 1 \\ 1 & 8 & 12 & 5 \\ 1 & 12 & 1 & 12 \\ 1 & 5 & 12 & 8 \end{pmatrix}, \qquad V([5]) = \begin{pmatrix} 1 & 1 & 1 & 1 \\ 1 & 5 & 12 & 8 \\ 1 & 12 & 1 & 12 \\ 1 & 8 & 12 & 5 \end{pmatrix},$$

and

$$V([8])V([5]) = \begin{pmatrix} 4 & 0 & 0 & 0 \\ 0 & 4 & 0 & 0 \\ 0 & 0 & 4 & 0 \\ 0 & 0 & 0 & 4 \end{pmatrix},$$

in accordance with Theorem 2. $\square$

Finally, let us consider the "Fourier interpolation" problem of size N, that of computing the (unique) interpolating polynomial $a(x) = \sum_{i=0}^{N-1} a_i x^i$ that satisfies $a(\omega^k) = b_k$ for arbitrarily specified b_k $(k = 0, \ldots, N-1)$, where $[\omega]$ is a set of N Fourier points.

Denote $V([\omega])$ by V and $V([\omega^{-1}])$ by V'. The coefficients a_i of $a(x)$ are then given by

$$\mathbf{a} = V^{-1}\mathbf{b} \qquad \text{Proposition 1'}$$

$$= N^{-1}V'\mathbf{b} \qquad \text{Theorem 2.}$$

Let $\mathbf{c} = V'\mathbf{b}$. Since $V' = V([\omega^{-1}])$ is a Vandermonde matrix, Proposition 1 tells us that c_k is the value of $b(x) = \sum_{i=0}^{N-1} b_i x^i$ at $x = (\omega^{-1})^k$, which is to say that the c_k's can be computed by *Fourier evaluation* of $b(x)$ at the N Fourier points $[\omega^{-1}]$. For this latter problem we have an efficient algorithm—the FFT. Thus the problem of Fourier interpolation can be essentially solved by Fourier evaluation in accordance with

Algorithm 1   (*Fast Fourier interpolation—FFI*).

Input.
   integer $N = 2^m$,
   primitive Nth root of unity ω,
   sample values $\mathbf{b} = (b_0, \ldots, b_{N-1})$.

Output. $a(x) = \sum_{i=0}^{N-1} a_i x^i$ where $a(\omega^k) = b_k$ $(k = 0, \ldots, N-1)$.

The following procedure uses the FFT procedure of Algorithm 2, Section 1.1.

> **procedure** FFI($N, \mathbf{b}, \omega, a(x)$);
> **begin**
> 1. $b(x) := \sum_{i=0}^{N-1} b_i x^i$;
> 2.   FFT($N, b(x), \omega^{-1}, \mathbf{c}$);
> 3. $a(x) := \sum_{i=0}^{N-1}(N^{-1}c_i)x^i$
> **end** $\square$

Clearly the above Fourier interpolation algorithm operates in FFT time $O(N \log N)$—i.e., it really is a *fast* Fourier interpolation algorithm as advertised. This we record as the companion to Theorem 4 of Section 1.1.

Theorem 3. The solution to a Fourier interpolation problem of size N can be computed in time $O(N \log N)$ (assuming $O(1)$ time for operations over the underlying field).

Example 3. Let us use Algorithm 1 to compute the solution $a(x) = \sum_{i=0}^{N-1} a_i x^i$ to the following $N = 4$ point Fourier interpolation problem over $\mathbf{Z}_{13}$ with respect to the Fourier points $[\omega = 8]$.

k	$\alpha_k = 8^k$	$b_k = a(\alpha_k)$
0	1	7
1	8	5
2	12	10
3	5	12

Line 1 of Algorithm 1 forms the polynomial $b(x) = 7 + 5x + 10x^2 + 12x^3$. Line 2 evaluates $b(x)$ via the FFT at the Fourier points $[\omega^{-1} = 5]$, returning $c_0 = b(1) = 8$, $c_1 = b(5) = 1$, $c_2 = b(12) = 0$, $c_3 = b(8) = 6$. Line 3 then multiplies these c_i's by $4^{-1} = 10$, returning $a(x) = 2 + 10x + 8x^3$ as the required interpolating polynomial. $\square$

1.3 Feasibility of mod p FFTs

To compute the solution of a Fourier evaluation-interpolation problem of size $N = 2^m$ over a field F, the FFT requires that F have a primitive Nth root of unity ω. Over the complex field (the field of traditional "analytic" applications —see Note 1), this requirement can always be satisfied: for any N, $e^{2\pi i/N}$ is a primitive Nth root of unity in $\mathbf{C}$. Over finite (modular) fields $\mathbf{Z}_p$, on the other hand, the situation is not nearly so cut and dry, but as we intend to show, it is still quite favorable: mod p FFTs are indeed computationally feasible. In applications to algebraic computing we are interested in primes p near computer wordsize, with a view to using the Chinese remainder and evaluation-interpolation techniques of Chapter VIII. Thus we are interested in very large primes, say on the order of 10^9.

This section, then, is devoted to answering two questions in the affirmative:

1. Do fields $\mathbf{Z}_p$ exist (p very large) having primitive $N = 2^m$th roots of unity (m in the 10–20 range, say)? Indeed, do such fields exist in abundance?

2. Given a field $\mathbf{Z}_p$ having a primitive Nth root of unity, can we find it efficiently (keeping in mind that $\mathbf{Z}_p$ may have billions of elements)?

A key result towards answering both of these questions is

Theorem 1. $\mathbf{Z}_p$ has a primitive Nth root of unity if and only if $N \mid (p - 1)$.

Proof.   By Lagrange's Theorem, the order of a group element divides the order of the group (as proved explicitly in Corollary 2 of Theorem 3, Section III.3.4). Since $\mathbf{Z}_p^*$ has order $p - 1$, we obtain the "only if" direction.

As for the "if" direction, let $N \mid (p - 1)$. Now $\mathbf{Z}_p$ contains a primitive element (Theorem 2 of Section VI.2.1), call it α. It is trivially verified (do it) that

$$\beta = \alpha^{(p-1)/N}$$

has order N in $\mathbf{Z}_p^*$, making β a primitive Nth root of unity. $\square$

Thus to compute $\bmod\, p$ FFTs of size $N = 2^m$, we require primes p such that $2^m \mid (p - 1)$, i.e., primes of the form $p = 2^e k + 1$ ($e \geq m$). We call a prime p of the form

$$p = 2^e k + 1 \qquad (k \text{ odd})$$

a *Fourier prime having (binary) exponent* e. Evidently any such prime can be used to compute FFTs of size $N = 2^m$ for $m \leq e$.

The assurance that Fourier primes exist in abundance rests on a result from analytic number theory (Note 2):

Generalized Prime Number Theorem   (Cf. the "ordinary" Prime Number Theorem of Appendix 1 to Section VIII.3). Let integers a and b be relatively prime. The number of primes $\leq x$ in the arithmetic progression $ak + b$ ($k = 1, 2, \ldots$) is approximately (and somewhat greater than)

$$(x/\log x)/\phi(a) \qquad (\phi = \text{Euler's phi function}).$$

Consequence.   The number of Fourier primes $p = 2^f k + 1 \leq x$ is approximately

(∗) $$(x/\log x)/2^{f-1}.$$

This, then, is our assurance that Fourier primes exist in reasonable abundance.

Example 1.   Let us take $x = 2^{31}$ (corresponding to the wordsize of a certain popular class of machines) and $f = 20$ in (∗). We conclude that there are approximately 180 Fourier primes $p = 2^e k + 1$ (k odd) with exponent $e \geq f = 20$. Any such Fourier prime could be used to compute FFTs of size 2^{20} (very large transforms indeed). $\square$

Now for the second question that we posed: Can we efficiently determine primitive Nth roots of unity in a very large finite field $\mathbf{Z}_p$?

In the proof of Theorem 1, we noted that if α is primitive in $\mathbf{Z}_p$, then $\beta = \alpha^{(p-1)/N}$ is a primitive Nth root of unity. Even if the exponent $(p-1)/N$ is very large (as it may well be), the powering of α can still be efficiently carried out (in log exponent time) using the fast algorithm developed in Example 2 of the Appendix to Chapter VII. So our problem of finding primitive Nth roots of unity boils down to that of *finding a primitive element in large finite fields* $\mathbf{Z}_p$.

What proportion of elements in $\mathbf{Z}_p$ are primitive? (Are we looking for a needle in a haystack?) From finite field theory (Theorem 3 of Section VI.2.1) the number of primitive elements in $\mathbf{Z}_p$ is given by $\phi(p-1)$ ($\phi =$ Euler's phi function). Analytic number theory tells us that the average value of $\phi(n)$ over all integers n is $6n/\pi^2$; it is readily argued that its average value over even integers is greater than $3n/\pi^2$ (Note 3). We conclude, therefore, that primitive elements abound in finite fields; on the average (over p) we would expect better than three of every π^2 elements to be primitive, or, in probabilistic terms, we would expect an element drawn at random from $\mathbf{Z}_p$ to be primitive with probability greater than $3/\pi^2 \approx 0.3$.

Thus we propose to find a primitive element of $\mathbf{Z}_p$ by testing the elements $2, 3, \ldots$ in turn until a primitive one is found, armed with the above probabilistic assurance that even for very large values of p we should not have to test many elements.

But how should we test an element α of $\mathbf{Z}_p$ for primitivity? We momentarily consider the obvious method of raising α to successive powers, checking whether or not $\alpha^n \neq 1$ for $1 < n < p - 1$. Thankfully enough, the need for carrying out this terribly inefficient test is obviated by the following elegant and completely algebraic result:

Theorem 2. $\alpha \in \mathbf{Z}_p$ is primitive $\Leftrightarrow$ $[\alpha^{(p-1)/q} \neq 1$ in $\mathbf{Z}_p$ for any prime factor q of $p - 1]$.

Proof. ($\Rightarrow$) Trivial.

($\Leftarrow$) Assume α has order $n < p - 1$. Then n divides $p - 1$ by Lagrange, so that $p - 1 = kn$. Let $k = qr$, where q is a prime in k's prime factorization. But then $p - 1 = qrn$, so that q is also a prime in $(p - 1)$'s (unique) prime factorization. We then have

$$\alpha^{(p-1)/q} = \alpha^{rn} = (\alpha^n)^r = 1. \quad \square$$

Example 2. Let us use Theorem 2 to find a primitive element in $\mathbf{Z}_{41}$. Here we have $p - 1 = 40 = 2^3 5$, so that $\alpha \in \mathbf{Z}_{41}$ is primitive if and only if $\alpha^{20}, \alpha^8 \neq 1$.

Computing over $\mathbf{Z}_{41}$, we see that $2^{20} = 1$, $3^8 = 1$, $4^{20} = (2^{20})^2 = 1$, $5^{20} = 1$, so none of $2, 3, 4, 5$ are primitive. But $6^8 = 10$ and $6^{20} = 40$, so that 6 is the smallest primitive element of $\mathbf{Z}_{41}$ ("smallest" in the usual integer ordering sense, of course). $\square$

We propose the following algorithm, based on Theorem 2, for computing a list of Fourier primes.

Algorithm 1   (*Computing a list of Fourier primes*).

Input.
  List L of primes $\leq W$ (where W is typically our computer's wordsize);
  Auxiliary list L' of primes $\leq \sqrt{W}$ (if the sieve method of Appendix 1 to
Section VIII.3 is used to compute L, then this list L' is already available);
  Positive integer f.

Output.
  List of Fourier primes of the form $p = 2^e k + 1$ (k odd) with $e \geq f$,
together with a primitive element of $\mathbf{Z}_p$.

For each prime p in L:

Step 1   {Is p a Fourier prime of required exponent?}
  Determine the exponent e of 2 in the prime factorization of $p - 1$. If
$e < f$, then discard p.

Step 2   {Prime power factorization of $p - 1$}.
  Determine the distinct primes a_i in the prime power factorization of
$p - 1$,

$$p - 1 = a_1^{e_1} a_2^{e_2} \cdots a_r^{e_r},$$

by cancelling out powers of primes in the auxiliary table L' (Note 4).

Step 3   {Find a primitive element}.
  Test $\alpha = 2, 3, \ldots$ for primitivity, using Theorem 2: α is primitive $\Leftrightarrow$
$\alpha^{(p-1)/a_i} \neq 1$ for all a_i. (Use fast powering for the computation $\alpha^{(p-1)/a_i}$.)

$\square$

In Fig. 2 are presented the results of executing Algorithm 1 to find the 10
largest Fourier primes $\leq 2^{31} - 1$ having exponent $e \geq f = 20$. (Such Fourier

p	e	α = least primitive element of $\mathbf{Z}_p$
2130706433	24	3
2114977793	20	3
2113929217	25	5
2099249153	21	3
2095054849	21	11
2088763393	23	5
2077229057	20	3
2070937601	20	6
2047868929	20	13
2035286017	20	10

Fig. 2.   Fourier primes $p = 2^e k + 1 \leq 2^{31} - 1$ ($e \geq 20$).

primes could be used to compute FFTs of size $2^{20} > 10^6$ using a 32 bit word (31 bits + sign) machine. Perhaps we should call these primes "IBM Fourier primes.")

From the results of this section, we conclude that $\bmod p$ FFTs *are* computationally viable: Fourier primes—primes p for which FFTs over $\mathbf{Z}_p$ are possible —exist in plenty. Moreover, we have a reasonably efficient algorithm (Algorithm 1) for determining such primes along with primitive elements in the associated fields. From these primitive elements, the required primitive roots of unity can be efficiently computed.

2. FAST ALGORITHMS FOR MULTIPLYING POLYNOMIALS AND INTEGERS

With the FFT in hand, it is an easy matter to give a faster than $O(n^2)$ algorithm for multiplying degree n polynomials (faster = faster in an asymptotic sense, i.e., for sufficiently large problems). With a little more effort, we can derive a faster than $O(n^2)$ algorithm for multiplying n digit integers. In view of the venerable character of the classical $O(n^2)$ integer and polynomial multiplication algorithms ("school algorithms"), perhaps the reader will agree that the new FFT-based algorithms deserve to be called "surprisingly fast." In any case, if the results of this section make the reader question the tried and true (at least in algorithm design), then fine.

2.1 Fast Polynomial Multiplication

We return to our polynomial multiplication algorithm by evaluation-interpolation (Example 1 of Section VIII.3.2). There we failed to do better than $O(n^2)$ time due to the time required by classical multipoint evaluation and interpolation. But now we have the FFT.

> **Algorithm 1**　(*FFT multiplication over $F[x]$*).
>
> *Input.*　　Polynomials $a(x), b(x) \in F[x]$ having degrees $\leq n$.
> *Output.*　$c(x) = a(x)b(x)$.
>
> Choose $N = 2^m$ to be $> 2n$. Let ω be a primitive Nth root of unity in F. (The following algorithm requires the existence of such an element ω.)
>
> **Step 1**　{Evaluation via FFT}.
> 　(a) Invoke FFT$(N, a(x), \omega, \mathbf{A})$,
> 　　　　　FFT$(N, b(x), \omega, \mathbf{B})$;
> 　(b) Compute $C_k = A_k B_k$ $(k = 0, \ldots, N - 1)$.
>
> **Step 2**　{Fourier interpolation—Algorithm 1 of Section 1.2}.
> 　Invoke FFI$(N, \mathbf{C}, \omega^{-1}, U(x))$.
>
> **Step 3.**
> 　Return $c(x) = U(x)$. □

Since ω has been chosen to be a primitive $N = 2^m$ th root of unity, the use of procedures FFT and FFI for solving the N-point evaluation and interpolation problem of steps 1(a) and 2 is valid. The overall validity of Algorithm 1—that $c(x)$ returned in step 3 is in fact $a(x)b(x)$— is then a consequence of our choice of N to be $> 2n$, appealing only to the validity of the general evaluation-interpolation scheme for polynomial multiplications over $F[x]$ (Example 1 of Section VIII.3.2).

The computing time required by Algorithm 1 is clearly dominated by the FFT steps 1(a) and 2, hence is $O(N \log N)$, assuming as always that operations over F require $O(1)$ time. If N is chosen to be the least power of 2 that is $> 2n$, then N is $\leq 4n$. Thus in terms of the original size parameter n, the computing time required by Algorithm 1 is $O(4n \log 4n) = O(n \log n)$. This we record as

Theorem 1. Multiplication of two polynomials of degree n over $F[x]$ can be carried out in time $O(n \log n)$ (provided F has the requisite primitive root of unity).

Thus over the complex field Theorem 1 holds unconditionally, since **C** contains primitive Nth roots of unity for any N. Over subfields of **C**, notably **R** and **Q**, Theorem 1 does not hold, strictly speaking, unless we are willing to perform Algorithm 1 over **C**. For example, FFTs involving real polynomials generate complex values, because the Fourier points are necessarily complex. Over finite fields $\mathbf{Z}_p$, Theorem 1 holds provided p is a Fourier prime $p = 2^e k + 1$ with $2^e > 2n$. (In applications of finite fields, we would of course restrict our attention to Fourier primes of high exponents as discussed in Section 1.3.)

Some concluding remarks about the relative speeds of FFT-based and classical multiplication algorithms as a function of the possibly different degrees of the operands. If $a(x)$ and $b(x)$ have degrees m and n, with $m \leq n$ say, then our FFT-based multiplication algorithm requires $O(n \log n)$ time while classical multiplication requires $O(mn)$ time. For $m \approx n$, the FFT-based algorithm is clearly at its best relative to the classical algorithm: $O(n \log n)$ versus $O(n^2)$. For $m \ll n$, on the other hand, the performance of the FFT-based algorithm becomes relatively poor due to the fact that it cannot effectively exploit the smallness of m relative to n. As an extreme case in point, let $m = 1$. Then the FFT-based algorithm is still $O(n \log n)$, whereas the classical algorithm becomes $O(n)$.

Out of this simple analysis emerges the following *balancing principle* which transcends the details of successful applications of fast (FFT-based) polynomial multiplication: polynomial multiplication, wherever it occurs, should involve polynomials of roughly the same size (degree). We shall encounter this principle at work in Section 3.

2.2 Fast Integer Multiplication

We have already exploited the observation that the conventional positional notation for integers is essentially a polynomial-based representation (see Example 4 of Section VIII.1.1): If $a = (a_{n-1} \cdots a_0)_B$ is a base B integer, then a represents the value of its *associated polynomial* $a(x) = \sum_{i=0}^{n-1} a_i x^i$ at $x = B$; i.e., $a = a(B)$. In this section we exploit that same observation, this time to achieve an integer multiplication algorithm that works in faster than the $O(n^2)$ time required by the venerable school algorithm.

Let us consider a sample product $c = ab$ of decimal ($B = 10$) integers:

$$a = 329$$

$$b = 617$$

$$c = 202993.$$

This result can also be achieved by the polynomial multiplication $c(x) = a(x)b(x)$ followed by the evaluation $c(10)$. For $c(10) = a(10)b(10)$, because polynomial evaluation (in this case at $x = 10$) is a morphism for multiplication. Thus $c = c(10)$ yields the product $a(10)b(10) = ab$. To illustrate with $a = 329$, $b = 617$:

$$a(x) = 3x^2 + 2x + 9$$

$$b(x) = 6x^2 + x + 7$$

$$c(x) = 18x^4 + 15x^3 + 77x^2 + 23x + 63.$$

Then $c(10) = 202993 = 329 \times 617$.

Now for a computer implementation of this polynomial multiplication-evaluation algorithm for multiplying two n-digit base B integers $a = (a_{n-1} \ldots a_0)_B$ and $b = (b_{n-1} \ldots b_0)_B$, where B is chosen to be $< W = $ the wordsize of our computer. An especially convenient choice for B, assuming that the multiplicands are presented to the computer as decimal integers, is the largest power of ten $< W$. Then the base B digits of the multiplicands are obtained by simply grouping consecutive decimal digits (e.g., the base 10^3 digits of 46709834 are 046, 709, 834); moreover these base B digits, which constitute the coefficients of the associated polynomials $[a = (a_{n-1} \ldots a_0) \leftrightarrow \sum_{i=0}^{n-1} a_i x^i]$, are all single-precision.

We now propose to compute the polynomial product $c(x) = a(x)b(x)$ by our polynomial MHI scheme of Section VIII.3.3; i.e., we propose to compute $c^{(k)}(x) = a(x)b(x) \bmod p_k$ for a sufficient number K of large, but less than word-size, primes p_k ($B \le p_k \le W$) in order to obtain $c(x)$ by the CRA. Choosing these p_k's to be Fourier primes of the form $p = 2^e l + 1$ for sufficiently large exponent e, we can use FFT-based polynomial multiplication to compute the $c^{(k)}(x)$'s. This, then, is the crux of our proposed algorithm: the replacement of the $O(n^2)$ base B digit calculations of classical long multiplication by a number (K) of *fast* polynomial multiplications.

For reasons that will become clear in the analysis phase of our development, we call the resulting algorithm the "three primes" algorithm.

Algorithm 1   (*"Three primes" algorithm for integer multiplication*).

Input. $a = (a_{n-1} \ldots a_0)_B$, $b = (b_{n-1} \ldots b_0)_B$.
Output. $c = ab$.
   The algorithm requires K Fourier primes $p = 2^e l + 1 \leq W$ with sufficiently large exponent e (K to be determined).

Step 1   {Multiplication of associated polynomials $a(x) = \sum_{i=0}^{n-1} a_i x^i$, $b(x) = \sum_{i=0}^{n-1} b_i x^i$ by the MHI scheme for $\mathbf{Z}[x]$ (Algorithm 1 of Section VIII.3.3)}.

   1.1 For K Fourier primes p_k ($B \leq p_k \leq W$): compute $c^{(k)}(x) = a(x)b(x)$
       over $\mathbf{Z}_{p_k}[x]$ using FFT-based polynomial multiplication.
   1.2 Solve the polynomial CRP

$$u(x) \equiv c^{(k)}(x) \ (\mathrm{mod}\, p_k) \qquad (k = 0, \ldots, K-1)$$

   for the least-positive coefficient solution $U(x)$.
   1.3 Return $c(x) = U(x)$.

Step 2   {Evaluation at radix}. Return $c = c(B)$. $\square$

   The algorithm and its analysis hinge on the determination of K (how big must K be?). Each coefficient $c_k = \sum_{i+j=k} a_i b_j$ of $c(x)$ is seen to be $< nB^2$. Hence step 1 correctly computes $c(x)$ provided

$$(*) \qquad\qquad p_0 p_1 \cdots p_{K-1} \geq nB^2.$$

Since each p_k is $\geq B$, ($*$) is satisfied provided B^K is $\geq nB^2$, i.e., provided K is $\geq (\log_B n) + 2$. *Therefore over the range $n \leq B$, three primes are sufficient* (thus the name "three primes" algorithm).

   Choosing $K = 3$, we now have a tacit restriction on the size of problem our algorithm can handle: n must be $\leq B$. But keeping in mind that B is a huge integer, on the order of 10^9 say, we can regard this restriction as being totally innocuous.

   We have one other restriction on the size of problem our algorithm can handle. Step 1.1 requires three Fourier primes $p = 2^e l + 1$ ($B \leq p \leq W$) with $2^e \geq$ length of $c^{(k)}(x) = 2n - 1$ (recall: length = degree + 1). Thus if E is the largest integer for which we can find three Fourier primes each having exponent $e \geq E$, then n must satisfy $2n - 1 \leq 2^E$, which will be the case if n is $\leq 2^{E-1}$.

   In summary, our algorithm imposes two constraints on n: $n \leq B$ and $n \leq 2^{E-1}$.

   As for the computing time analysis of our algorithm, step 1.1 requires $O(n \log n)$ time, while step 1.2 requires $O(n)$ time ($2n - 1$ CRPs each involving three integer congruences). Thus step 1 requires $O(n \log n)$ time overall. We leave it as a not too difficult exercise to show that the polynomial evaluation of step 2 requires $O(n \log n)$ time (Exercise 4). This gives the following

Theorem 1. Let our computer have (fixed) wordsize W, so that $\bmod p$ operations for $p \leq W$ can be carried out in $O(1)$ time. Then this computer can multiply two n digit base B integers, $B < W$, in $O(n \log n)$ time provided:

1. $n \leq B$;
2. $n \leq 2^{E-1}$ where three Fourier primes $p = 2^e l + 1$ ($B \leq p \leq W$) can be found with $e \geq E$, E the largest such integer.

Algorithm 1 thus has a curious *subasymptotic* property: Although it multiplies *n*-digit numbers in $O(n \log n)$ time (versus the $O(n^2)$ time required by the classical method), it is operational only over a finite range of n. Of course this range can in principle be extended by increasing W, the computer wordsize. But W cannot be extended indefinitely with increasing n, for then the assumption that $\bmod p$ operations can be carried out in $O(1)$ time—time bounded by a constant independent of n—would become untenable. This, then, is why Algorithm 1 is *sub*asymptotic in character.

With such an algorithm it is obligatory to ask what kind of range of application it has. After all, if the algorithm turns out to be applicable only for small n (say $n \leq 100$), then it would have to be dismissed as uninteresting; table lookup would provide a much better subasymptotic algorithm. But as we now show, Algorithm 1 is "practically asymptotic"; for all intents and purposes its range of application for n is effectively infinite.

To establish this result we confine our attention to a specific "typical" wordsize, namely $W = 2^{31} - 1$ (corresponding to the same 31 bit + sign wordsize that we used for illustrative purposes in Section 1.3).

With $W = 2^{31} - 1$, we choose $B = 10^9$, the largest power of ten $< 2^{31} - 1$. The "$n \leq B$" constraint means that in Algorithm 1, n must be $\leq 10^9$.

From Fig. 3 we see that there are three Fourier primes $B \leq p \leq W$ having exponents ≥ 24. Thus E in the "$n \leq 2^{E-1}$" constraint can be taken as 24 (the largest possible value, as it turns out, for this particular word size), which means that n must be $\leq 2^{23} \approx 8.38 \times 10^6$. The latter, therefore, is the determining constraint.

$p = 2^e k + 1$ (k odd)	e	α = least primitive element of $\mathbf{Z}_p$
2013265921	27	31
2113929217	25	5
2130706433	24	3

Fig. 3 Three Fourier primes having exponent ≥ 24 (for a 32 bit word computer).

Thus Algorithm 1, *employed on a 32 bit word machine, is capable of multiply-ing integers having in excess of eight million decimal digits.* (Perhaps the reader concurs that Algorithm 1 is indeed "practically asymptotic"!)

In Note 1 we have gathered together some mainly historical remarks about fast integer multiplication algorithms.

3. FAST ALGORITHMS FOR MANIPULATING FORMAL POWER SERIES

Although the FFT will be vital to our achieving fast power series algorithms, it is Newton's method—that most venerable of numerical algorithms—that is the highlight of this concluding section.

3.1 Truncated Power Series Revisited

A (formal) power series $a(t) = \sum_{i=0}^{\infty} a_i t^i$ is in general an infinite mathematical object (because there are infinitely many coefficients a_i), unlike a polynomial or an integer, but very much like the infinite decimal expansion of a real number. For computational purposes it is usually both desirable and necessary to represent a power series $a(t) = \sum_{i=0}^{\infty} a_i t^i$ by its first (say) n terms $\sum_{i=0}^{n-1} a_i t^i$. This is the *truncated* power series $T_n[a(t)] = a(t) \bmod t^n$ introduced in Section V.2.2. (For convenience we are now writing $T_n[a(t)]$ rather than $T_n(a(t))$.) We regard $T_n[a(t)]$ as a mod t^n *approximation* to $a(t)$ in that $T_n[a(t)]$ agrees with $a(t)$ in its first n terms, much as we regard 3.14159 as a six figure approximation to π.

Let us now review the truncated power series ring $F[[t]]_n$ (Example 3 of Section V.2.2) from the viewpoint of the complexity of its operations. We make the usual assumption that operations over the coefficient field F require $O(1)$ time.

Let $a(t), b(t) \in F[[t]]_n$. Then

$$(1) \qquad\qquad a(t) \oplus b(t) = T_n[a(t) + b(t)]$$
$$= a(t) + b(t),$$

$$(2) \qquad\qquad a(t) \odot b(t) = T_n[a(t)b(t)],$$

where the right-hand side operations $+$, $\cdot$ are polynomial operations. As for inversion over $F[[t]]_n$, $a(t)^{\ominus}$, we have the general ring morphism result $\phi(a^{-1}) = \phi(a)^{\ominus}$ (Proposition 7 of Section IV.2.1). Here the morphism is $T_n : F[[t]] \to F[[t]]_n$, so we have

$$(3) \qquad\qquad a(t)^{\ominus} = T_n[a(t)^{-1}],$$

which states: to compute $a(t)^{\ominus}$, compute the first n terms of $a(t)^{-1}$ over $F[[t]]$ (for example, using the algorithm of Theorem 4, Section IV.3.1).

From (1), (2), and (3) we immediately conclude

Proposition 1. Over $F[[t]]_n$:

(1) $a(t) \oplus b(t)$ can be computed in $O(n)$ time;

(2) $a(t) \odot b(t)$ can be computed
   (i) in time $O(n^2)$ using classical polynomial multiplication,
   (ii) in time $O(n \log n)$ using FFT polynomial multiplication (provided
        F supports the FFT);
(3) $a(t) \ominus$ can be computed in $O(n^2)$ time.

Now suppose it is desired to compute $T_n[a(t) + b(t)]$, $T_n[a(t)b(t)]$, $T_n[a(t)^{-1}]$ for specified power series $a(t)$, $b(t) \in F[[t]]$—call these operations T_n-addition, T_n-multiplication, and T_n-inversion. Since T_n is a morphism $F[[t]] \rightarrow F[[t]]_n$, we have

$$T_n[a(t) + b(t)] = T_n[a(t)] \oplus T_n[b(t)],$$

$$T_n[a(t)b(t)] = T_n[a(t)] \odot T_n[b(t)],$$

$$T_n[a(t)^{-1}] = T_n[a(t)] \ominus.$$

These equations tell us that to compute T_n-sums, products, or inverses, the power series operands need only be specified mod t^n, and these operations can then be interpreted over $F[[t]]_n$; i.e., mod t^n. Henceforth we do not distinguish between mod t^n operations and T_n-operations.

* * *

The gap between multiplication time and inversion time begs the question, can we find a faster inversion algorithm? For an affirmative answer to this question we look to numerical computing.

3.2 Fast Power Series Inversion; Newton's Method

Our objective is to derive a fast algorithm for power series inversion. *Newton's method*, of numerical computing fame, provides just the tool we need. (This section, together with the next, might well have been entitled "Newton's method: a great *algebraic* algorithm"—Note 1.)

Let us briefly review Newton's method in its familiar numerical setting, as an algorithm for solving $f(x) = 0$ for an approximation to a numerical (say real) root $\bar{x}$. The method consists of computing a sequence of so-called *iterates* $x_1, x_2, \ldots$ (approximants to $\bar{x}$) according to *Newton's iteration*,

(1) $x_{k+1} = x_k - f(x_k)/f'(x_k),$

starting from some specified initial approximation x_0 to the desired root $\bar{x}$.

The geometry of Newton's method is illustrated in Fig. 4. The tangent to the curve at $(x_k, f(x_k))$, where x_k is the current iterate, is seen to intercept the x-axis at a point that provides a closer approximation to the root $\bar{x}$. This point, then, is taken to be the next iterate x_{k+1}. The forementioned tangent has the equation

$$\frac{y - f(x_k)}{x - x_k} = f'(x_k)$$

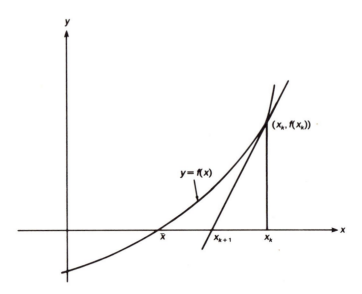

Fig. 4   Newton's method.

and passes through the point $(x_{k+1}, 0)$. Solving for x_{k+1} yields Newton's iteration (1).

The convergence properties of Newton's method are well-known (Note 2). If the sequence of iterates x_k converges at all to some root $\bar{x}$, then it converges *quadratically*, meaning that $\epsilon_{k+1} = C\epsilon_k^2$, where $\epsilon_k = \bar{x} - x_k$ is the error of the kth iterate and C is some real constant. For the case of real root-finding, this quadratic convergence property implies that the number of correct decimal places in an iterate roughly doubles with each successive iteration. The other side of the convergence coin, however, is a bad one. Newton's method is only *locally* convergent, meaning that convergence to a root $\bar{x}$ is assured only if the initial approximation, x_0, is sufficiently close to $\bar{x}$. In summary: Newton's method, *if* it works, works very well.

Let us now apply Newton's method to compute the reciprocal of a real number a. For $f(x)$ we take $1/x - a$, noting that $f(\bar{x}) = 0$ implies $\bar{x} = 1/a$. Newton's iteration (1) becomes

$$(2) \qquad x_{k+1} = x_k - (1/x_k - a)/(-1/x_k^2)$$
$$= 2x_k - ax_k^2.$$

This iteration, happily enough, does not involve division. It is the basis for virtually all computer algorithms (usually hardware but occasionally software) for performing real inversion or division.

For the case of (2) we can explicitly establish quadratic convergence. With $\epsilon_k = \bar{x} - x_k$, we have

$$
\begin{aligned}
\epsilon_{k+1} &= \bar{x} - x_{k+1} \\
&= \bar{x} - 2(\bar{x} - \epsilon_k) + a(\bar{x} - \epsilon_k)^2 \\
&= \bar{x} - 2\bar{x} + 2\epsilon_k + a\bar{x}^2 - 2a\bar{x}\epsilon_k + a\epsilon_k^2 \\
&= a\epsilon_k^2 \quad (\text{using } a\bar{x} = 1).
\end{aligned}
$$

So much for Newton's method in its familiar numerical setting. The numerical ideas that make Newton's method go through all have sharp analogs in a formal power series setting.

We have already introduced the concept of *approximation* in $F[[t]]$: If $b(t) = T_n[a(t)]$, then we regard $b(t)$ as a $\bmod t^n$ (polynomial) approximation to $a(t)$. And we can regard $\epsilon(t) = a(t) - b(t)$ as the *error* in the approximation $b(t)$ to $a(t)$. (But note that this error is a formal power series.) The following lemma is an immediate consequence of the definition of order of a power series (Section IV.3.1).

Lemma 1. Let $a(t) \in F[[t]]$, $b(t) \in F[[t]]_n$, and let $\epsilon(t) = a(t) - b(t)$. Then

$$
b(t) = T_n[a(t)] \Leftrightarrow \text{ord } \epsilon(t) \geq n.
$$

The notion of *convergence* is also a natural one in $F[[t]]$. We say that the sequence of power series $\{(x^{(k)}(t)\}_{k=0}^{\infty}$ *converges* to $\bar{x}(t)$ if for each $J \in \mathbf{N}$ there exists $K \in \mathbf{N}$ such that

$$
\text{ord}(\bar{x} - x^{(k)}) \geq J \quad \text{for } k \geq K.
$$

The following theorem, the key result of this section, underlies the efficacy of Newton's iteration (2) for power series inversion.

Theorem 1 (*Quadratic convergence of Newton's iteration for power series inversion*). Let $a(t)$ be a unit in $F[[t]]$ ($a_0 \neq 0$), so that $b(t) = a(t)^{-1}$ exists. If $u(t)$ is a $\bmod t^n$ approximation to $b(t)$, then $v(t)$ defined over $F[[t]]_{2n}$ by

$$
(3) \qquad\qquad v(t) = 2u(t) - T_{2n}[a(t)] \odot u(t)^2
$$

is a $\bmod t^{2n}$ approximation to $b(t)$.

Proof (Cf. the above proof of quadratic convergence of Newton's iteration for numerical inversion). Let $\epsilon(t) = b(t) - u(t)$, $\delta(t) = b(t) - v(t)$. By assumption, $u(t) = T_n[b(t)]$; hence ord $\epsilon(t) \geq n$ (Lemma 1). If we can show that ord $\delta(t) \geq 2n$, then $v(t) = T_{2n}[b(t)]$ (again by Lemma 1) and we will be done.

Now $\delta(t) = b(t) - v(t)$. Let us first examine $v(t)$. Since T_{2n} is a morphism $F[[t]] \to F[[t]]_{2n}$, we can appeal to our Fundamental Morphism Theorem of Section VIII.1.1 to rewrite (3) as

$$v(t) = T_{2n}\left[2u(t) - a(t)u(t)^2\right]$$

$$= 2u(t) - a(t)u(t)^2 + t^{2n}k(t)$$

for some $k(t) \in F[[t]]$.

Writing u, v, etc., for $u(t)$, $v(t)$, etc., we have

$$\delta = b - v$$

$$= b - 2u + au^2 - t^{2n}k$$

$$= b - 2(b - \epsilon) + a(b - \epsilon)^2 - t^{2n}k$$

$$= a\epsilon^2 - t^{2n}k \quad \text{(using } ab = 1\text{)}.$$

From our results on order (Proposition 1(b) and the corollary to Theorem 2 of Section IV.3.1), we have that $\operatorname{ord} a\epsilon^2 = \operatorname{ord} a + 2\operatorname{ord} \epsilon$ is $\geq 2n$ as is $\operatorname{ord} t^{2n}k$, whence $\operatorname{ord} \delta$ is also $\geq 2n$. $\square$

Thus we have shown that Newton's iteration for inversion yields a $\operatorname{mod} t^{2n}$ approximation (to the true power series inverse) from a $\operatorname{mod} t^n$ approximation. This is to say that the number of correct terms of the approximate inverse is doubled (thus the term "*quadratically* convergent"). Now it is an easy matter to obtain the $\operatorname{mod} t^1$ approximation to $a(t)^{-1}$; this is simply a_0^{-1}. We can then apply Newton's iteration (3) repeatedly to compute the $\operatorname{mod} t^2$, $\operatorname{mod} t^4$, $\operatorname{mod} t^8$, etc., approximations to $a(t)^{-1}$. Thus we obtain

Algorithm 1 (*Newton's method for power series inversion*).

Input. $a(t) \operatorname{mod} t^{2^n} = \sum_{i=0}^{2^n-1} a_i t^i$, $a_0 \neq 0$.

Output. $x^{(n)}(t) = a(t)^{-1} \operatorname{mod} t^{2^n}$.

> **begin**
> $\quad x^{(0)}(t) := a_0^{-1}$;
> $\quad$ **for** $k := 0$ **until** $n - 1$ **do**
> $\quad\quad x^{(k+1)}(t) := 2x^{(k)}(t) - T_{2^{k+1}}[a(t)] \odot x_k(t)^2$
> $\quad\quad\quad$ (computed over $F[[t]]_{2^{k+1}}$)
> **end** $\square$

Example 1. Let us compute $a(t)^{-1} \operatorname{mod} t^8$ for the $Z_{11}[[t]]$ power series

$$a(t) = 1 + 7t + 10t^2 + 4t^3 + 3t^4 + 3t^5 + 8t^6 + 2t^7.$$

With the initial approximation $x^{(0)}(t) = a(t)^{-1} \operatorname{mod} t = 1$, Newton's method computes

$$x^{(1)}(t) = a(t)^{-1} \operatorname{mod} t^2 = 1 + 4t$$

$$x^{(2)}(t) = a(t)^{-1} \operatorname{mod} t^4 = 1 + 4t + 6t^2 + 2t^3$$

$$x^{(3)}(t) = a(t)^{-1} \operatorname{mod} t^8 = 1 + 4t + 6t^2 + 2t^3 + 6t^4 + 9t^5 + 7t^6 + 10t^7. \quad \square$$

The quadratic convergence of Newton's method, illustrated above, is to be contrasted with the linear convergence of the classical algorithm which computes one only new term per iteration. Although quadratic convergence means that Newton's method requires only $\log_2 n$ iterations to compute $\mod t^n$ inverses (rather than the n iterations required by the classical algorithm), we need to check that the increasing complexity of successive Newton iterations does not override the advantages accruing from the small number of iterations that are required.

Computing time analysis. Let $T(N)$ be the time required for Newton's method to compute $a(t)^{-1} \mod t^N$. The iteration of Algorithm 1 computes $x^{(k+1)}$ (the first 2^{k+1} terms of $a(t)^{-1}$) in terms of $x^{(k)}$ (the first 2^k terms of $a(t)^{-1}$). Hence $T(N)$ satisfies the recursion

(4) $$T(2N) = T(N) + C(N)$$

where $C(N)$ is the time required to compute the first $2N$ terms of $a(t)^{-1}$ from the previously computed first N terms. Charging some constant cost c for the computation of a_0^{-1} gives $T(1) = c$.

Lemma 2. Let γ be a positive real number, $T(N)$ as above.

(a) If $C(N) = O(N^\gamma)$, then $T(N) = O(N^\gamma)$;
(b) If $C(N) = O(N^\gamma \log N)$, then $T(N) = O(N^\gamma \log N)$.

Proof. Let $N = 2^n$. From (4) we have

$$T(2^n) = T(2^{n-1}) + C(2^{n-1}).$$

Iterating this relationship n times and using $T(1) = c$ give

$$T(2^n) = \sum_{k=0}^{n-1} C(2^k) + c.$$

For $C(N) \le \sigma N^\gamma$, σ a constant, we have

$$T(2^n) = \sum_{k=0}^{n-1} \sigma(2^\gamma)^k + c$$

$$= \sigma\left(\frac{2^{n\gamma} - 1}{2^\gamma - 1}\right) + c$$

$$= O(2^{n\gamma}) = O(N^\gamma),$$

which establishes (a). The proof of (b) is similar. $\square$

$C(N)$ for Algorithm 1 is essentially the time required to compute the two $\mod t^{2N}$ products of Newton's iteration, hence is $O(N \log N)$ if FFT multiplication is used, $O(N^2)$ if classical multiplication is used. From Lemma 2 we immediately conclude the main result of this section:

Theorem 2. Mod t^n inverses over $F[[T]]$ can be computed in time $O(n \log n)$, provided F supports the FFT.

We note that if classical multiplication is used, then Newton's method requires $O(n^2)$ time to compute $\bmod t^n$ inverses, just as the classical algorithm does.

Corollary. Mod t^n division by units over $F[[t]]$ can be carried out in $O(n \log n)$ time.

Proof. If $c(t) = a(t)/b(t)$ with $b(t)$ a unit, then $T_n[c(t) = a(t)b(t)^{-1}]$ can be computed according to $T_n[b(t)] \odot T_n[a(t)^{-1}]$; the $\bmod t^n$ inversion and multiplication each requires $O(n \log n)$ time. $\square$

Interesting enough, fast power series division can be utilized to achieve fast polynomial (quotient-remainder) division (Exercise 1). See Note 3 for historical remarks on fast division algorithms.

3.3 Polynomial Root-finding over Power Series Domains

The polynomial root-finding problem for an integral domain D is simply this: *given a polynomial $f(x) \in D[x]$, find a root $\bar{x} \in D$, or at least a "good approximation" $\hat{x}$ (in some precise sense) to $\bar{x}$.*

Taking D to be the real or complex field yields the familiar and well-studied problem of computing approximate roots of real or complex polynomials. Taking D to be a power series domain $F[[t]]$ (F an arbitrary field) yields the essentially algebraic problem of computing a root $\bar{x}(t) = \sum_{i=0}^{\infty} \bar{x}_i t^i$ of a polynomial $f(x) = \sum_{i=0}^{r} f_i x^i$ with power series coefficients $f_i = f_i(t) \in F[[t]]$. It is this algebraic root-finding problem that we are interested in.

The concept of approximation that we need for computational purposes is precisely the one that we have already introduced and used in Section 3.2: *a "good approximation" to a root $\bar{x}(t)$ is $\bar{x}(t) \bmod t^n$ for n sufficiently large.*

Example 1. Consider $f(x) = x^2 - a \in D[x]$. Then a root $\bar{x}$ of $f(x)$ yields a square root of $a \in D$.

(i) Take $D = \mathbf{Q}[[t]]$ and $a(t) = 1 + t + 2t^3 + 3t^3 + \cdots$. Then $\sqrt{a(t)} \bmod t^4$ $= 1 + \frac{1}{2}t + \frac{7}{8}t^2 + \frac{17}{16}t^3$.

(ii) Take $D = \mathbf{Z}_7[[t]]$ and $a(t) = 1 + t$. Then $\sqrt{a(t)} \bmod t^4 = 1 + 4t + 6t^2 + 4t^3$. (Both (i) and (ii) can be verified directly.) $\square$

In the previous section we employed Newton's method (suitably algebraicized) to solve $f(x) = 1/x - a(t)$ for $\bar{x}(t) = a(t)^{-1}$. Unsurprisingly we intend to do the same to solve our polynomial root-finding problem $f(x) = 0$ over $F[[t]]$. Although polynomial root-finding has much of the flavor of the general case (i.e., general $f(x)$), we observe that the $f(x)$ of inversion, $1/x - a(t)$, does not fall within its scope: $1/x - a(t)$ is not a polynomial.

We require a couple of preliminary results, the first and most interesting of which is an algebraic analog of *Taylor's Theorem* from the calculus. We have already defined the concept of a (formal) derivative of a polynomial $f(x) = \sum_i f_i x^i \in F[x]$ (Section IV.3.3):

$$f'(x) = \sum_i (i+1)f_{i+1}x^i.$$

This definition conforms with the analytic derivatives of the calculus, and we established (algebraically) that the well-known properties of analytic derivatives hold for formal derivatives (Proposition 3 of Section IV.3.3).

Lemma 1 (*Formal Taylor series*). Let $f(x)$ be a polynomial in $D[x]$. Then in $D[x][y]$

$$f(x+y) = f(x) + f'(x)y + \rho(x,y)y^2$$

for some $\rho(x,y) \in D[x][y]$.

Proof. We note first that $x+y$ is a polynomial in $D[x][y]$. Therefore the value of $f(x)$ evaluated at $x+y$ is a polynomial that is also in $D[x][y]$, which can be expressed as

$(*)$ $\qquad\qquad f(x+y) = f_0(x) + f_1(x)y + \rho(x,y)y^2$

for some $f_0(x), f_1(x) \in D[x]$ and $\rho(x,y) \in D[x][y]$.

Letting $y = 0$ in $(*)$ yields $f(x) = f_0(x)$. Taking formal derivatives in $D[x][y]$ of both sides of $(*)$ (appealing to the properties of formal derivatives contained in Proposition 3 of Section IV.3.3) yields

$(**)$ $\qquad\qquad f'(x+y) = f_1(x) + 2y\rho(x,y) + \rho'(x,y)y^2$

($\rho'(x,y) =$ the derivative of $\rho(x,y)$ in $D[x][y]$). Letting $y = 0$ in $(**)$ yields $f'(x) = f_1(x)$. $\square$

Example 1. Consider $f(x) = x^4 \in \mathbf{Z}[x]$. Then in $\mathbf{Z}[x][y]$ we have

$$
\begin{aligned}
f(x+y) &= (x+y)^4 \\
&= x^4 + 4x^3y + 6x^2y^2 + 4xy^3 + y^4 \\
&= f(x) + f'(x)y + \rho(x,y)y^2 \\
&\qquad (\text{for } \rho(x,y) = 6x^2 + 4xy + y^2). \quad \square
\end{aligned}
$$

Let $\phi: D \to E$ be a morphism of domains. Then ϕ induces a natural morphism $\hat{\phi}: D[x] \to E[x]$ defined by: for $f(x) = \sum_i f_i x^i \in D[x]$,

(1) $\qquad\qquad \hat{\phi}(f(x)) = \sum_i \phi(f_i)x^i.$

Lemma 2.   Let $\phi\colon D \to E$ be a morphism of domains, and let $\hat{f}(x) = \hat{\phi}(f(x))$. Then for $a \in D$,

(a) $\phi(f(a)) = \hat{f}(\phi(a))$;
(b) $\phi(f'(a)) = \hat{f}'(\phi(a))$ ($f', \hat{f}' = $ derivatives of $f, \hat{f}$ with respect to x).

Proof.   Part (a) follows routinely by the Fundamental Morphism Theorem. Part (b) is reduced to (a) by showing that $\hat{f}'(x) = \widehat{f'}(x)$. We leave the details to Exercise 2. $\square$

Example 2.   Consider $f(x) = 2x^2 - 7x - 19 \in \mathbf{Z}[x]$ and $\phi = r_5\colon \mathbf{Z} \to \mathbf{Z}_5$. Then $\hat{f}(x) = 2x^2 + 3x + 1 \in \mathbf{Z}_5[x]$. Also, $f'(x) = 4x - 7$ and $\hat{f}'(x) = 4x + 3$ $[= \widehat{f'}(x)]$. With $a = 6$, for example, we see that

(a) $\phi(f(6)) = \phi(11) = 1 = \hat{f}(\phi(6))$,
(b) $\phi(f'(6)) = \phi(17) = 2 = \hat{f}'(\phi(6))$,

in accordance with Lemma 2. $\square$

With $f(x)$ a polynomial over $F[[t]]$, let us define $f_{(n)}(x)$ per eq. (1) by

$$(2) \qquad\qquad f_{(n)}(x) = \hat{T}_n[f(x)].$$

Thus $f_{(n)}(x)$ is obtained from $f(x)$ by reducing $\bmod\, t^n$ the coefficients f_i of $f(x)$. For example, if $f(x) = (1 + 2t + 3t^2 + \cdots)x^2 + 6t^2 x + (5 + 7t)$, then $f_{(1)}(x) = x^2 + 5$, $f_{(2)}(x) = (1 + 2t)x^2 + (5 + 7t)$, etc. Note in particular that $f_{(1)}(x)$ is obtained by setting $t = 0$ in $f(x)$.

Lemma 2 now specializes to

Lemma 3.   Let $f(x)$ be a polynomial over $F[[t]]$. Then for $a = a(t) \in F[[t]]$,

(a) $T_n[f(a)] = f_{(n)}(T_n[a])$;
(b) $T_n[f'(a)] = f_{(n)}'(T_n[a])$.

(The right-hand sides are of course evaluated over $F[[t]]_n$.)

Remark.   Lemma 3 shows that in the computation of $f(a)$ [or $f'(a)$] $\bmod\, t^n$, the coefficients $f_i = f_i(t)$ and argument $a = a(t)$ can all be reduced $\bmod\, t^n$ and the subsequent evaluation carried out over $F[[t]]_n$ — a computationally vital state of affairs.

We now investigate Newton's iteration

$$x^{(k+1)} = x^{(k)} - f(x^{(k)})f'(x^{(k)})^{-1}$$

in the context of our polynomial root-finding problem $f(x) = 0$ over a power series domain $F[[t]]$. The following key theorem is the analog of Theorem 1 of the previous section.

Theorem 1   (*Quadratic convergence of Newton's iteration for solving $f(x) = 0$ over $F[[t]]$*). Let $f(x)$ be a polynomial over $F[[t]]$, and let $\bar{x} = \bar{x}(t) = \sum_{i=0}^{\infty} \bar{x}_i t^i \in F[[t]]$ be a root of $f(x)$ which further satisfies $f_{(1)}'(\bar{x}_0) \neq 0$. For any $n > 0$, if $u = u(t)$ is a mod t^n approximation to $\bar{x}$, then $v = v(t)$ defined according to the following Newton's iteration is a mod t^{2n} approximation to $\bar{x}$: with $g(x) = f_{(2n)}(x) \; (= \hat{T}_{2n}[f(x)])$,

(3) $$v = u - g(u) \odot g'(u)^{\ominus} \qquad \text{(over } F[[t]]_{2n}).$$

Proof. Let us first check that v is defined over $F[[t]]_{2n}$—that $g'(u)$ is invertible over $F[[t]]_{2n}$. By Lemma 3(b), we have

$$T_{2n}[f'(u)] = f_{(2n)}'(T_{2n}[u]) = g'(T_{2n}[u]) = g'(u).$$

From ring theory (Proposition 7 of Section IV.2.1) we have

(4) $$g'(u)^{\ominus} = T_{2n}[f'(u)^{-1}]$$

provided $f'(u)^{-1}$ exists in $F[[t]]$.

The condition for the existence of $f'(u)^{-1}$ is that $T_1[f'(u)]$ be $\neq 0$. Checking this out,

$$T_1[f'(u)] = f_{(1)}'(T_1[u]) \qquad \text{Lemma 3(b)}$$
$$= f_{(1)}'(T_1[\bar{x}]),$$

the last equality holding because by assumption u is a mod t^n approximation $(n > 0)$ to $\bar{x}$. But $f_{(1)}'(T_1[\bar{x}]) = f_{(1)}'(\bar{x}_0)$ is $\neq 0$, also by assumption, so that $g'(u)$ is indeed invertible over $F[[t]]_{2n}$, with inverse given by (4).

The proof now proceeds much as the proof of Theorem 1, Section 3.2. Let $\epsilon = \bar{x} - u$, $\delta = \bar{x} - v$. By assumption, $u = T_n[\bar{x}]$, so that ord ϵ is $\geq n$. The proof is complete if we can show that ord δ is $\geq 2n$, for then $v = T_{2n}[\bar{x}]$.

By Lemma 1 we can express $f(x + y) \in D[x][y]$ as

$$f(x + y) = f(x) + f'(x)y + \rho(x, y)y^2.$$

We now evaluate $f(x + y)$ over $F[[t]]$ by substituting $u = u(t)$ and $\epsilon = \epsilon(t)$ for the indeterminates x and y, respectively. This gives

$$0 = f(\bar{x})$$
$$= f(u + \epsilon)$$
(*) $$= f(u) + f'(u)\epsilon + \rho\epsilon^2$$

(where $\rho = \rho(u, \epsilon) \in F[[t]]$).

We have already shown that $f'(u)$ is a unit in $F[[t]]$. Therefore we can multiply through (*) by $f'(u)^{-1}$ to obtain

$$\epsilon = -f(u)f'(u)^{-1} - \epsilon^2\rho f'(u)^{-1}.$$

Substituting $\epsilon = \bar{x} - u$ gives

(5) $$\bar{x} = u - f(u)f'(u)^{-1} - \epsilon^2\rho f'(u)^{-1}.$$

Let us now examine v as defined by (3). Since $T_{2n}[f(u)] = g'(u)$ (Lemma 3(a)), $T_{2n}[f'(u)^{-1}] = g(u)^{\ominus}$ (eq. (4)), and $T_{2n}[u] = u$ (trivially), it follows that the Fundamental Morphism Theorem can be applied to (3) to obtain

$$v = T_{2n}[u - f(u)f'(u)^{-1}],$$

whence

$$v = u - f(u)f'(u)^{-1} + t^{2n}k \qquad (k \in F[[t]]).$$

From (5) we then obtain

$$\bar{x} = v - t^{2n}k - \epsilon^2 \rho f'(u)^{-1},$$

so that

$$\delta = -t^{2n}k - \epsilon^2 \rho f'(u)^{-1}.$$

Now $\operatorname{ord}(-t^{2n}k)$ is $\geq 2n$, and $\operatorname{ord}(-\epsilon^2 \rho f'(u)^{-1})$ is $\geq 2n$ (since $\operatorname{ord}\epsilon$ is $\geq n$). Hence $\operatorname{ord}\delta$ is $\geq 2n$. $\square$

For a given polynomial $f(x)$ over $F[[t]]$, suppose that we obtain just the constant term α of a root $\bar{x} \in F[[t]]$. Thus $\alpha = \bar{x} \bmod t$ satisfies $f_{(1)}(\alpha) = 0$. Suppose further that $f_{(1)}'(\alpha) \neq 0$. Then one application of Newton's iteration (3) yields $\bar{x} \bmod t^2$, the next application $\bar{x} \bmod t^4$, the next $\bar{x} \bmod t^8$, and so on. Thus we have obtained the following analog of Algorithm 1 of the previous section.

Algorithm 1   (*Newton's method for solving* $f(x) = 0$ *over* $F[[t]]$).

Input.   Polynomial $f(x)$ over $F[[t]]$,
$\alpha = $ constant term of root $\bar{x}$, satisfying $f_{(1)}'(\alpha) \neq 0$ (in addition to $f_{(1)}(\alpha) = 0$).

Output. $x^{(n)} = \bar{x} \bmod t^{2^n}$.

begin
$\quad x^{(0)} := \alpha;$
$\quad$ **for** $k := 0$ **until** $n - 1$ **do**
$\quad$ **begin**
$\qquad g(x) := f_{(2^{k+1})}(x)$
$\qquad x^{(k+1)} := x^{(k)} - g(x^{(k)}) \odot g'(x^{(k)})^{\ominus}$
$\qquad\qquad$ (computed over $F[[t]]_{2^{k+1}}$)
$\quad$ **end**
end $\square$

Remark.   On input, $f(x)$ clearly need only be specified up to $f_{(2^n)}(x)$.

Suppose that for a given polynomial $f(x)$ over $F[[t]]$ we obtain $\alpha \in F$ such that
$(*)$ $\qquad\qquad\qquad f_{(1)}(\alpha) = 0, \qquad f_{(1)}'(\alpha) \neq 0$

(which means that α is a *simple* root of $f_{(1)}(x) = 0$). Although it is "intuitively obvious" that successive convergents of Newton's method, starting from $x^{(0)} = \alpha$, converge to a root $\bar{x}$ of $f(x)$, Theorem 1, which assumes the *a priori* existence of a root $\bar{x}$ (having constant term α), does not in itself prove it. We leave it as an exercise (Exercise 3) to show that if α satisfies ($*$), then $f(x)$ has a (unique) root $\bar{x}$ with constant term α. (Newton's method then converges quadratically to this root, as we have shown.)

Now for a couple of examples of Newton's method (over $F[[t]]$) in action.

Example 1.   Take $f(x) = x^2 + tx - 4$ considered as a polynomial over $\mathbf{Q}[[t]]$. Now $\alpha = 2$ is a simple root of $f_{(1)}(x) = x^2 - 4$: $f_{(1)}(2) = 0, f_{(1)}'(2) = 4 \neq 0$. Newton's method with $\alpha = 2$ then converges quadratically to the root $\bar{x}$ of $f(x)$ with $\bar{x}_0 = 2$. The first two iterations yield

$$\bar{x} \bmod t^2 = 2 - \tfrac{1}{2}t,$$

$$\bar{x} \bmod t^4 = 2 - \tfrac{1}{2}t + \tfrac{1}{16}t^2. \quad \square$$

Example 2   (*Solution of a cubic*).   Let us compute a root $\bar{x}$ of

$$f(x) = x^3 - \frac{2}{1-t}x + \beta$$

for (i) $\beta = 1$ and (ii) $\beta = 7/8$.

(i) $\beta = 1$: $\alpha = 1$ is evidently a simple root of $f_{(1)}(x)$. The first two iterations of Newton's method yield

$$\bar{x} \bmod t^2 = 1 + 2t,$$

$$\bar{x} \bmod t^4 = 1 + 2t - 6t^2 + 58t^3.$$

(ii) $\beta = \tfrac{7}{8}$: $\alpha = \tfrac{1}{2}$ is a simple root of $f_{(1)}(x)$. Two iterations of Newton's method yield

$$\bar{x} \bmod t^2 = \tfrac{1}{2} - \tfrac{4}{5}t,$$

$$\bar{x} \bmod t^4 = \tfrac{1}{2} - \tfrac{4}{5}t + \tfrac{156}{125}t^2 - \tfrac{13508}{3125}t^3. \quad \square$$

Computing time analysis.   Let $f(x)$ be a polynomial of degree m over $F[[t]]$ and let $\bar{x}$ be a root of $f(x)$. Let $T(N)$ $[= T_m(N)]$ be the time required for Newton's method, Algorithm 1, to compute $\bar{x} \bmod t^N$ starting from the initial approximation $\alpha = \bar{x}_0$. As with our power series inversion algorithm of the previous section, the iteration of Algorithm 1 computes $x^{(k+1)}$ (the first 2^{k+1} terms of $\bar{x}$) in terms of $x^{(k)}$ (the first 2^k terms of $\bar{x}$). Hence $T(N)$ satisfies the same recursion

$$T(2N) = T(N) + C(N)$$

where $C(N)$ in this case is the time to compute an expression of the form

$$u - g(u) \odot g'(u)^{\textstyle\ominus^{-1}}$$

over $F[[t]]_{2N}$, with u a power series in $F[[t]]_{2N}$ (actually in $F[[t]]_N$) and $g(x)$ a polynomial of degree m over $F[[t]]_{2N}$. The basis for the recursion is

$$T(1) = c$$

where the constant c (independent of N) is the time required to compute the root α of $f_{(1)}(x) = 0$ over the ground field F.

The crucial quantity is $C(N) = C$. Let us write $C = C_1 + C_2$ where

C_1 = the time to compute $g(u), g'(u) \in F[[t]]_{2N}$,

C_2 = the time to compute the operations $\ominus$, $\odot$, $-$ over $F[[t]]_{2N}$.

C_1 is the interesting component. For the moment, let $g(x)$ be a polynomial over an arbitrary commutative ring R, with $u \in R$. Dividing by $x - u$ gives

$$g(x) = (x - u)h(x) + g(u)$$

according to the Factor Theorem. Dividing the quotient $h(x)$ by $x - u$ gives

$$h(x) = (x - u)k(x) + g'(u)$$

according to our result of Exercise 14, Section IV.3 (do this exercise!). If $g(x)$ has degree m, then the above two divisions evidently require a total of $m + (m - 1) = 2m - 1$ multiplications over R. (Remark: The algorithm which implements these two divisions is called *Horner's rule for the simultaneous evaluation of a polynomial and its first derivative*.)

In our case, the ring R is $F[[t]]_{2N}$. Hence

$$C_1 = O(mN \log N) \quad \text{or} \quad O(mN^2)$$

according to whether fast (FFT) or classical mod t^{2N} power series multiplication is used.

Also,

$$C_2 = O(N \log N) \quad \text{or} \quad O(N^2)$$

according to whether fast or classical mod t^{2N} multiplication and inversion are used. (Fast inversion is via Algorithm 1 of the previous section.)

Thus we have

$$C(N) = O(mN \log N) \quad \text{or} \quad O(mN^2)$$

according to whether fast or classical algorithms are used for mod t^{2N} multiplication and inversion.

From Lemma 2 of the previous section we conclude

Theorem 2. The time $T_m(N)$ for Newton's method to compute the first N terms of a root $\bar{x} \in F[[t]]$ of a degree m polynomial over $F[[t]]$ for specified $\bar{x}_0$ is

(a) $O(mN \log N)$ if fast power series multiplication and division are used;
(b) $O(mN^2)$ if classical power series multiplication and division are used.

To summarize these last two sections, we have seen that the analytic concepts of approximation and convergence, differentiation and Taylor series expansion can all be interpreted in the algebraic setting of a power series domain $F[[t]]$ over an arbitrary field F. This permitted us to establish in completely algebraic terms the applicability of Newton's method for computing inverses of power series (Section 3.2) and roots of polynomials over power series domains (Section 3.3). Newton's method in a power series setting was found to be quadratically convergent (the number of correct terms doubles with each successive iteration) subject to the initial approximation being "sufficiently close" to the desired root $\bar{x}$. Our "sufficiently close" condition is sharp—we require the constant term only of $\bar{x}$. (Note 4.)

EXERCISES

Section 1

1. (*b-ary FFT.*)
   (a) Derive a b-ary FFT algorithm based on splitting a polynomial $a(x)$ into b "pieces," each piece being a polynomial in x^b. (Thus $b = 2$ corresponds to the ordinary (= binary) FFT of Section 1.1.)
   (b) Show that the number of field multiplications required by the b-ary FFT to compute an $N = b^m$ point transform is $(b-1)N\log_b N$.
   (c) For b a power of 2, show that $b = 2$ is always best (with respect to the required number of multiplications to compute an N point transform). Is a nonbinary b ever better than binary?

2. (a) Devise a better than $O(n^3)$ algorithm for solving a *Vandermonde* system of n linear equations in n unknowns over a field F:

   $$V\mathbf{x} = \mathbf{b} \qquad (v_{ij}, b_i \in F)$$

   where V is a Vandermonde matrix $V(t_0, \ldots, t_{n-1}) = (t_i^j)$ (t_i's distinct).
   (b) Illustrate your algorithm of (a) by solving for $\mathbf{x}$ over $\mathbf{Z}_7$:

   $$V = V(1, 2, 5), \qquad \mathbf{b} = (0, 2, 6).$$

Section 2

1. Let $\alpha_0, \ldots, \alpha_{n-1}$ be n elements in a field F that supports the FFT. Derive and analyze a fast algorithm for computing the coefficients a_i of $a(x) = \prod_{i=0}^{n-1}(x - \alpha_i)$.

2. (*Fast multivariate polynomial multiplication.*) Let F be a field that supports the FFT.
   (a) Derive an evaluation-interpolation based algorithm for multiplying polynomials $a(x, y), b(x, y) \in F[x, y]$ having degree m in x, degree n in y
   (i) using classical evaluation-interpolation,
   (ii) using FFT evaluation-interpolation.
   Give bounds on the required computing times $T_{(i)}(m, n)$, $T_{(ii)}(m, n)$.
   (b) Same as (a) only for polynomials $a(x_1, \ldots, x_r), b(x_1, \ldots, x_r) \in F[x_1, \ldots, x_r]$ having degree n in each of r variables.

3. (*Fast exact division over $F[x]$.*) Let $a(x) = \sum_{i=0}^m a_i x^i$, $b(x) = \sum_{i=0}^n b_i x^i$ be polynomials over a field F that supports the FFT. Derive an algorithm for testing whether $a(x)$ divides $b(x)$ and if so returning $q(x) = b(x)/a(x)$. (Does your algorithm impose any restrictions on $a(x)$ or $b(x)$?)

4. Show that step 2 of Algorithm 1, Section 2.2, can be carried out in $O(n \log n)$ time.

5. Devise a "one prime" large integer multiplication algorithm and compare its range of application for the case of a 32 bit computer wordsize with that of the three primes algorithm of Section 2.2.

Section 3

1 (*Reduction of polynomial division to power series division.*) Let $a(x), b(x) \in F[x]$ have degrees m, n, respectively, with $m \geq n$. Our goal here is to reduce the problem of computing $q(x), r(x)$ satisfying the Division Property

$$a(x) = b(x)q(x) + r(x), \qquad \deg r(x) < \deg b(x)$$

to power series division. To this end we define the notion of the *reciprocal* of a polynomial $a(x)$ which we denote by $\tilde{a}(x)$: if $a(x) = a_n x^n + a_{n-1} x^{n-1} + \cdots + a_0$, then

$$\tilde{a}(x) = a_0 x^n + a_1 x^{n-1} + \cdots + a_n$$

(thus the reciprocal of a polynomial is obtained by writing down the coefficients in reverse order).

(a) Prove that the quotient $q(x)$ when $a(x)$ is divided by $b(x)$ satisfies

$$\tilde{a}(x) \equiv \tilde{b}(x)\tilde{q}(x) \pmod{x^{m-n+1}}.$$

[*Hint:* Note that $\tilde{a}(x) = x^n a(1/x)$; argue and exploit that $\phi_{1/x} : a(x) \mapsto a(1/x)$ is a morphism $F[x] \rightarrow F\langle x \rangle$.]

(b) Conclude that $\tilde{q}(x) = T_{m-n+1}[\tilde{a}(x)/\tilde{b}(x)]$.

(c) Further conclude that if F supports the FFT, then $q(x)$ and $r(x)$ can be computed in time $O(n \log n)$ when $n \approx m/2$. (This compares with the $O(n^2)$ time required by the classical algorithm.)

2. Prove Lemma 2 of Section 3.3.

3. Let $f(x)$ be a polynomial over $F[[t]]$. Prove: If $\alpha \in F$ is a simple root of $f_{(1)}(x)$ ($f_{(1)}(\alpha) = 0$, $f_{(1)}'(\alpha) \neq 0$), then $f(x)$ has a unique root $\bar{x} \in F[[t]]$ having constant term $\bar{x}_0 = \alpha$.

4. (*r-th roots of formal power series.*)
(a) Devise an algorithm for computing an rth root of a power series $a(t) \in F[[t]]$.

(b) What is the time $T_r(n)$ required by your algorithm to compute $\sqrt[r]{a(t)} \bmod t^n$ (i) using fast power series operations (assuming F supports the FFT), (ii) using classical power series operations?

(c) Illustrate your algorithm of (a) in computing $\sqrt{a(t)} \bmod t^4$ for $a(t) = 1 + t + 2t^2 + 3t^3 + \cdots \in Q[[t]]$.

5. (*Reversion.*)
(a) Given t as a polynomial in x over a field F,

$$t = a(x) \quad \text{for } a(x) \in F[x],$$

devise an algorithm for finding x in terms of t; i.e., for computing $x(t) \in F[[t]]$ such that $t = a(x(t))$. (The process of finding x in terms of t given t in terms of x is called *reversion*.)

(b) What is the time $T(n)$ required by your algorithm to compute $x(t) \bmod t^n$, assuming that F supports the FFT?

(c) Illustrate your algorithm by computing $x(t) \bmod t^8$ for $a(x) = x - x^3/3 + x^5/5 - x^7/7$. (Since $a(x) = \tan x \bmod x^8$, what you are asked to compute is $\arctan t \bmod t^8$ — Note 4.)

NOTES

Section 1

1. (*Analytic setting for the FFT.*) The *Fourier transform* of a continuous (say time) function $a(t)$ is given by

(1)
$$A(f) = \int_{-\infty}^{\infty} a(t) e^{2\pi i f t} \, dt \qquad (i = \sqrt{-1}).$$

Analogous to this (continuous) Fourier transform is the *discrete* Fourier transform which applies to samples $a_0, a_1, \ldots, a_{N-1}$ of $a(t)$:

(2)
$$A_j = \sum_{k=0}^{N-1} a_k e^{2\pi i j k / N} \qquad (j = 0, \ldots, N-1).$$

The continuous Fourier transform is useful because many phenomena of engineering, physics, and other areas of applied mathematics are more easily analyzed and understood in the transform domain. The discrete Fourier transform is useful because it effectively approximates the continuous Fourier transform.

The connection between the above discrete Fourier transform and our multipoint evaluation problem becomes obvious once we observe that each so-called Fourier coefficient A_j of (2) is nothing more than the value of the polynomial

$$a(x) = \sum_{k=0}^{N-1} a_k x^k$$

at $x = \omega^j$, where $\omega = e^{2\pi i/N}$. We have already noted in Example 1 of Section 1.1 that $e^{2\pi i/N}$ is a primitive Nth root of unity in $\mathbf{C}$. Thus the problem of computing the discrete Fourier transform — the A_j's of (2) — is nothing more than the problem of evaluating a polynomial of length N at N Fourier points. (Thus our terminology "Fourier.")

It was to this important problem of computing discrete Fourier transforms that J. W. Cooley and J. W. Tukey addressed their now-celebrated paper "An algorithm for the machine calculation of complex Fourier series" [*Math. Comp.* **19** (1965), 297–301], in which they essentially discovered the FFT — an $O(N \log N)$ algorithm for computing an N point discrete Fourier transform. (Actually, the FFT has a long and involved history predating the Cooley-Tukey paper [see J. W. Cooley, P. A. W. Lewis, and P. D. Welch, *Proc. IEEE* **55** (1967), 1675–1677].) In short time the Cooley-Tukey FFT completely revolutionized the economics of computing numerical transforms. What previously took several minutes (dollars) of computer time using classical $[O(N^2)]$ methods now required only a few seconds (pennies) using the FFT. It is small wonder that the FFT has exerted a powerful fascination on its beholders. Probably in response to this "powerful fascination," a number of papers were written to explain the workings and applications not only of the FFT but of the discrete Fourier transform which it computes. See, for example,

G. D. Bergland, "A guided tour of the fast Fourier transform," *IEEE Spectrum* **6** (1969), 41–52.

W. T. Cochran et al., "What is the fast Fourier transform?," *Proc. IEEE* **55** (1967), 1664–1674.

J. W. Cooley, P. A. W. Lewis, and P. D. Welsh, "The fast Fourier transform and its applications," *IEEE Trans. on Education* **12** (1969), 27–34.

W. M. Gentleman and G. Sande, "Fast Fourier transforms—for fun and profit," *AFIPS Proc.* **29** (1966), 536–568.

A concluding "programming" remark: The power of recursion as a control structure is evident on comparing our recursive FFT procedure with the usual iterative ones reported in the literature such as in any of the papers cited above, including the Cooley-Tukey one. For example see p. 29 of the Cooley-Lewis-Welch paper for an iterative FFT program written in Fortran. Nearly half the program is devoted to a rather complicated data accessing scheme, while the other half (the transform proper) defies easy comprehension.

2. See T. Estermann, *Introduction to Modern Prime Number Theory*, (Cambridge University Press, 1952), Chapter 2.

3. For the $6n/\pi^2$ result for the average value of $\phi(n)$ over all integers, see Hardy and Wright, *The Theory of Numbers*, 4th ed. (Oxford, 1960), Section 18.5.

    The argument that the average value of $\phi(n)$ over even integers is greater than $3n/\pi^2$ goes as follows. For n even, write $n = 2^e k$, k odd. We then have $\phi(n) = \phi(2^e)\phi(k) = 2^{e-1}\phi(k)$ (Exercise 17 of Section III.3). Now the average value of $\phi(k)$ over *odd* k must be greater than $6k/\pi^2$, since its average value over *even* k is $\leq k/2 < 6k/\pi^2$. Thus $\phi(2^e k)$ has average value greater than $2^{e-1}6k/\pi^2 = 3(2^e k)/\pi^2 = 3n/\pi^2$.

4. Only primes $\leq \sqrt{p-1}$ need be cancelled. For $p - 1$ can have at most one prime factor $> \sqrt{p-1}$, clearly. Thus, if $p - 1 = a_1^{e_1} \cdots a_k^{e_k} x$, where $a_1, \ldots, a_k$ are all of the prime factors of $p - 1$ that are $\leq \sqrt{p-1}$, then x must either be unity or a prime.

Section 2

1. A comprehensive account of pre-FFT fast integer multiplication algorithms is to be found in Knuth (1969), Section 4.3.3. It was not until 1962 that a faster than classical $[< O(n^2)]$ integer multiplication algorithm appeared. This was an order $n^{\log_2 3} \approx n^{1.57}$ algorithm due to A. Karatsuba, a simplified version of which is presented in Knuth. (This algorithm can be readily adapted for polynomial multiplication.)

    By 1966 there had been discovered two asymptotically faster algorithms, one due to S. A. Cook, the other to A. Schönhage. The fact that these two ostensibly quite different algorithms had the same order of magnitude running times—$O(n\alpha^{\sqrt{\log n}})$ for α some real constant—lead Knuth to conjecture [Knuth (1969), p. 274] that an *optimal* multiplication algorithm would have such a time bound. Post-FFT developments, however, proved this conjecture false.

    In 1971, A. Schönhage and V. Strassen published an FFT-based algorithm with a time bound of $O(n \log n \log\log n)$ [Computing **7** (1971), 281–292]. This is the current best (at time of writing); optimality conjectures abound. A noteworthy feature of the FFTs in the Schönhage-Strassen algorithm is that they are applied over *Fermat rings:* rings of integers modulo *Fermat numbers* $2^{2^n} + 1$ for appropriately chosen n. The FFTs of the Schönhage-Strassen algorithm do not require that these Fermat numbers be prime (which is just as well, since for certain smaller values of n they are known to be composite—e.g., $n = 5, 6, 7, 8$. For larger n the situation is mainly unknown.)

Our subasymptotic "three primes" algorithm of Section 2.2, superior over its astronomically large but finite range of application to the current asymptotic best, is based on an algorithm due to J. M. Pollard. [This is just one of several interesting ideas to be found in his important paper "The fast Fourier transform in a finite field," *Math. Comp.* **25** (1971), 365–374; see pp. 368–369 in particular.] This author owes the apt term "practically asymptotic" (to describe the behavior of such an algorithm) to his colleague A. Borodin.

Section 3

1. This is the title of this author's paper in *SYMSAC 76*, pp. 260–270, on which Sections 3.2 and 3.3 are based.

2. See, for example, P. Henrici, *Elements of Numerical Analysis* (New York: Wiley, 1964), Section 6.

3. (*History of fast polynomial division and power series inversion.*) As we have seen, the classical division algorithm for dividing a degree m polynomial by a degree n polynomial requires $O(n^2)$ time, as does the classical algorithm for computing power series inverses mod t^n. The first speedup of either of these problems was an $O(n \log^2 n)$ polynomial division algorithm due to A. Borodin and R. Moenck [*J. Comput. System Sci.* **8** (1974), 366–386]. Independently, M. Sieveking [*Computing* **10** (1972), 153–156] devised an $O(n \log n)$ algorithm for power series inversion. By reducing polynomial division to power series inversion, V. Strassen [*Numer. Math.* **20** (1973), 238–251] was able to apply Sieveking's algorithm to obtain an improved $[O(n \log n)]$ polynomial division algorithm. (Exercise 1 is essentially Strassen's result.) These algorithms all exploited fast (FFT-based) polynomial multiplication.

   An important conceptual simplification was made by H. T. Kung [*Numer. Math.* **22** (1974), 341–348] who showed that Sieveking's power series inversion algorithm was essentially Newton's method for computing reciprocals, only applied over a power series domain. Kung's paper is the cornerstone for both Sections 3.2 and 3.3.

   But interestingly enough, the integer analog of these power series developments, including the use of Newton's method to compute the reciprocals of large integers, had been carried out several years earlier by S. Cook in his 1966 Ph.D. thesis (Harvard Univ.).

4. For further results about Newton's method as a tool for computer algebra see J. F. Traub and H. T. King, "All algebraic functions can be computed fast," *J. ACM* **25** (1978), 245–260; D. Y. Y. Yun, "Hensel meets Newton—algebraic constructions in an analytic setting," in J. F. Traub (ed.), *Analytic Computational Complexity* (New York: Academic Press, 1976), 205–215; also D. Y. Y. Yun, "Algebraic Algorithms using *p*-adic constructions," *SYMSAC 76*, 248–250.

   The reversion operation of Exercise 5 is an important topic unto itself. See P. Henrici, *Applied and Computational Complex Analysis* **1** (New York: Wiley, 1974), Section 1.7, for an algebraic treatment of reversion of power series. For an ingenious reversion algorithm (cf. your result in Exercise 5) see R. P. Brent and H. T. Kung, "$O((n \log n)^{3/2})$ algorithms for composition and reversion of power series," in *Analytic Computational Complexity* (cited above), 217–225.

SELECTED BIBLIOGRAPHY

Two classics of modern algebra are

G. BIRKHOFF AND S. MAC LANE, *A Survey of Modern Algebra*, 4th ed. (New York: Macmillan, 1977).

B. L. VAN DER WAERDEN, *Algebra* (Vols. I and II), 7th ed. (New York: Ungar, 1970).

In addition to the above there are many excellent texts on (pure) modern algebra. A short list (consisting of five of this author's favorites) might include

R. A. DEAN, *Elements of Abstract Algebra* (New York: Wiley, 1966).

R. GODEMENT, *Algebra* (Boston: Houghton Miffin, 1968).

L. J. GOLDSTEIN, *Abstract Algebra* (Englewood Cliffs, N.J.: Prentice-Hall, 1973).

I. N. HERSTEIN, *Topics in Algebra* (New York: Blaisdell, 1964).

S. LANG, *Algebraic Structures* (Reading, Mass.: Addison-Wesley, 1967).

The area of *applied* algebra has received increasing attention in recent years. The following texts contain a variety of computer-related applications of algebra.

G. BIRKHOFF AND T. C. BARTEE, *Modern Applied Algebra* (New York: McGraw-Hill, 1969).

J. L. FISHER, *Application-oriented Algebra* (New York: Crowell, 1977).

John D. Lipson, Elements of Algebra and Algebraic Computing. ISBN 0-201-04115-4.

A. Gill, *Applied Algebra for Computer Sciences* (Englewood Cliffs, N.J.: Prentice-Hall, 1976).

F. P. Preperata and R. T. Yeh, *Introduction to Discrete Structures* (Reading, Mass.: Addison-Wesley, 1973).

The following computer science texts contain a wealth of material on algebraic computing:

A. V. Aho, J. E. Hopcroft, and J. D. Ullman, *The Design and Analysis of Computer Algorithms* (Reading, Mass.: Addison-Wesley, 1974).

A. Borodin and I. Munro, *The Computational Complexity of Algebraic and Numeric Algorithms* (New York: American Elsevier, 1975).

D. E. Knuth, *The Art of Computer Programming, Vol. 2: Seminumerical Algorithms* (Reading, Mass.: Addison-Wesley, 1969).

Extensive collections of research and survey papers in computer algebra—dealing with both algorithms and systems—are to be found in the following symposia proceedings:

(*SIGSAM 2*) S. R. Petrick (ed.), *Proceedings of the Second Symposium on Symbolic and Algebraic Manipulation* (ACM, 1971).

(*SYMSAC 76*) R. D. Jenks (ed.), *Proceedings of the 1976 ACM Symposium on Symbolic and Algebraic Computation* (ACM, 1976).

NASA CP-2012, Proceedings of the 1977 MACSYMA User's Conference (held at Berkeley, Calif., July 27–29, 1977).

Also recommended are the following survey papers on computer algebra. (Each author, as it turns out, is the principal designer of a major computer algebra system.)

W. S. Brown and A. C. Hearn, "Applications of symbolic algebraic computation," *Bell Laboratories Computing Science Technical Report* #66, 1978.

G. E. Collins, "Computer algebra of polynomials and rational functions," *Amer. Math. Monthly* **80** (1973), 725–755.

J. Moses, "Algebraic structures and their algorithms," in J. F. Traub (ed.), *Algorithms and Complexity* (New York: Academic Press, 1976), 301–319.

INDEX TO NOTATION

INDEX TO NOTATION

Notation	**Meaning**	**Page**
$\check{\rho}$	converse of ρ	12
$\rho \circ \sigma$	composition of ρ and σ	12
$[a]$	equivalence class containing a	13
A/E	quotient set A mod E	13
$a \equiv_m b$	a is equivalent mod m to b	14
$x \wedge y$	meet of x and y	20
$x \vee y$	join of x and y	20
$A \to B$	function from A to B	21
$a \mapsto b$	assignment of a to b	22
B^A	all functions from A to B	22
$f(S)$	image of S under f	22
$f^{-1}(T)$	inverse image of T under f	22
$\mathrm{Im}\, f$	image (range) of f	23
c_S	characteristic function of S	26
E_f	kernel relation of f	28

Chapter II

$m \mid a$	m divides a	44
$\left.\begin{array}{c} r_m(a) \\ a \bmod m \end{array}\right\}$	remainder when a is divided by m	44
$\mathbf{Z}_m$	integers mod m: $\{0, 1, \ldots, m-1\}$	45
$\left.\begin{array}{c} \gcd(a, b) \\ (a, b) \end{array}\right\}$	g.c.d. of a and b	46
$\left.\begin{array}{c} \mathrm{lcm}(a, b) \\ [a, b] \end{array}\right\}$	l.c.m. of a and b	52

Prologue to Algebra

$[A; \Omega]$	algebraic system with carrier A and operations Ω	57
$\lvert A \rvert$	number of elements of set A; order of algebraic system A	58
$a \oplus b, a \odot b$	addition, multiplication mod m	58

Chapter III

$\Sigma(A)$	finite strings over A	69
S_X	symmetric group on X	72
S_n	symmetric group on n letters	72
$A \le B$	A is a subalgebra of B	77
$[H]$	subalgebra generated by H	78
$gS\,(Sg)$	left (right) coset of S by g	81
$g + S$	additive coset of S by g	81

Notation	**Meaning**	**Page**
Chapter VI		
$F(\alpha)$	F adjoined by α	174
$F(\alpha_1, \ldots, \alpha_r)$	F adjoined by $\alpha_1, \ldots, \alpha_r$	174
$m_\alpha(x)$	minimum polynomial of α	175
GF(q)	Galois field of order q	182
Chapter VII		
$O(f(m,n))$	order ("big oh" of) $f(m,n)$	194
var := expr	assignment command (Algol)	199, 230
$\neg \gamma$	not γ	204, 220
$\{p\}\,\sigma(q)$	assertion: if p holds before execution of σ then q holds afterwards	204, 218
Chapter VIII		
$\mathcal{E}(X_s)$	formal expressions in the s indeterminates of X_s	236
$e(x)$	formal expression $e(x_1, \ldots, x_s)$	236
$e(a)$	value of $e(x_1, \ldots, x_s)$ at $x_i = a_i$	237
$\det A$	determinant of A	243
adj A	adjoint of A	283
Chapter IX		
P_N	multipoint evaluation problem of size N	294
F_N	Fourier evaluation problem of size N	296
$[\omega]$	$\{1, \omega, \ldots, \omega^{N-1}\}$: set of N Fourier points	296
$V(\alpha_0, \ldots, \alpha_{N-1})$	Vandermonde matrix associated with $\alpha_0, \ldots, \alpha_{N-1}$	300
$a(t)^{\ominus 1}$	inverse of $a(t)$ in $F[[t]]_n$	312
$\hat{\phi}$	morphism $D[x] \to E[x]$ induced by morphism $\phi: D \to E$	319

INDEX

INDEX